THE THEORY
OF
RELATIVITY

THE THEORY
OF
RELATIVITY

SECOND EDITION

R.K. PATHRIA

Distinguished Professor Emeritus
University of Waterloo
and
Visiting Scholar
University of California at San Diego

DOVER PUBLICATIONS, INC.
Mineola, New York

Bibliographical Note

This Dover edition, first published in 2003, is an unabridged reprint of the second edition, published by Pergamon Press (Oxford and New York) in 1974, of the work originally published by Hindustan Publishing Corporation (India) in 1963.

Library of Congress Cataloging-in-Publication Data

Pathria, R.K.
    The theory of relativity / R.K. Pathria.
        p. cm.
    Originally published: 2nd ed. Delhi : Hindustan Pub. Corp., 1963.
    Includes bibliographical references and index.
    ISBN 0-486-42819-2 (pbk.)
        1. Relativity (Physics) I. Title.

QC173.55.P37 2003
530.11—dc21

                                                            2003043452

Manufactured in the United States of America
Dover Publications, Inc., 31 East 2nd Street, Mineola, N.Y. 11501

CONTENTS

PREFACE TO THE FIRST EDITION

The present book arose primarily out of the notes of lectures that I have been giving to the post-graduate students at the University of Delhi. Quite inevitably, the subject matter, in its finer details, got considerably modified during the process of preparing the manuscript; the style of presentation, however, remains the same as followed in the lectures.

I strongly feel that one of the essential aspects of teaching is to arouse the curiosity of the students in their subject; seemingly, the most effective way of doing this is to discuss with them (in a reasonable proportion, of course) the salient features of the circumstances that led to the emergence of the subject and of the course of development it subsequently followed. In the case of a subject like relativity, these features are all the more exciting because, in the main, they have been carved out by men of the highest calibre in physics; one is, therefore, tempted to stop time and again in order to see rather minutely as to *how* such and such developments actually came about. At the same time, one may not be prepared to let the flow of a pedagogical text, such as the one in hand, get hampered by unwholesome discontinuities arising from the intermittent addition of historical material. Accordingly, I have included in the present account an historical introduction to the subject which stands separate from the main text.

As for the material contents of the book, there is hardly anything that is not customary. The presentation, however, is, to a large extent, according to my own taste. This is reflected, rather conspicuously, in the treatment of those parts of the subject which come in direct contact with experiment. Here, I have discussed in a considerably greater detail than is ordinarily done the extent to which the various conclusions of the theory have been verified by observation; in fact, during the course of these discussions, I have attempted to present to the reader a report as up-to-date as possible.

Generally speaking, I have devoted a respectable part of the available space to the elaboration of physical ideas underlying various arguments rather than being contented with mathematical derivations alone; in other words, I have avoided the use of the so-called telegraphic language which, unfortunately, is becoming a practice with a growing number of authors writing on theoretical subjects. Further, to provide the reader with an opportunity to gauge the extent to which he has grasped what he has read, a variety of unsolved problems has been included in the text.

I am of the opinion that for a student at the graduate level it is the right time to get introduced to periodical scientific literature which, in the years to come, is possibly going to be his major resort. With this in view, and also for reasons of documental propriety, I have been quite particular about quoting original references; further, to enhance their impact and utility, I have given them in the form of on-the-spot footnotes rather than piling them up at the end of the book.

Of course, the extent to which I have succeeded in doing justice to this project can be gauged only by the readers. I, on my part, will be grateful to receive any suggestions that are intended to improve the warp and woof of this book.

The completion of this task has left me indebted to many. First of all, I am thankful to Professor R. C. Majumdar for his encouraging interest in my plan of writing this book. Thanks are also due to my colleagues Dr. K. K. Singh and Mr. M. P. Kawatra for going through the first draft of the Historical Introduction and making suggestions which helped me unkink the exposition at a number of points. Mr. Kawatra also assisted me in the compilation of the Indexes; this, too, is gratefully acknowledged.

I earnestly appreciate the unfailing cooperation of my publishers who always respected my exacting demands, and the untiring efforts of the printing staff who strove to tap all possible resources in order to make the best of their job.

I cannot express adequately my indebtedness to my wife, the contribution of whose considerate cooperation, through all stages of this project and against all odds, is, I sincerely feel, beyond my sense of assessment.

Delhi, March 1963   R. K. Pathria

PREFACE TO THE SECOND EDITION

The main consideration in preparing this edition was to update the contents of the first edition which appeared ten years ago. This has been done by adding new sections to the text, by rewriting the existing ones and by augumenting the list of references. In doing this I have benefited considerably from the opinions and suggestions of the various referees who reviewed the first edition of this book and of others who used it as a text or as a supplementary reference. As a result of these changes and additions, the book should satisfy its intended purpose, laid down in the Preface to the first edition, more adequately than before.

A special feature of the new edition is the inclusion of a large number of problems, at the end of each chapter, which should enhance the usefulness of the book as a text for graduate or senior undergraduate students in physics and mathematics.

The preparation of this edition has again left me indebted to many—most importantly, to my students and colleagues whose interest and criticism have helped me in remaining attached to this wonderful subject, despite commitments elsewhere. On the technical side I am thankful to MRS. JEAN SPOWART for her skillful typing of the manuscript and to MRS. ANMAY CHEN for assistance in proof-reading. Once again, the unreserved cooperation of my wife has been a source of strength and inspiration throughout the arduous task of writing and rewriting.

Waterloo, November 1973   R. K. PATHRIA

HISTORICAL INTRODUCTION

Although the influence of the theory of relativity has by now permeated practically all branches of physics, it is to the fields of optics and electromagnetism that its origin is chiefly due. And the one single concept, which has played far greater part than any other in bringing about this new development, is that of the 'aether'. This mysterious entity had been figuring quite frequently in the deliberations of the early philosophers, but it was brought into science only by RENE DESCARTES (1596-1650) who, for the first time, postulated that the aether be regarded as endowed with mechanical properties rendering it capable of transmitting force. Subsequently, it came into prominence through the work of ROBERT HOOKE (1635-1703) on the propagation of light waves, in which this all-pervading, imponderable medium was employed as a carrier of optical disturbances—hence the title of being 'luminiferous'. For ISAAC NEWTON (1642-1727), according to whom light consisted not of waves but of corpuscles, this medium was essentially unwanted so long as it was required to do nothing else but undulate. However, it *did* find a respectable place in his considerations as well—at least in the role of an intermediary for mutual interaction between light and the ponderable matter.

It was AUGUSTIN FRESNEL (1788-1827) who, in the twenties of the last century, developed a systematic, mathematical theory of light based on a set of pre-assigned dynamical properties of the aether. This theory made such remarkable achievements in the realm of physical optics—a subject under vigorous pursuit during those times—that it not only overthrew the corpuscular hypothesis but also placed the aether on (so to say!) a sound basis. Later authors, notably GEORGE GREEN (1793-1841), AUGUSTINE-LOUIS CAUCHY (1789-1857), JAMES MACCULLAGH (1809-1847) and WILLIAM THOMSON (later LORD KELVIN; 1824-1907), contributed substantially towards elucidating the various aspects of the role that aether could be made to play in the field of optics.[1]

It was MICHAEL FARADAY (1791-1867) who, for the first time, suggested that if the existence of a luminiferous aether were to be admitted, then it might as well be the vehicle of the magnetic force. He wrote in 1851, "It is not at all unlikely that if there be an aether, it *should* have other uses than simply the conveyance of radiations."[2] In 1853, BERNHARD RIEMANN (1826-1866), attempting to realize an aspiration of his teacher KARL FRIEDRICH GAUSS (1777-1855), proposed another set of properties for the aether, notably the power of resisting compression, which could enable it to transmit gravitational and electric effects as well as optical and magnetic ones.

However, a systematic theory of the transmission of electric and magnetic influences through aether, or for that matter through any continuous medium, did not emerge until 1861-62 when JAMES CLERK MAXWELL (1831-1879) propounded his famous theory of the electromagnetic field. All the salient concepts and conclusions of his theory (for instance, the displacement current, the stresses and strains in the field, the propagation of electromagnetic waves with a finite velocity and the transmission of energy) seemed to point strongly towards the active role of aether in various manifestations of the electromagnetic field. And, finally, the identification of electromagnetic waves with light made it all the more plausible that there existed a medium (in the nature of the aether), which could be regarded as the seat of both optical and electromagnetic phenomena, irrespective of what one may think of its multifarious, and not completely self-consistent, properties. Consequently, much thought was expended by MAXWELL, his contemporaries and their successors (WILLIAM THOMSON, GUSTAV KIRCHHOFF, 1824-1887; GEORGE FRANCIS FITZGERALD, 1851-1901; JOSEPH LARMOR, 1857-1942, etc.) in order to evolve suitable mechanical models of the aether. This hypothetical medium thus assumed a central importance in the theoretical physics of the last century.[3]

---

[1] For details, see E. T. WHITTAKER, *A History of the Theories of Aether and Electricity*, Vol. I (Thomas Nelson and Sons Ltd., Edinburgh, 1951), Chap. V. Reprinted by Harper and Brothers, New York, 1960.

[2] M. FARADAY, *Experimental Researches in Electricity* (London, 1839), Sec. 3075. FARADAY's attitude, however, amounts to replacing aether by the 'lines of force' which, likewise, fill all space and make material bodies mutually interact.

[3] See E. T. WHITTAKER, *loc. cit.*, Vol. I, Chap. IX. It is not advisable for us to dwell here, at length, on the *pros* and *cons* of the various models developed, because we know very well what ultimate fate the very concept of aether had to meet in the wake of subsequent developments.

One must, however, admit that all such attempts as were directed at evolving a mechanical interpretation of the electromagnetic phenomena actually gave little satisfaction—so much so that finally it appears most reasonable to regard the MAXWELL field vectors themselves as the real basic data and not seek for them an interpretation in terms of mechanical images. But things did not appear so simple to the successors of MAXWELL, for they felt it necessary to retain aether in view of another fundamental difficulty, viz. the failure of MAXWELL's field equations to satisfy the principle of 'mechanical' relativity. According to this principle, which had played a decisive role in the development of the fundamental laws of mechanics, at the hands of GALILEO GALILEI (1564-1642), CHRISTIAN HUYGENS (1629-1695) and ISAAC NEWTON, the whole discipline of mechanics held equally valid in all those systems of reference which conformed to the Galilean law of inertia. Such systems are generally referred to as *Galilean* or *inertial* systems of reference; they differ from one another only through a uniform (translational) relative motion.

Now, the very fact that the electromagnetic field equations did not satisfy the Galilean principle of relativity made it imperative for the theorists who succeeded MAXWELL, notably HENDRIK ANTOON LORENTZ (1853-1928), to assume an aether *in a state of absolute rest,* the so-called 'stationary' aether, which could be set apart as *the* system of reference in which the field equations were valid as such.[1] This attitude gave rise to a new topic of investigation—the so-called electrodynamics of 'moving' bodies—to which LORENTZ himself proved to be the most prolific contributor. However, the aether, even after having been reduced to a pretty modest role of being simply a reference stage, did not cease to create theoretical as well as practical difficulties.

It is quite clear that in the totality of the inertial systems of reference—called *The Body Alpha* by CARL NEUMANN (1832-1925)—the ones rigidly attached to the aether would stand singled out and may possibly claim a special privilege over the others; for instance, all laws of Nature may assume an especially simplified and elegant form when referred to *these*, rather than to other, systems of reference. It is then tempting to inquire if one could carry out a measurement of the velocity of any particular body, say the

---

[1] It could then also serve as the 'absolute space' of NEWTON which he had introduced in order to distinguish the so-called inertial systems of reference from others. According to NEWTON, this particular class of systems would be such that their motion with respect to the absolute space system is unaccelerated; in such systems, he asserted, no *fictitious* forces would arise.

Earth, with respect to the aether, which might be regarded as its *absolute velocity,* and thus establish this medium as Nature's *really* privileged system of reference (with respect to which the questions of absolute rest and absolute motion could be straightaway settled).

The lead had to come now from the experimental side and was provided by the pioneering work of ALBERT ABRAHAM MICHELSON (1852-1931) who, in the eighties of the last century, performed a series of experiments[1] designed to detect, and even measure, the relative motion of the Earth and the aether. These experiments, however, failed to yield positive results, which made physicists cast serious doubts on the very existence of the aether and, for that matter of any 'privileged' system of reference.[2]

Struck by these circumstances, HENRI POINCARE (1854-1912) observed in 1899 that "it was very probable that the optical phenomena depend only on the *relative* motions of the material bodies, luminous sources and optical apparatus concerned", with the implication that 'absolute' motion was, in principle, indetectable so far as optical means were concerned. He further felt that *any* physical means whatsoever, not only the optical ones, were necessarily going to fail in this sort of an attempt. He said in 1900, "Our aether, does it really exist? I do not believe that more precise observations could ever reveal anything more than *relative* displacement."[3]

In 1904, POINCARE enunciated a new principle—calling it *The Principle of Relativity*—according to which '*the laws of physical phenomena must be the same for a 'fixed' observer as for an observer who possesses a uniform relative motion of translation*'. Thus, one could not hope to discern, in any manner whatsoever, whether one is, or is not, being carried along in 'uniform motion'. In the light of this principle, the explanation for the failure of MICHELSON's experiments is quite straightforward, viz. that they were intended to achieve something which, in principle, is impossible. Making a brilliant use of his intense originality and penetrating intuition, POINCARE declared, "From all these results there must arise an entirely new kind of dynamics, which will

---

[1] A. A. MICHELSON, *Am. J. Sci.* (3), **22**, 20(1881); A. A. MICHELSON and E. W. MORLEY, *ibid.*, **34**, 333(1887). See also A. A. MICHELSON, *Am. J. Sci.* (4), **3**, 475(1897). For details of these experiments, see Sec. 1.3 of the text.

[2] Efforts to save the aether were made by FITZGERALD and LORENTZ by introducing, in 1892, an *ad hoc* hypothesis—the so-called contraction hypothesis (see Sec. 1.3). To start with, it did appear to work; however, its purely artificial character could not prolong the life of the aether by more than about a decade.

[3] For original sources, whence these extracts have been taken, refer to E. T. WHITTAKER, *A History of the Theories of Aether and Electricity*, Vol. II (Thomas Nelson and Sons Ltd., Edinburgh, 1953), Chap. II. Reprinted by Harper and Brothers, New York, 1960.

be characterized above all by the rule that no velocity can exceed the velocity of light."[1]

It became, therefore, impelling to devise an analytical scheme which could enable the gamut of physics to be formulated in accordance with POINCARE's principle of relativity. The first to be subjected to such an analysis were the laws of the electromagnetic field, which were already being examined by LORENTZ in connection with his studies on the electrodynamics of moving bodies. During the period 1892-1904, LORENTZ carried out extensive investigations into the question of the interrelationship between the electromagnetic measures pertaining to two observers, one 'at rest' and the other in uniform rectilinear motion. In his preliminary work,[2] quantities of orders higher than the first in v/c, where v is the magnitude of the velocity of the 'moving' observer and c the velocity of the electromagnetic waves in vacuum, were neglected. In 1900, LARMOR[3] extended this analysis to include quantities of the second order in v/c, while in 1904 LORENTZ reattacked[4] the problem on a wider scale and obtained results in a form exact to all orders in v/c.

It was in these investigations that LORENTZ arrived at the well known space-time transformations that go after his name; it was here that he found it mathematically convenient to introduce what he called a *local* time t' for a 'moving' system of reference,[5] in terms of which the field equations of MAXWELL were found to remain invariant under a transformation from the observer 'at rest' to the observer 'in motion'.[6]

There remained, however, one serious deficiency in LORENTZ's work,

[1] The extracts in this paragraph are from an address delivered by POINCARE at the Congress of Arts and Sciences at St. Louis, U.S.A., on September 24, 1904; the original version appeared in *Bull. des Sc. Math.* **28**, 302 (1904) and an English translation by G. B. HALSTED was published in the January 1905 issue of *The Monist*.

[2] H. A. LORENTZ, *Versuch einer Theorie der electrischen und optischen Erscheinungen in bewegten Körpern* (E. J. Brill, Leiden, 1895); reprinted by Teubner of Leipzig in 1906.

[3] J. LARMOR, *Aether and Matter* (Cambridge, 1900), pp. 167-177.

[4] H. A. LORENTZ, 'Electromagnetic phenomena in a system moving with any velocity less than that of light', *Proc. Acad. Sci.*, Amsterdam 6, 809 (1904); also available in *The Principle of Relativity* : a collection of the original papers of A. EINSTEIN, H. A. LORENTZ, H. MINKOWSKI and H. WEYL (Methuen and Co., London, 1923 ; reprinted by Dover Publications, New York).

[5] The time t' was first introduced by W. VOIGT, in 1887, in connection with the study of the equations of vibratory motion.

[6] It is interesting to note that, in spite of the invariance of the field equations and the consequent equivalence of the various inertial systems of reference, LORENTZ could not abandon, even towards the end of his career, the notions of 'absolute space' and 'absolute time'.

viz. that he could establish the invariance of MAXWELL's equations only in *charge-free* space (because he failed to transform the charge density and current density correctly). This deficiency was removed by POINCARE[1] who hit upon the correct transformation equations for these two quantities and thus established the invariance of MAXWELL's equations quite generally. Of course, even in the general case, the space-time transformations obtained earlier by LORENTZ applied *as such*. The aggregate of all transformations so obtained, combined with the aggregate of rotations in ordinary space, constitutes a *group*, which POINCARE called the '*group of Lorentz transformations*'.[2]

It was at this stage that ALBERT EINSTEIN (1879-1955) appeared on the scene. His epoch-making paper[3] of 1905, which, on all accounts, appears to have been written without prior knowledge of the advances made by POINCARE and LORENTZ, not only contained all the essential results deduced by these two authors but also afforded an entirely novel and profound understanding of the problem.[4]

Pointing out a formal asymmetry in the description of electric currents generated by relative motion between a magnet and a conductor, EINSTEIN felt that the phenomena of electrodynamics, as well as of mechanics, possess no properties corresponding to the concept of 'fixity' in space; it appears that in all systems of reference, in which the fundamental laws of mechanics hold good, exactly the *same* laws of electrodynamics and optics also apply.[5] EINSTEIN chose to adopt this conjecture as a postulate. Next, he introduced a second postulate, viz. the propagation of light in empty space with a *definite* speed c, which is independent of the relative motion of the source and the observer.[6]

[1] H. POINCARE, 'Sur la dynamique de l'electron', *C. R. Acad. Sci.*, Paris **140**, 1504 (1905); *R. C. Circ. mat.*, Palermo **21**, 129 (1906).

[2] This terminology first appeared in the 1905 paper of POINCARE, *loc. cit.*

[3] A. EINSTEIN, 'On the Electrodynamics of Moving Bodies', *Ann. der Phys*. **17**, 891 (1905); English translation available in *The Principle of Relativity, loc. cit.*

[4] In the same issue of the *Annalen* appeared his other two memorable papers, one on the photoelectric effect, p. 132, and the other on the Brownian motion, p. 549. A closer scrutiny of these three papers, dealing with apparently unrelated problems, shows that they probably arose from the study of a single general problem, viz. *the fluctuations in the pressure of radiation*. This reveals, to a significant extent, the inner continuity (and outer independence!) of EINSTEIN's work of 1905. For an admirable account on this point, reference may be made to G. HOLTON, *Am. J. Phys.* **28**, 627 (1960).

[5] Here, EINSTEIN does refer (though not citing it categorically) to the 1892-95 work of LORENTZ.

[6] Here too, EINSTEIN's reference to the failure of all attempts to discover any motion of the Earth relative to the 'light medium' is significant, though he makes no specific mention of the work of MICHELSON *et al.*

Based on these postulates, EINSTEIN was able to develop a simple and consistent theory, the so-called special theory of relativity, which successfully formulated the electrodynamics of moving bodies on the basis of MAXWELL'S theory (for stationary bodies). Accordingly, he dismissed aether as a mere *superfluity* and stressed the need of developing physics independently of it.[1]

It goes to EINSTEIN's credit that he brought out clearly the *relativity* of length and time measurements, thus giving physical meaning to the mathematical formulae of the LORENTZ transformations. In the light of his approach to the problem, t', the *local* time introduced by LORENTZ for a moving system of reference, had to be regarded as the real *physical* time appropriate to that system and not merely a mathematical artifice. EINSTEIN's derivation of space-time transformation equations, on the basis of the aforementioned postulates, was amazingly straightforward; as a result, the interwoven character of space and time revealed itself with admirable clarity. A new dynamics was thereby born, which satisfied, as best as one could imagine, the aspirations of POINCARE.

EINSTEIN's work, and the spirit behind it, provided strong stimulus to detailed investigations into the very fundamentals of physics; these, as is well known, resulted in an almost complete overhauling of this branch of science.[2]

A contribution of great importance to the theory of relativity was made in 1908 by HERMAN MINKOWSKI[3] (1864-1909), which proved vital to the subsequent development of the subject. The salient feature of this contribution

---

[1] Later on, in 1920, EINSTEIN suggested that one could think of aether as the totality of those physical quantities which are to be associated with matter-free space. In this wider sense, one could say that there does exist an aether; of course, it is conceptually very different from the one conceived by EINSTEIN's predecessors.

It is interesting to view this suggestion of EINSTEIN in the light of the quantum field theory, where one obtains physically observable effects arising from the polarisation of the vacuum; see, for instance, S. S. SCHWEBER, *An Introduction to the Relativistic Quantum Field Theory* (Row, Petterson and Company, Illinois, 1961), Sec. 15e.

[2] It seems advisable that further discussion on this aspect of the theory be reserved for the main text.

[3] H. MINKOWSKI: (I) 'Das Relativitätsprinzip' a lecture delivered in Göttingen on November 5, 1907; published posthumously in *Jber. Deutsch Math. Ver.* 24, 372 (1915).

(II) 'Die Grundgleichungen für die elektromagnetischen Vorgange in bewegten Körpern', *Nachr. Ges. Wiss.*, Göttingen (1908), p. 53, and *Math. Ann.* 68, 472 (1910).

(III) 'Raum und Zeit', a lecture delivered at the 80th assembly of German Natural Scientists and Physicians, at Cologne, on September 21, 1908 and published in *Phys. ZS.* 10, 104 (1909); English translation entitled 'Space and Time' available in *The Principle of Relativity*, *loc. cit.*

These works will be referred to as I, II and III, respectively.

was a reformulation to the theory in terms of a four-dimensional space-time continuum, which prompted the use of tensors in this theory.[1] This amplification not only resulted in making the structure of the various formulae more compact and elegant but also provided a tool for ascertaining the covariance of the basic equations describing physical laws. The latter aspect of the tensor formalism played a pivotal role in the development of the general theory of relativity to which we now turn.

It was first pointed out by POINCARE in 1906 that NEWTON's law of gravitation, which requires an *instantaneous* action at a distance, was incompatible with the special theory of relativity, according to which one could not admit signals propagating faster than light.[2] He suggested that the Newtonian theory be modified, so that (i) the gravitational influence from one body to another propagated with a speed equal to that of light and (ii) the law of gravitation itself assumed a form which was LORENTZ-invariant. His ideas were further developed by MINKOWSKI and others, as a result of which some light was thrown on the astronomical significance of introducing these modifications into the Newtonian theory.[3]

It is, however, obvious that once we realize the *finiteness* of the speed of propagation of a certain physical influence, a logical theory can be developed only in terms of a set of continuously varying functions of space and time. The foremost consideration then should be given to the *field* of that influence (and to the *differential equations* governing that field) rather than to the *law of force* operating therein; the latter would actually be one of the many conclusions obtainable by carrying out the integration of the field equations under appropriate boundary conditions. In the present context, the modifications of the Newtonian theory of gravitation could be regarded as physically complete only when one discovered a covariant substitute for the classical POISSON equation of the gravitational field. This is precisely what EINSTEIN

---

[1] MINKOWSKI had been, to some extent, anticipated by POINCARE who, in 1906, made use of the imaginary coordinate $x_4 = ict$ and interpreted as point coordinates, in a four-dimensional continuum, the quantities which we now call the 'components of a four-vector'; the invariant interval ds also plays a role in his considerations. See his paper of 1906, *loc. cit.*

It appears of interest to mention here that in their monumental four-volume work on *The Theory of the Top*, 1897-1910, FELIX KLEIN (1849-1925) and ARNOLD SOMMERFELD (1868-1951) made use of a four-dimensional non-Euclidean space in which time figured as the fourth dimension; however, this space was introduced solely for mathematical convenience and no physical significance was intended.

[2] H. POINCARE, *loc. cit.*

[3] H. MINKOWSKI, II, *loc. cit.*; W. DE SITTER, *Mon. Not. Roy. Astr. Soc.* **71**, 388 (1911).

achieved through his monumental work carried out during the period 1907-1916.

EINSTEIN's starting point in this direction was his determination to extend the principle of relativity to systems of reference in *arbitrary* motion.[1] He postulated that physical laws should be covariant with respect to *all* systems of reference; to this postulate he gave the name *The General Principle of Relativity*, in contrast to *The Special Principle* that applied only to inertial systems of reference.

In the process of examining the physical significance of the foregoing extension, EINSTEIN invoked the so-called *Principle of Equivalence*, according to which a homogeneous gravitational field is regarded as physically *equivalent* to a field 'arising' in a suitably accelerated frame of reference.[2] This innovation was based on a remarkable property of the gravitational fields, first discovered by GALILEO, viz. that in such a field *all bodies fall alike*— a property which holds for the field 'generated' in a non-inertial system as well. At a deeper level it rested on the equality of the 'inertial' mass and the 'gravitational' mass of a body—in line with the celebrated concepts of ERNST MACH (1858-1916); this result had been established experimentally by ROLAND VON EÖTVÖS[3] (1848-1919) and was emphasized *in time* by MAX PLANCK[4] (1858-1947).

The practical importance of the principle of equivalence lies in the fact that it enables us to apply the results obtained by calculating the development of physical events in an accelerated system of reference to phenomena taking place in a homogeneous gravitational field. In this way, EINSTEIN showed, in 1911, that a ray of light passing close to a massive body, such as the Sun, would get deflected from its rectilinear path; moreover, the frequency of a spectral line would depend on the gravitational field and so must the speed of light.

The next advance came with a paper published by HARRY BATEMAN[5] (1882-1946) who showed that the introduction of a gravitational field (that

---

[1] A. EINSTEIN, *Jahrb. für Radioakt. und Elekt.* **4**, 411 (December 4, 1907).

[2] A. EINSTEIN, *Ann. der Phys.* **35**, 898 (1911); English translation available in *The Principle of Relativity, loc. cit.* For details, see Sec. 7.1 of the text.

[3] R. VON EÖTVÖS, *Math. und Naturw. Ber. aus Ungarn* **8**, 65 (1890); *Ann. der Phys.* **59**, 354 (1896). Also R. VON EÖTVÖS, D. PEKAR and E. FEKETE, *Ann. der Phys.* **68**, 11 (1922).

[4] M. PLANCK, *Berlin Sitzungsberichte* (June 13, 1907), p. 542. It was clearly pointed out by PLANCK that 'since all energy has inertial properties, so *it must gravitate*'.

[5] H. BATEMAN, *Proc. London Math. Soc.* (2), **8**, 223(1910); see also *Am. J. Math.* **34**, 325(1912).

is, passage from an inertial system of reference to a non-inertial one) implied a transition from the (quasi-) Euclidean space-time of the special relativity to a Riemannian manifold, characteristically endowed with a 'curvature' in the space-time continuum. BATEMAN noted a close connection of his work with the absolute differential calculus[1] developed by GREGORIO RICCI (1853-1925), in collaboration with his pupil T. LEVI-CIVITA, and recognized that the ten quantities g_{ik}, which were characteristic of the Riemannian metric (or of the gravitational field), constituted a symmetric covariant tensor of rank 2.

These ideas were subjected to a more profound treatment by EINSTEIN[2] who, working in collaboration with MARCEL GROSSMANN during the years 1912-14, introduced the striking prescription that *the path of a freely moving particle in a gravitational field would be a 'geodesic' in the four-dimensional curved space-time (with a metric of the Riemannian type)*; this prescription may be regarded as an adaptation of the famous principle of PIERRE DE FERMAT (1601-1665) that *Nature always acts by the 'shortest' course*.

Accordingly, the Newtonian single-parameter theory of gravitation got replaced by a ten-parameter theory of EINSTEIN; however, the task of discovering field equations, the covariant counterpart of the pre-relativistic POISSON equation, which, in accordance with MACH's principle (as adopted by EINSTEIN), would determine the g_{ik} in terms of the mass (or mass-energy) distribution in the continuum, yet remained to be done. This was accomplished by EINSTEIN in a series of papers[3] published in 1915 in which, drawing heavily upon the elements of the absolute differential calculus (or tensor[4] calculus, as EINSTEIN henceforth called it), he set up the desired, covariant field equations for the g_{ik} which met all the basic physical requirements. It was in these papers that EINSTEIN calculated, *by the method of successive approximations*, (i) the advance of the perihelion of the planet Mercury and (ii) the bending of light rays in the gravitational field of the Sun; the amount of bending now turned out to be exactly double the value derived previously, in 1911, on the basis of the principle of equivalence alone. Soon after, his concluding paper[5] entitled 'Die Grundlagen der allgemeinen Relativitäts-

---

[1] G. RICCI and T. LEVI-CIVITA, *Math. Ann.* **54**, 125(1900).

[2] A. EINSTEIN and M. GROSSMANN, *Zeits. für Math. und Phys.* **62**, 225 (1913); **63**, 215(1914) ; also A. EINSTEIN, *Phys. ZS.* **14**, 1249(1913).

[3] A. EINSTEIN, *Berlin Sitzungsberichte* (1915), pp. 778, 799, 831, 844.

[4] The word *tensor* had been used earlier by VOIGT, in 1898, in connection with the elasticity of crystals.

[5] A. EINSTEIN, *Ann. der Phys.* **49**, 769(1916); English translation entitled 'The Foundations of the General Theory of Relativity' is available in *The Principle of Relativity, loc. cit.*

theorie' appeared, which represented the culmination of one of the most brilliant adventures in theoretical physics—*the geometrization of gravitation*.

In 1916, KARL SCHWARZSCHILD (1873-1916) made an important advance in the theory of gravitation by discovering an *exact* solution of the field equations for the space-time occupied by a single material particle.[1] In the years following 1916, several other particular solutions of these equations were obtained, for an account of which reference may be made to other sources.[2]

In 1917, EINSTEIN[3] modified his field equations by introducing a new term— the so-called *cosmological* term—in the hope of making these equations compatible with MACH's principle; for details, see Chapter 9 of the text. The modified equations were applied to the universe as a whole in order to see how far the theory of relativity could account for the type of universe we actually have. Different assumptions about the sign and magnitude of the cosmological term, the sign of the overall curvature of the universe and the nature of the universe as regards its being static or non-static, lead to a variety of cosmological models;[4] however, only some of these conform reasonably to the actual situation.

The great achievement of EINSTEIN, which was in *formal* accord with an idea expressed in 1894 by FITZGERALD that 'gravity is probably due to a change of structure of the aether produced by the presence of matter', was bound to urge physicists and mathematicians alike to strive to evolve a still more general theory—the so-called *unified field theory*—which could possibly account for all fundamental fields in terms of the properties of the space-time continuum. This has been a challenging problem throughout the intervening years and still remains an open question. The difficulties that creep up in the course of such attempts arise mainly because of the revolutionized nature of the fields characterizing elementary particles. With the present

---

[1] K. SCHWARZSCHILD, *Berlin Sitzungsberichte* (1916), p. 189.

[2] See, for instance, R. C. TOLMAN, *Relativity, Thermodynamics and Cosmology* (Clarendon Press, Oxford, 1934), Chap. VII, Pt. II; see also references to a number of original sources in E. T. WHITTAKER, *loc. cit.*, Vol. II, pp. 175-183.

[3] A. EINSTEIN, *Berlin Sitzungsberichte* (1917), p. 142; English translation entitled 'Cosmological Considerations on the General Theory of Relativity' is available in *The Principle of Relativity, loc. cit.*

[4] For a detailed account of these, see H. BONDI, *Cosmology* (University Press, Cambridge, 1952); G. C. McVITTIE, *General Relativity and Cosmology* (Chapman and Hall Ltd., London, 1956) ; H. P. ROBERTSON and T. W. NOONAN, *Relativity and Cosmology* (W. B. Saunders, Philadelphia, 1968); D. W. SCIAMA, *Modern Cosmology* (University Press, Cambridge, 1971).

state of knowledge, it is too early to say as to when and how the desired unification is going to be brought about.[1]

Among the most modern problems in general relativity,[2] a specific mention may be made of the *quantization* of the gravitational field; this, too, is not yet completely solved and is engaging the attention of several groups of workers all over the world.[3]

[1] For a review of the various attempts in this direction (till 1955), see W. PAULI, *Theory of Relativity* (Pergamon Press, London, 1958), supplementary note No. 23, pp. 224-232. A detailed account of this subject, with a wealth of references to earlier works, is given by C. W. MISNER and J. A. WHEELER, *Annals of Physics* 2, 525(1957); J. A. WHEELER, *ibid.* 2, 604(1957).

[2] For information on these, refer to the Proceedings of the 1957 Conference on the Role of Gravitation in Physics, published in the *Revs. Mod. Phys.* 29, 255-546 (1957). See also H. Y. CHIU and W. F. HOFFMANN, *Gravitation and Relativity* (W. A. Benjamin, New York, 1964); B. K. HARRISON, K. S. THORNE, M. WAKANO and J. A. WHEELER, *Gravitation Theory and Gravitational Collapse* (University of Chicago Press, 1965); M. CARMELI, S. I. FICKLER and L. WITTEN. *Relativity* (Plenum Press, New York, 1969); C. G. KUPER and A. PERES, *Relativity and Gravitation* (Gordon and Breach, New York, 1971).

[3] For a review of this problem, see P. G. BERGMANN and A. B. KOMAR in *Recent Developments in General Relativity* (Pergamon Press, London, 1962), pp. 31-46; S. MANDELSTAM, *Proc. Roy. Soc. London* A 270, 346(1962); also several articles in CARMELI, FICKLER and WITTEN. *loc. cit.*

PART I

THE SPECIAL THEORY

CHAPTER 1

SPACE-TIME TRANSFORMATIONS

1.1. Reference systems and the physical world

In our study of the physical world we encounter a multitude of phenomena which have to be, first of all, properly specified and then subjected to a systematic analysis aimed at discovering fundamental relationships among them. For carrying out the first part of this project unambiguously, each of these phenomena has to be specified by expressing numerical measures of the various parameters involved; for this purpose, it is necessary to refer our measures to a certain *system of reference* adopted for the investigation concerned. The basic equipment for any system of reference consists, so far as kinematical investigations are concerned, of an arrangement to measure the three space coordinates of the point at which an event takes place (this would, in general, involve a measurement of distances and angles) and the time of its happening.

It is obvious that for any physical observation the choice of a system of reference is quite unlimited. However, an observer might decide to choose one particular system of reference in preference to another if he feels that the description of the physical phenomenon under study would be simpler in *that* system than in the other one. For instance, in most of the terrestrial experiments the system of reference adopted is the one at rest relative to the Earth—the so-called *laboratory system*—while in hydrodynamical studies one often employs a system with respect to which the fluid is at rest—the so-called *rest system* of the fluid. Further, in the realm of atomic physics it is customary to adopt a system of reference in which the centre of mass of the group of particles under study is at rest—the so-called *centre-of-mass system—*, while in some of the astronomical work the choice of a system attached to the Sun, the centre of the solar system, may appear more natural.

Further, even in a given system of reference one is free to adopt any of the several systems of coordinates; the choice, of course, will be governed by the requirement that the resulting mathematical relationships are not unduly involved; this can generally be anticipated by noting the types of symmetry involved in the phenomenon under study. Actually, there are about a dozen systems of coordinates in use,[1] the most popular ones being the Cartesian and the spherical polar. For most fundamental situations in physical space, the Cartesian system of coordinates is found to be the simplest. According to this, the position of any event in space is described by specifying three numbers, x, y and z, which denote perpendicular distances of the site of the event from three mutually perpendicular planes, the x-plane, the y-plane and the z-plane, respectively; the lines of intersection of these planes, taken in pairs, constitute the *axes* of the system (x-axis being the one normal to the x-plane, etc.), while the point where the axes intersect defines the *origin* ($x=y=z=0$) of the system. Throughout our considerations here, we shall employ the Cartesian system of coordinates, unless specified otherwise.

Going back to the question of choice of a system of reference, we pose the following question: 'Granted that from the point of view of simplicity and convenience of description one system of reference may be preferable to others, can this be taken to mean that the fundamental laws governing physical phenomena would assume an especially simple form in any particular class of systems, while in others they would appear as complicated mathematical structures'? For studying this important question, which forms the very backbone of the theory of relativity, let us first confine ourselves to the field of mechanics, primarily because this field has always been regarded as the best founded discipline in physics and also because the question we have posed here was given serious consideration by the founders of this discipline, viz. GALILEO and NEWTON.

As is well known, the fundamental laws of mechanics are embodied in the famous *trio* of NEWTON, the first one of which is the so-called *law of inertia*. According to this law, a material body continues to be in its state of rest or of uniform rectilinear motion unless it is acted upon by an external force. It is important to note that the law holds only in a restricted class of systems of reference, which, by definition, are referred to as the *inertial*

---

[1] For details of these, see H. MARGENAU and G. M. MURPHY, *The Mathematics of Physics and Chemistry* (D. Van Nostrand Company, New York, 1943), Chap. 5.

systems of reference.[1]   Once this particular class of systems is recognized, it is easy to demonstrate that the whole scheme of Newtonian mechanics applies *with equal validity* to each member of this class.

It appears worthwhile to discuss briefly the question of recognizing the inertial systems of reference.   For NEWTON, and for most of his successors, the solution lay firstly in the identification of an *absolute* system of reference (the absolute space, later identified with the aether), in which the laws of mechanics are supposed to be valid *as such,* and secondly in the recognition that the same would be true for any other system of reference which, with respect to the absolute one, is either at rest or moving with an unaccelerated motion.   Whereas the second part of the statement presents no serious difficulty, the first part is quite intriguing.   One suggestion, based on experience and plausibility, is to regard the system of 'fixed' stars as an alternative to the absolute system.   From a practical point of view, this suggestion can indeed be recommended for adoption.[2]   From a theoretical point of view, however, it appears more cogent to define an inertial system of reference as one for which *space is homogeneous and isotropic* and *time is homogeneous.*[3]   A little consideration will show that a uniform relative motion of a system will not impair the foregoing properties of space and time.   Moreover, as pointed out in Chapter 3, the laws of conservation of momentum and energy for a closed system of bodies are direct consequences of these fundamental properties of space and time; consequently, these laws hold equally well in all inertial systems of reference.

From the foregoing it follows that all inertial systems of reference are *completely equivalent,* at least from the point of view of mechanics.   Hence, in the formulation of mechanical laws there is no reason whatsoever to prefer one inertial system to another.   This constitutes *the principle of relativity* of Newtonian mechanics or, as one normally puts it, the *Galilean* principle of relativity.

[1] EINSTEIN used for these systems the term '*Galilean* systems of reference'.

[2] For an excellent discussion on the identification of a system of reference in which distant stars are in unaccelerated motion with the inertial system of NEWTON's theory, and on the cosmological evidence for this identification, see H. BONDI, *Cosmology* (University Press, Cambridge, 1952), Chap. IV; see also F. HOYLE and J. V. NARLIKAR, *Proc. Roy. Soc. London* A. **270**, 334 (1962), especially the ensuing remarks by P. A. M. DIRAC and H. BONDI.

[3] L. D. LANDAU and E. M. LIFSHITZ, *Mechanics* (Pergamon Press, Oxford, 1960), Sec. 3.   This view is fully consistent with Machian concepts; see, in this connection, H. BONDI, *Rep. Prog. Phys.* **22**, 97 (1959), Sec. 5.5.

At this stage it appears quite natural to ask: 'Why should the validity of the principle of relativity be restricted to mechanics alone? Should it not be extended to other domains of physics as well ?' From a purely aesthetic point of view, the answer to this question must be in the affirmative. From a matter-of-fact point of view too, it is not difficult to visualize that curious asymmetries would arise in physical phenomena if one inertial system were regarded as a privileged one in that physical laws, as ordinarily formulated, hold only in that system and not in others (or if one of the two given inertial observers is regarded on a different footing from the other so far as the formulation of fundamental laws is concerned). Actually, the opening paragraph of EINSTEIN's first paper on the subject is devoted exactly to pointing out such an asymmetry in the realm of electromagnetism.[1] One is, therefore, inclined to demand that the principle of relativity be valid *quite generally* and, hence, the form of *any* given physical law be the same in all inertial systems of reference.

In carrying out the foregoing extension of the principle of relativity, our first concern would be electrodynamics, as developed by MAXWELL. One of the well known phenomena in electrodynamics is the 'propagation of electromagnetic waves' which is intimately connected with the 'propagation of light'. In view of the role it plays in bringing electromagnetism and optics under the same roof, we start our considerations of relativity with this particular phenomenon.

1.2. Propagation of light in free space

One of the most outstanding conclusions of MAXWELL's theory of electromagnetism was the propagation of electromagnetic waves in accordance with the ordinary laws of wave motion. In free space, this propagation takes place with a fixed velocity c, which is independent of the direction of propagation of the waves. Originally, the quantity c had entered into the theory as a ratio between the measures of various physical quantities on the electrostatic system of units and the electromagnetic system of units; its numerical magnitude, obtained by WEBER and KOHLRAUSCH[2] by comparing the measures of the charge of a Leyden jar determined both electrostatically and electro-

[1] A. EINSTEIN, *Ann. der Phys.* **17**, 891 (1905); English translation is available in *The Principle of Relativity, loc. cit.*

[2] W. WEBER and R. KOHLRAUSCH, *Ann. der Phys.* **99**, 10 (1856). WEBER was really following up the work on absolute measurements begun by himself and GAUSS in connection with terrestrial magnetism.

magnetically, turned out to be nearly 3.1×10^{10} cm/sec which, within the limits of the experimental error, was the same as the speed of light obtained a few years earlier by FIZEAU.[1] This close agreement between the two results impressed MAXWELL strongly. He, therefore, did not hesitate to identify the two phenomena and assert :

"We can scarcely avoid the inference that light consists in the transverse undulations of the same medium which is the cause of electric and magnetic phenomena."

By now this identification has taken the strongest possible roots, so that the propagation of electromagnetic waves and the propagation of light are now regarded as synonymous.

Now the question arises as to which system of reference we have in mind when we say that 'light propagates through free space with a velocity c, *independent* of the direction of propagation'. Obviously, this cannot be true for *all* inertial systems of reference together, because the (Newtonian) law of addition of velocities has to come in somewhere; this will not only affect the magnitude of the velocity of propagation but will also make it direction-dependent. We should, therefore, be able to pick out one particular inertial system out of the whole set and say, 'Well, this is *the* inertial system we had in mind when we made the foregoing statement about the propagation of light'. This system, and all others *at rest* with respect to this one, will then form a *privileged* subset of inertial systems such that in these, *and only in these*, the propagation of light would display the afore-mentioned simplicity. At the same time, it is only for the members of this subset that MAXWELL's field equations would hold *as such*.

The foregoing situation is, however, quite the opposite of the principle of relativity which we had set out to extend to the whole of physics. The dilemma, of course, arises from the fact that if we require MAXWELL's field equations to hold *as such* for the whole set of inertial systems, then all the conclusions of these equations must also do so; consequently, light would propagate in free space with a *fixed* velocity c, *independent of the direction of propagation*, in respect of *all* the inertial systems of reference. But these systems are in motion with respect to one another. How can we understand this peculiar situation in terms of the ordinary concepts of space and time ? On the other hand, if we feel obliged towards a generalization of the principle of relativity we cannot shirk from the foregoing type of propagation of light; this, of course, will necessitate a revision of our age-old concepts of

---

[1] H. FIZEAU, *Compt. Rend.* **29**, 90 (1849). The present recommended value of this constant is $(2.997925 \pm 0.000004) \times 10^{10}$ cm/sec.

space and time in such a way that the Newtonian law of addition of velocities gets replaced by a new one which leaves c unchanged on composition with any other velocity. Clearly, in that case, the formalism of mechanics will have to be drastically revised!

Though the pressure of physical facts has forced us to adopt this line of approach, it did not come about so directly. Lots of considerations were given to find out *the* medium with respect to which light propagated with equal speed in all directions; some of these considerations deserve a brief mention.[1]

(i) First of all we have the 'emission hypothesis' of RITZ, according to which it is with respect to the *source,* or a system attached to the source, that the velocity of light is isotropically equal to c. However, COMSTOCK[2] pointed out the case of double stars which causes serious trouble to the RITZ hypothesis; related problems were discussed quantitatively by DE SITTER.[3]

Since the two components of a double-star system are revolving around their centre of mass, their velocities with respect to a terrestrial observer are, at each instant, different. The velocity of light emitted by a star being equal to c with respect to *that* star, the velocity of light waves, starting simultaneously from the two stars, would be affected differently; consequently, the two wave trains would arrive at the Earth at different instants of time. In other words, at any *given* instant of time the terrestrial observer views the two stars at positions corresponding to two different (earlier) instants. And the difference between the two instants would itself be a periodic function of time. The net result would, then, be that the over-all picture of the binary system would appear a lot distorted. Sometimes we may observe the same component simultaneously at more than one place, i.e. 'ghost stars' may appear and disappear in the course of motion. However, such an effect has never been observed; consequently, the emission hypothesis is ruled out.

It was shown by DE SITTER that this hypothesis would also have the effect of introducing an apparent eccentricity into the otherwise (practically) circular orbits of a double-star system. Such an effect is also not observed; in fact, DE SITTER concluded that if one assumes that

$$v_{\text{light}} = c + \alpha v_{\text{star}} \,,$$

---

[1] For more details, see P. G. BERGMANN, *Introduction to the Theory of Relativity* (Prentice-Hall, Inc., New York, 1942), Chap. III.

[2] D. F. COMSTOCK, *Phys. Rev.* **30**, 267(1910).

[3] W. DE SITTER, *Proc. Acad. Sci.*, Amsterdam **15**, 1297(1913); **16**, 395(1913).

then α, instead of being unity, is no more than 2×10^{-3}.

More recently, experiments have been performed on rapidly moving terrestrial sources which support astronomical evidence against the RITZ hypothesis. In one such experiment with γ-ray photons, the source consisted of unstable particles (π°-mesons) travelling at 99.975% of the speed of the light.[1] The measured speed of the photons emitted in the forward direction was $(2.9977 \pm 0.0004) \times 10^{10}$ cm/sec, which is in excellent agreement with the value of c obtained for stationary sources. The corresponding value of α is less than 10^{-4}.

(ii) Next we have another hypothesis, according to which it is with respect to the *medium* of propagation that the velocity of light is isotropically equal to c. To assess this hypothesis, we consider two different media, side by side, in relative motion. Then, the velocity of light, as it passes from one medium into the other, would suffer a change of direction not only because of the difference in the refractive indices of the media but also because of the relative motion between them. Now, if we were to go over to the limit of zero density for both the media, the first cause would tend to vanish (because both refractive indices would tend to unity) while the second cause would persist. Finally, we would have two *vacua* in relative motion, leading to refraction of light. This is clearly absurd.

In this connection, we also have important experimental evidence on record. In 1851, FIZEAU determined quantitatively the effect of motion of the medium of propagation on the speed of light.[2] On the basis of the hypothesis under consideration, the speed of light in the laboratory system, c', would be given by

$$c' = \frac{c}{n} \pm v, \tag{1.1}$$

where c/n is the speed of light in the medium itself (n being the refractive index) while v is the speed of the medium in the laboratory; we assume that the two motions are either parallel or antiparallel. The actual result, however, was (see also Sec. 1.8)

$$c' = \frac{c}{n} \pm v \left[1 - \frac{1}{n^2} \right]. \tag{1.2}$$

---

[1] T. ALVÄGER, F. J. M. FARLEY, J. KJELLMAN and I. WALLIN, *Phys. Letters* **12**, 260(1964); for a fuller account by the same authors, with J. M. BAILEY, see *Arkiv für Fysik* **31**, 145(1966).

[2] H. FIZEAU, *Compt. Rend.* **33**, 349(1851).

It should be noted that this result is free from the objection levelled in the preceding paragraph for, in the limit of zero density ($n \to 1$), it does yield the desired result, $c' = c$, irrespective of the state of motion of the 'evacuated' medium.

Yet another phenomenon which invalidates this hypothesis is the phenomenon of *aberration*. This was first discovered by BRADLEY who, in 1727, observed that the distant 'fixed' stars execute elliptical paths in the sky, the period of the path being one year and the major axis of the ellipse being *in each case* 41″ of arc. The explanation of this phenomenon, given correctly by BRADLEY himself, is quite straightforward, viz. that it arises due to the orbital motion of the Earth around the Sun, which necessitates a continual adjustment of the direction of the telescope in order to keep it pointed towards a particular star. The maximum range of this adjustment would be given by

$$2\theta = 2\tan^{-1}\frac{v_{\text{Earth}}}{c} \simeq 2 \times 10^{-4} \text{ radian}, \tag{1.3}$$

in agreement with the value actually observed.

Now, the foregoing explanation assumes that the speed of light is equal to c, irrespective of the changes in the velocity of the Earth in its orbit. On the other hand, according to the hypothesis under study, the medium of propagation being the decisive factor, light rays, on entering the Earth's atmosphere, should be 'swept along' with the atmosphere (analogously to the phenomenon of *refraction* mentioned above) and hence no aberration should result.

Thus, we find that this hypothesis is also untenable.

(iii) Finally, we are left with the most crucial hypothesis, namely that there exists an 'absolute' reference system, say the aether, which is the *real* privileged system being looked for. In order to see how far this hypothesis may be true, we discuss in the following section some of the important experiments which have been conducted with a view to deciding whether such a system *really* exists. It is on the results of these experiments that our further considerations of the problem will have to be based.

1.3.   Search for the aether

We start our discussion with the famous series of experiments, performed by MICHELSON and MORLEY,[1] which were designed to 'observe' the motion

[1] A. A. MICHELSON, *Am. J. Sci.* (3), **22**, 20(1881); A. A. MICHELSON and E. W. MORLEY, *ibid.* **34**, 333(1887). For an appraisal of the situation existing prior to the experiments of MICHELSON and MORLEY, see E. T. WHITTAKER, *loc. cit.*, Vol. I, Chap. XIII.

of the Earth with respect to the privileged medium—the aether. If the velocity of light is supposed to be equal to c only in this particular medium, light would (be expected to) propagate with different speeds in different directions, as viewed in the system of the Earth. It is this difference of speeds that the MICHELSON-MORLEY experiments were intended to detect.

The optical arrangement of a typical experiment of this type is shown in Fig. 1.1. Light from a terrestrial source S is separated into two mutually perpendicular beams by a half-silvered plate P. These beams are reflected from two mirrors M_1 and M_2 which are placed normal to their paths, at almost equal distances from the plate P. The reflected beams reunite and are observed in the field of view of the telescope T. The path difference between the two beams arises solely from their transit across the distances l_1 and l_2, respectively; the fringe pattern resulting from this arrangement, therefore, depends upon the relative magnitudes of l_1 and l_2.

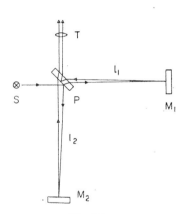

Fig. 1.1.

However, in view of the hypothesis under test, the speed of light would be different in different directions (unless the Earth, at the time of the experiment, happens to be at rest with respect to the aether, which we do not presently assume); consequently, the fringe pattern will be influenced by this difference as well. Let us suppose that the Earth, and with it the whole experimental set up, is moving with a velocity v with respect to the aether in the direction $P{\rightarrow}M_1$. Then, by the Newtonian law of addition of velocities, light would propagate with velocity $(c - v)$ in the forward direction $P{\rightarrow}M_1$, $(c + v)$ in the backward direction $M_1 \rightarrow P$ and $(c^2 - v^2)^{1/2}$ in the perpendicular direction $P \rightarrow M_2$ or $M_2 \rightarrow P$. The times t_1 and t_2 taken by the two beams for their return journeys between the plate and the respective mirrors would then be

$$t_1 = \frac{l_1}{c-v} + \frac{l_1}{c+v} = \frac{2l_1/c}{(1-v^2/c^2)} \, , \tag{1.4}$$

and

$$t_2 = 2 \cdot \frac{l_2}{\sqrt{c^2-v^2}} = \frac{2l_2/c}{\sqrt{1-v^2/c^2}} \, , \tag{1.5}$$

respectively; the difference $(t_1 - t_2)$, which is *not* proportional to $(l_1 - l_2)$, will govern the fringe pattern.

Now, let the whole assembly be given a 90° rotation in its own plane. The roles of the two beams would thereby be reversed. For the resulting situation, we would have

$$t_1' = \frac{2l_1/c}{\sqrt{1-v^2/c^2}} \quad \text{and} \quad t_2' = \frac{2l_2/c}{(1-v^2/c^2)} \, ; \tag{1.6}$$

the difference $(t_1' - t_2')$ would govern the new fringe pattern. The change in the time difference is given by

$$\delta(\Delta t) = (t_1' - t_2') - (t_1 - t_2)$$

$$= \frac{2(l_1+l_2)/c}{\sqrt{1-v^2/c^2}} \left[1 - \frac{1}{\sqrt{1-v^2/c^2}} \right], \tag{1.7}$$

which does not depend explicitly on the path difference $(l_1 - l_2)$; rather it exists by virtue of the *nonzero* value of v. From (1.7), we obtain for the fringe shift in the pattern

$$\delta n = |\, \delta(\Delta t)\,| \, \frac{c}{\lambda}$$

$$= \frac{l_1+l_2}{\lambda} \left[\frac{v^2}{c^2} \right] \quad \text{for} \quad (v^2/c^2) \ll 1 \, ; \tag{1.8}$$

here, λ is the wavelength of the light.

For the orbital motion of the Earth, $(v/c) \simeq 10^{-4}$. Hence, we can write (1.8) as

$$\delta n \simeq \frac{(l_1+l_2) \text{ in cm}}{\lambda \text{ in } \mathring{A}} \, . \tag{1.9}$$

Now, in the actual experiment the factor $(l_1 + l_2)$, by means of multiple reflections, could be made several metres, say 20 m. Since the value of λ for light ordinarily used is about 6000 Å, the expected shift would be about one-third of a fringe. The design of the experiment was such that it could detect a shift of even one-hundredth of a fringe. However, *no effect was observed*!

It is, of course, likely that at the time of the experiment the Earth was at rest with respect to the aether. In that case, obviously, no effect could be expected. To check against this possibility MICHELSON and MORLEY, in the second round of their experiment, performed it twice, at an interval of six months. Now, if the afore-mentioned situation held at the time of one of the performances, it could not hold for the other. Hence, the effect should have been observed on *at least* one of the two occasions. Actually, the experiment gave 'null' result on both the occasions!

MICHELSON tried to explain the absence of the effect by assuming that the aether was being constantly dragged along with the Earth in its orbital motion. In that case, there would be no aether wind at the surface of the Earth; however, it could possibly exist at high altitudes. He, therefore, repeated his experiment on a high mountain, but still without any detectable effect.[1]

A different suggestion for the explanation of the null result of the MICHELSON-MORLEY experiments, which drew considerable attention, was made by FITZGERALD[2] in 1892. He proposed that the dimensions of all material bodies get altered when they are in motion with respect to the aether. He postulated that dimensions in the direction of motion get contracted by a factor of $\sqrt{1 - v^2/c^2}$ while those perpendicular to it remain unaffected; a sphere, for instance, would become an oblate spheroid. As can be seen from Eqs. (1.4) and (1.5), the amount of contraction, postulated by FITZGERALD, is just sufficient to nullify the effect of the relative motion of the Earth and the aether.[3] Consequently, the MICHELSON-MORLEY experiment would not be expected to show any fringe shift.

A few months later, LORENTZ[4] adopted the same hypothesis whereupon it began winning favor in a gradually widening circle, until eventually it came

---

[1] A. A. MICHELSON, *Am. J. Sci.* (4), 3, 475 (1897).

[2] See O. LODGE, *Nature* 46, 165 (1892) where the author mentions the relevant communication from FITZGERALD to him.

[3] This is, however, true only if $l_1 = l_2$. For unequal arms of the interferometer, a fringe shift would be expected *in spite of* the contraction hypothesis. See, in this connection, the discussion of the KENNEDY-THORNDIKE experiment in Sec. 1.7.

[4] H. A. LORENTZ, *Amst. Verh. Akad. van Wetenschappen* 1, 74 (1892).

to be regarded as the *basis* of all theoretical investigations concerning motion of ponderable bodies through the aether.

Although the LORENTZ-FITZGERALD hypothesis explains the null result of MICHELSON-MORLEY experiments reasonably well, its *ad hoc* nature and hypothetical character prevent it from being convincing enough. In fact, we shall find that in the theory of relativity there is no place at all for such a hypothesis; the explanation of the null results of MICHELSON-MORLEY experiments is actually contained in the postulate on the 'constancy of the speed of light for *all* inertial observers'.

A number of investigators have elaborated the pioneering work of MICHELSON and MORLEY; in consequence, there is a general agreement now that motion with respect to aether, if it exists at all, has always been *indedectable*.[1] A brief reference to some of these investigations may be made here.

First of all we notice that the null result of the MICHELSON-MORLEY experiments is quite compatible with the hypothesis of RITZ, according to which light propagates with velocity *c* in *all* directions with respect to its source. Now, in the experimental set up discussed here, every part of the apparatus is at rest with respect to the (terrestrial) source employed (the whole apparatus being at rest with respect to the Earth); consequently, both beams would have a common velocity *c*, and, hence, on giving rotation to the assembly no fringe shift would occur. One would, however, expect that if light came from an extra-terrestrial source (the Sun or the stars) and were subsequently subjected to the MICHELSON-MORLEY experiment, a positive result might be obtained. However, if the result once again happened to be null, then both the aether hypothesis and the RITZ hypothesis would fall to the ground.

Such an experiment has indeed been performed,[2] and again the result has turned out to be *null*.

Another confirmation of the foregoing conclusions has come from an

---

[1] This conclusion has been contested by D. C. MILLER, *Revs. Mod. Phys.* 5, 203 (1933), on the basis of a statistical study of the results obtained in his experiments over a long period of years. However, his considerations have not been accepted generally as valid; refer to the excellent critique by R. S. SHANKLAND, S. W. McCUSKEY, F. C. LEONE and G. KUERTI, *Revs. Mod. Phys.* 27, 167 (1955). For an illustrative physical assessment of the various experimental results, see G. W. STROKE's article entitled 'Michelson-Morley Experiment' in *Encyclopaedic Dictionary of Physics* (Pergamon Press Ltd., Oxford, 1961), Vol. 4, p. 635.

[2] R. TOMASCHEK, *Ann. der Phys.* 73, 105 (1924).

experiment, carried out by CEDARHOLM, BLAND, HAVENS and TOWNES,[1] who employed two *maser* (microwave amplification by stimulated emission of radiation) beams in place of the customary light beams. The precision with which the frequencies of the masers could be compared was about one part in 10^{12}. The results showed that the effect, if it exists at all, is less then one-thousandth of what one would expect on the basis of the known orbital velocity of the Earth.

Before we close this section, mention must be made of another set of experiments carried out by TROUTON,[2] in collaboration with NOBLE, which were designed to study the effect of the motion of the Earth with respect to the aether on some of the basic electrical phenomena. Calculation shows[3] that the field of a charged condenser, which is in motion, should exert a turning moment on the material parts of which the system is constructed. However, experiments demonstrated that there is *no* tendency at all for a charged condenser, moving with the Earth, to turn about its axis. This again showed that the motion of a physical system does not possess an absolute character that might be observable.

1.4. Postulates of the special theory of relativity and their basic consequences

Discussions in the preceding sections have shown that the assumption of a privileged system of reference is not only a negation of our basic expectations regarding the nature of physical phenomena but is also observationally untenable. It is, therefore, tempting to suggest that this assumption is physically unsound and deserves an outright rejection. This is precisely the stand taken by EINSTEIN who summed up these conclusions in the form of two postulates, which have become basic pillars for the structure of the *special theory of relativity*.

POSTULATE I. *It is impossible to measure, or detect, the unaccelerated trans-*

---

[1] J. P. CEDARHOLM, G. F. BLAND, B. L. HAVENS and C. H. TOWNES, *Phys. Rev. Letters* 1, 342 (1958); J. P. CEDARHOLM and C. H. TOWNES, *Nature* 184, 1350 (1959). For a more recent confirmation of these results, see T. S. JASEJA, A. JAVAN, J. MURRAY and C. H. TOWNES, *Phys. Rev.* 133, A1221 (1964).

[2] F. T. TROUTON, *Trans. Roy. Dub. Soc.* 7, 379 (1902), F. T. TROUTON and H. R. NOBLE, *Phil. Trans. Roy. Soc. London* 202, 165 (1903); *Proc. Roy. Soc. London* 72, 132 (1903).

[3] See, for instance, W. K. H. PANOFSKY and M. PHILLIPS, *Classical Electricity and Magnetism* (Addison-Wesley Publishing Company, Inc., Reading, Mass., U.S.A., 1955), Sec. 14-2.

latory motion of a system through free space or through any aether-like medium which might be assumed to pervade it.

POSTULATE II.   *The speed of light in free space is the same for all inertial observers, independent of the relative velocity of the source and the observer.*

The first postulate is just another way of asserting that the laws of physics are *covariant* in respect of all inertial systems of reference. For, if this were not so it would be possible to distinguish between different systems on the basis of the *differences* in the fundamental features of the various physical phenomena; the state of motion would then become an absolute property of a system. The assertion in this postulate is, therefore, in the nature of a *principle of covariance*, as applied to the family of inertial observers.

The second postulate, on the other hand, is not easy to comprehend (though it has been forced upon us by the pressure of experimental facts). It requires the velocity of light to be constant and direction-independent for *all* inertial observers who, with respect to one another, may be in motion. This is incompatible with the customary concepts of space and time. The only way to admit this postulate into the realm of physics would be to give up the commonplace notions about space and time and adopt, in their place, certain new ones. Its primary result is the introduction of the element of 'relativity' into the measures of space and time, which we now discuss briefly.

A.   *Relativity of simultaneity*

We consider two inertial observers S and S' in uniform relative motion, such that at some instant of time the origins of their coordinate systems coincide. Let, at that very instant of time, a light signal be emitted from the site of their common origin. Then, according to the second postulate, each observer will find that the locus of the space points reached *simultaneously* by the signal, at some later instant of time, is a sphere (of appropriate radius) with *his* origin as the centre. However, their origins are no longer coincident. Therefore, if any such locus is defined by one of the observers it cannot correspond to any of the loci defined by the other. This means that the set of points reached (by the signal) *simultaneously* according to one observer is not reached *simultaneously* according to the other.

It follows that the concept of 'simultaneity' of events (in our example, the arrival of the signal at a set of space points) is not an *absolute* concept. Rather, it is a *relative* concept, dependent upon the state of motion of the observer.

B. Relativity of time measurements

As a consequence of the 'relativity of simultaneity', we find ourselves unable to define an 'absolute' universal time which would flow alike for all inertial observers, viz. the one for which $(dt/dt') \equiv 1$. This can be understood by reversing the argument, viz. if we could define an absolute universal time, then the time of occurrence of any physical event could be specified in a unique manner; this, in turn, would make simultaneity an absolute concept. This, however, is not the case. We, therefore, conclude that the measure of time must itself be a 'relative' concept, dependent upon the state of motion of the observer.

C. Relativity of length measurements

Finally, we notice that the foregoing considerations affect length measurements as well, for the length of a rod is determined by the difference of the coordinates of its ends, measured *simultaneously*. The last qualification is absolutely necessary, for otherwise it would not be a consistent definition. Now, since simultaneity is itself a relative concept, so would be the measurement of the coordinates of the two points and, hence, the distance between them. Thus, the measure of length is also a relative concept, dependent upon the state of motion of the observer.

Through these considerations we arrive at the conclusion that the concepts of 'absolute space' and 'absolute time' are no longer admissible. We are rather confronted with a novel character of space and time, viz. the relativity of their measures. Our next problem then consists in determining mathematical formulae which govern the relativistic character of the space-time measurements.

1.5. Special Lorentz transformations

In order to derive the fundamental relationships connecting *spatial* and *temporal* measurements, made by inertial observers in relative motion, we consider in detail the example touched upon in the preceding section. For simplicity, we assume that: (i) the Cartesian coordinate systems employed by the observers S and S' are such that the axes of one are parallel to the corresponding axes of the other, and (ii) their relative motion is confined in the direction of one of the axes, say the x- (or x'-) axis, the velocity of S' with respect to S being equal to v; see Fig. 1.2. We may also assume that the *x*-axis of *S* and the *x'*-axis of *S'* are not only parallel but also coincident; accordingly, their origins 0 and 0' would, at some instant of time, coincide. Let that instant of time be taken as the origin of the time measurements, *t* and *t'*, of the two observers, with the result that

$$x = y = z = t = 0 \qquad (1.10)$$

and

$$x' = y' = z' = t' = 0 \qquad (1.11)$$

hold together. At this very instant of time and from this very space point, let a light signal be emitted. For each observer, this signal will develop into a *spherical* wave-front, of ever-increasing radius, with his *own* origin as the centre. Suppose that, according to the observer *S*, the wave-front passes through a given point *P(x, y, z)* at time *t*. Then, we must have

$$x^2 + y^2 + z^2 - (ct)^2 = 0. \qquad (1.12)$$

Now, if the coordinates of the point *P*, according to the observer *S'*, are *x'*, *y'* and *z'* and the wave-front reaches there at time *t'*, then we must also have

$$x'^2 + y'^2 + z'^2 - (ct')^2 = 0. \qquad (1.13)$$

Our problem, then, consists in determining the transformation formulae connecting *x'*, *y'*, *z'* and *t'* with *x*, *y*, *z* and *t*, such that Eqs. (1.12) and (1.13) are satisfied together.

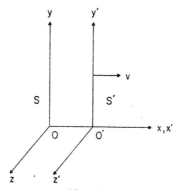

Fig. 1.2.

Before we set out to solve this problem, let us reflect a little closely on the nature of the formulae we expect to obtain here. In view of the fact that the law of inertia must be valid in both S and S', a uniform rectilinear motion in one system must appear as a uniform rectilinear motion in the other system as well, i.e. a linear relationship among x, y, z and t must go over to a linear relationship among x', y', z' and t'. It follows then that the transformation formulae must themselves be linear. This is further suggested by the fact that for both S and S' space and time are homogeneous; therefore, if the numbers x, y, z and t are each multiplied by a factor m, the numbers x', y', z' and t' must also get multiplied by the same factor m, which can only be guaranteed by a *linear, homogeneous transformation.*

Next, we note some of the features of this transformation which arise due to the *special* restriction introduced right in the beginning, viz. that the relative motion of the two systems is confined to the x- (or x'-) direction.[1] Imagine the totality of points constituting the plane $y = y_1$ which is parallel to, and is located at a distance y_1 (as measured by S) from the primary plane $y = 0$. It is obvious that these planes would be parallel for the observer S' too; however, the distance y_1' between them might not be the same as y_1. The difference between y_1 and y_1' cannot be a function of the other spatial coordinates (because of the parallelism of the two planes) nor can it be a function of time (because of the fact that this parallelism continues indefinitely in time); it can only be a function of v, the relative velocity of the two systems. We may, therefore, write

$$y' = f(v)\, y, \tag{1.14}$$

where the function $f(v)$ is yet to be determined. Suppose we consider another reference system S'' moving along the same line but with a velocity $-v$ with respect to S'. Then, the relation between y and y'' would be

$$y'' = f(-v)\, y' = f(-v)\, f(v)y, \tag{1.15}$$

where, in view of the equivalence of the various inertial systems, $f(-v)$ is the same function of $-v$ as $f(v)$ is of v. However, the system S'' is no different from the system S; consequently, y'' must be equal to y and, therefore, we must have

$$f(-v)\, f(v) = 1. \tag{1.16}$$

Now, since the measurement involved here is in a direction *orthogonal* to the

---

[1] It is because of this restriction that the transformations studied in this section are referred to as *special* LORENTZ transformations.

direction of relative motion, we have no reason to expect that it will be affected differently by a motion to the right in comparison with an equivalent motion to the left. Consequently, we must also have

$$f(-v) = f(v),\tag{1.17}$$

which, on combination with (1.16), gives

$$f(v) = \pm 1.\tag{1.18}$$

Since we are not interested in the *inversion* of the axes, the solution $f(v) = -1$ does not interest us. The function $f(v)$ must, therefore, be identically equal to unity; hence, Eq. (1.14) gives

$$y' = y.\tag{1.19}$$

By a similar reasoning,

$$z' = z.\tag{1.20}$$

In view of Eqs. (1.19) and (1.20), the compatibility of (1.12) and (1.13) now requires that the equality

$$x^2 - (ct)^2 = x'^2 - (ct')^2\tag{1.21}$$

must hold for all events observed in the systems S and S'. Let us introduce, for the sake of simplicity, a new coordinate

$$ict = x_4; \quad ict' = x'_4, \qquad (i = \sqrt{-1})\tag{1.22}$$

and at the same time replace x and x' by x_1 and x'_1, respectively. In terms of these coordinates, (1.21) takes the form

$$x_1^2 + x_4^2 = x_1'^2 + x_4'^2.\tag{1.23}$$

Now, if we regard x_1 and x_4 as the rectangular coordinates of a point on a (complex) plane, then the left-hand side of (1.23) would mean the square of the distance of this point from the origin $(0, 0)$. The right-hand side, being of the same form as the left-hand side, then suggests that x'_1 and x'_4 could also be regarded as the coordinates of the same point but referred to a new set of axes; moreover, these new axes must also be rectangular so that the equality (1.23) is naturally satisfied. Obviously, the new axes can be obtained from the original ones by means of a rotation in the representative plane,

$$\left.\begin{array}{l} x_1' = x_1 \cos \varphi + x_4 \sin \varphi \\ x_4' = - x_1 \sin \varphi + x_4 \cos \varphi, \end{array}\right\} \tag{1.24}$$

where φ, the angle of rotation in the plane, would be a function of the relative velocity v of the systems S and S'. The relation between φ and v can be determined as follows.

An object at rest in the system S' must have a velocity v in the system S; hence, for $(dx_1'/dx_4') = 0$, (dx_1/dx_4) should be equal to $-iv/c$. Now, Eqs. (1.24) give

$$\frac{dx_1}{dx_4'} = \frac{\dfrac{dx_1}{dx_4} \cos \varphi + \sin \varphi}{-\dfrac{dx_1}{dx_4} \sin \varphi + \cos \varphi}; \tag{1.25}$$

it then follows that

$$\tan \varphi = iv/c. \tag{1.26}$$

Accordingly,

$$\sin \varphi = \frac{iv/c}{\sqrt{1 - v^2/c^2}}; \quad \cos \varphi = \frac{1}{\sqrt{1 - v^2/c^2}}. \tag{1.27}$$

The transformation equations (1.24) then become

$$x' = \frac{x - vt}{\sqrt{1 - v^2/c^2}},$$

and

$$t' = \frac{t - \dfrac{v}{c^2} x}{\sqrt{1 - v^2/c^2}}. \left.\phantom{\begin{array}{c}a\\b\\c\\d\end{array}}\right\} \tag{1.28}$$

Equations (1.28), along with Eqs. (1.19) and (1.20), constitute the *special Lorentz transformation* formulae.

From the very structure of these formulae one can see how intimately the space and time coordinates are intertwined. A detailed study of these formulae should, therefore, enable us to appreciate quantitatively the effects of 'relativity' on the basic measures of space and time; this study will be taken up in Sec. 1.7.

Before we close this section, a remark appears called for. Looking at (1.23), it becomes evident that transformation from one set of coordinates to the other is such that the equation of a *circle* remains unchanged in form. This is effected by the transformation equations (1.24), which involve circular functions with *imaginary* argument φ and which arise from a rotation of the rectangular axes so that the new axes are inclined to the original ones at angle φ. Likewise, equality (1.21) suggests that the transformation involved (when both coordinates are real) must be such that the equation of a hyperbola remains unchanged in form. This would be effected by a set of transformation equations which involve hyperbolic functions with a *real* argument α and which arise from a transition from the original set of (rectangular) axes to a new set of (oblique) axes, which too are conjugate diameters of the hyperbola, inclined to the former at angle α. Mathematically, we have

$$\left.\begin{array}{l} x' = x \cosh \alpha - ct \sinh \alpha \\ ct' = - x \sinh \alpha + ct \cosh \alpha, \end{array}\right\} \tag{1.24'}$$

with

$$\tanh \alpha = v/c, \tag{1.26'}$$

$$\sinh \alpha = \frac{v/c}{\sqrt{1 - v^2/c^2}} ; \quad \cosh \alpha = \frac{1}{\sqrt{1 - v^2/c^2}}. \tag{1.27'}$$

Parameter α is referred to as the *'rapidity'* of the relative motion.[1] The final formulae (1.28), of course, remain the same.

1.6. The Lorentz group

In the preceding section we investigated the specialized, single-parameter,

---

[1] See V. KARAPETOFF, *Revs. Mod. Phys.* **16**, 33(1944); R. K. PATHRIA, *Am. J. Phys.* **24**, 411(1956); A. SHADOWITZ, *Special Relativity* (W. B. Saunders, Philadelphia, 1968).

LORENTZ transformations and obtained the following formulae:

$$x' = \frac{x - vt}{\sqrt{1 - v^2/c^2}},$$

$$y' = y, \qquad z' = z,$$

$$t' = \frac{t - \dfrac{v}{c^2} x}{\sqrt{1 - v^2/c^2}}.$$

(1.29)

First of all we note that these formulae hold only for $v < c$; for $v \geqslant c$, they yield unphysical results. Consequently, the relative velocity between S and S' must always be less than the velocity of light. This fundamental restriction will be assumed throughout the following discussion.[1]

Solving (1.29) for x, y, z and t, one obtains

$$x = \frac{x' + vt'}{\sqrt{1 - v^2/c^2}},$$

$$y = y', \qquad z = z',$$

$$t = \frac{t' + \dfrac{v}{c^2} x'}{\sqrt{1 - v^2/c^2}};$$

(1.30)

formulae (1.30) pertain to the inverse transformation $S' \to S$. A compari--son of (1.30) with (1.29) shows that the *inverse transformation is also a special Lorentz transformation*, the transformation parameter being $-v$ instead of v.

Next, we consider two successive transformations, $S \to S'$ and $S' \to S''$, both being single-parameter transformations, as are presently under discussion, the values of the respective transformation parameters being v_1 and v_2. Then, we have two sets of formulae:

$$x' = \frac{x - v_1 t}{\sqrt{1 - v_1^2/c^2}},$$

$$y' = y, \qquad z' = z,$$

$$t' = \frac{t - \dfrac{v_1}{c^2} x}{\sqrt{1 - v_1^2/c^2}},$$

(1.31)

[1] For transformations involving velocities exceeding the velocity of light, see L. PARKER, *Phys. Rev.* **188**, 2287(1969).

and

$$x'' = \frac{x' - v_2 t'}{\sqrt{1 - v_2^2/c^2}},$$

$$y'' = y', \qquad z'' = z',$$

$$t'' = \frac{t' - \dfrac{v_2}{c^2} x'}{\sqrt{1 - v_2^2/c^2}}.$$

$$\left.\vphantom{\begin{array}{c}1\\1\\1\\1\\1\\1\end{array}}\right\} \qquad (1.32)$$

Eliminating x', y', z' and t' from these equations, we obtain[1]

$$x'' = \frac{x - Vt}{\sqrt{1 - V^2/c^2}},$$

$$y'' = y, \qquad z'' = z,$$

$$t'' = \frac{t - \dfrac{V}{c^2} x}{\sqrt{1 - V^2/c^2}},$$

$$\left.\vphantom{\begin{array}{c}1\\1\\1\\1\\1\\1\end{array}}\right\} \qquad (1.33)$$

where

$$V = \frac{v_1 + v_2}{1 + v_1 v_2/c^2}. \qquad (1.34)$$

The form of Eqs. (1.33) shows that the *composite transformation* $S \to S''$ *is also a special Lorentz transformation* where the transformation parameter is given by the relation (1.34); accordingly, (1.34) becomes the *relativistic law of addition of (parallel) velocities*. Moreover, since v_1 and v_2 enter symmetrically into the expression for V, their ordering is not relevant to the process of composition; consequently, the process of combining successive *special Lorentz transformations* to obtain a composite *special Lorentz transformation* is a *commutative* process.

Finally, we notice that the identity transformation $S \to S$ is also a special LORENTZ transformation, the transformation parameter in this case being equal to zero; this transformation can also be looked upon as a composite transformation, as in (1.33) and (1.34), with $v_2 = -v_1$.

The foregoing properties of these transformations show that all mathematical conditions necessary for the formation of a *group of transformations*

---

[1] Analytically, it is simpler to carry out the process of elimination after writing Eqs. (1.31) and (1.32) in terms of hyperbolic functions, as in (1.24').

are fulfilled. We, therefore, conclude that special LORENTZ transformations, for all permissible values of the parameter v, constitute a group, which is commutative. However, this group is only a *subgroup* of a more general one which we shall now consider.

To examine this generalization, we remove the restriction that the relative velocity of the two observers is confined to the direction of one of the coordinate axes. Rather, let the relative velocity of S' with respect to S be $\mathbf{v}(v_x, v_y, v_z)$. However, we shall continue to confine ourselves to transformations *without rotation of the coordinate axes*. This qualification does not, however, mean that the (rectangular) axes of S' are parallel to the corresponding (rectangular) axes of S because, as will be seen in the following section, a vector parallel to the x'-axis, as judged by an observer in S', is not *in general* parallel to the x-axis, as judged by an observer in S; further, two vectors which are perpendicular in S' are not *in general* perpendicular in S. Under these circumstances, the only meaning that can be given to the aforementioned qualification is that the angles (measured in S and S', respectively) through which the Cartesian axes in S and S' must be turned so as to obtain the orientation shown in Fig. 1.2 are the same.[1]

In order to obtain transformation equations relevant to this situation, which are known as *general Lorentz transformations without rotation,* we split the spatial vector $\mathbf{r}(x, y, z)$ into two components, one parallel to $\mathbf{v}$ and the other perpendicular to $\mathbf{v}$:

$$\mathbf{r} = \mathbf{r}_{\|} + \mathbf{r}_{\perp}, \tag{1.35}$$

where

$$\mathbf{r}_{\|} = \frac{(\mathbf{r} \cdot \mathbf{v})\,\mathbf{v}}{v^2}, \tag{1.36}$$

and

$$\mathbf{r}_{\perp} = \mathbf{r} - \frac{(\mathbf{r} \cdot \mathbf{v})\,\mathbf{v}}{v^2}. \tag{1.37}$$

We apply transformation (1.29) to the two components separately and then combine them to form a single vector. The final result is

and

$$\left. \begin{aligned} \mathbf{r}' &= \mathbf{r} + \frac{(\mathbf{r} \cdot \mathbf{v})\,\mathbf{v}}{v^2}\,(\gamma - 1) - \gamma \mathbf{v} t, \quad \left[\gamma = \frac{1}{\sqrt{1 - v^2/c^2}} \right] \\ t' &= \gamma \left[t - \frac{(\mathbf{r} \cdot \mathbf{v})}{c^2} \right]. \end{aligned} \right\} \tag{1.38}$$

---

[1] For details, refer to C. MOLLER, *The Theory of Relativity* (Clarendon Press, Oxford, 1952), Secs. 18 and 19.

Transformations inverse to these can be obtained by making use of the obvious device, viz. interchanging the roles of the primed and unprimed coordinates and replacing $\mathbf{v}$ by $-\mathbf{v}$. One thereby gets

and
$$\left.\begin{aligned}
\mathbf{r} &= \mathbf{r}' + \frac{(\mathbf{r}' \cdot \mathbf{v})\,\mathbf{v}}{v^2}\,(\gamma - 1) + \gamma \mathbf{v} t', \\
t &= \gamma\left[t' + \frac{(\mathbf{r}' \cdot \mathbf{v})}{c^2}\right],
\end{aligned}\right\} \qquad (1.39)$$

γ being the same as in (1.38); clearly, the inverse transformation is again a general LORENTZ transformation without rotation.

However, if we consider two successive transformations of this type, the composite transformation is no longer without rotation.[1] It turns out that the transformation $S \rightarrow S''$ in general implies a change of orientation of the Cartesian axes, even though this was not the case with *any* of the component transformations;[2] the only exception which leads to composite transformation *without* rotation corresponds to the case when the relative velocity between S'' and S' is parallel (or antiparallel) to the relative velocity between S' and S.

We, therefore, conclude that the three-parameter family of the general LORENTZ transformations without rotation *does not* constitute a group.

Next, we consider the *most general Lorentz transformations,* viz. the ones in which a reorientation of the Cartesian axes is also included. This enhances the degrees of freedom of the transformation from three to six. One pleasant feature of this extension is that since a change of orientation of the Cartesian axes is now a part of the transformation program, any such change arising from a composition of successive LORENTZ transformations will not throw the composite transformations out of the (wider) family of transformations now under view. Hence, the family of the six-parameter general LORENTZ transformations *does* constitute a group. However, because of the rotations of the axes, this group will not be commutative.

[1] See, e.g. C. MOLLER, *loc. cit.*, Sec. 22.

[2] This leads to the phenomenon of *Thomas precession,* according to which a set of axes undergoing accelerated motion possesses, as observed from an inertial reference system, an angular velocity of precession $\omega = (\gamma - 1)\,(\dot{\mathbf{v}} \times \mathbf{v})/v^2$, where $\mathbf{v}$ and $\dot{\mathbf{v}}$ are the instantaneous velocity and acceleration of the moving system. See L. H. THOMAS, *Nature* 117, 514 (1926); *Phil. Mag.* (7), 3, I (1927); also J. D. JACKSON, *Classical Electrodynamics* (John Wiley and Sons, New York, 1962), Sec. 11.5. For a derivation making use of the formulae of spherical trigonometry, see A. SOMMERFELD, *Atomic Structure and Spectral Lines* (English translation of the 1931 German edition), mathematical appendix 12.

There are certain other degrees of freedom which could be made use of in further widening the scope of these transformations. For instance, one could allow for a displacement of the origin of the coordinate axes and also a shift in the zero of the time measurements. This would not allow conditions (1.10) and (1.11) to hold together and, in consequence, would make the otherwise homogeneous transformations *inhomogeneous* [by throwing constants into Eqs. (1.29)]. Physically, these changes would be trivial; nevertheless, this four-fold freedom, when admitted into the most general group already considered, results in a ten-parameter group of LORENTZ transformations. Since each of these degrees of freedom allows for an infinity of values for the respective parameter, the most generalized group consists of $(\infty)^{10}$ members.

We do not propose to consider at this stage further possibilities of widening our group of transformations, which can be realized by allowing for the *inversion* of the space axes and the *reversal* of the arrow of time. We shall touch upon these possibilities at a later stage; see Sec. 2.5.[1]

In closing, we note that the common feature of all the transformations discussed here is that while the *homogeneous* ones leave the quantity

$$s^2 = (x^2 + y^2 + z^2 - c^2 t^2) \tag{1.40}$$

invariant, the *inhomogeneous* ones leave the *infinitesimal* quantity

$$ds^2 = (dx^2 + dy^2 + dz^2 - c^2\, dt^2) \tag{1.41}$$

invariant; obviously, (1.41) remains invariant under homogeneous transformations as well. Transformations satisfying this invariance are referred to as *Lorentz transformations*.

1.7. Kinematic consequences of Lorentz transformations

In the preceding section we obtained for a general LORENTZ transformation without rotation

and

$$\mathbf{r}' = \mathbf{r} + \frac{(\mathbf{r} \cdot \mathbf{v})\,\mathbf{v}}{v^2}(\gamma - 1) - \gamma \mathbf{v} t, \quad \left[\gamma = \frac{1}{\sqrt{1 - v^2/c^2}}\right]$$
$$t' = \gamma\left[t - \frac{(\mathbf{r} \cdot \mathbf{v})}{c^2}\right]. \tag{1.38}$$

---

[1] For a detailed discussion of these transformations, see J. AHARONI, *The Special Theory of Relativity* (Clarendon Press, Oxford, 1959), Sec. 10.

For $v \ll c$, these formulae reduce to

$$\left.\begin{array}{l} \mathbf{r'} = \mathbf{r} - \mathbf{v}t, \\ \\ t' = t, \end{array}\right\} \qquad (1.38')$$

and

which correspond to the nonrelativistic mechanics of GALILEO and NEWTON. Following P. FRANK, we refer to Eqs. (1.38') as *'Galilean transformation equations'*. It is evident that any conclusions drawn from Eqs. (1.38) would, for $v \ll c$, reduce to the customary conclusions following from Eqs. (1.38'). However, for velocities comparable to c, it would be *incorrect* to use (1.38') even as an approximation.

We shall now make a comparative study of the spatial and temporal measurements carried out in reference systems S and S'.   First of all, let us consider a straight rod which is stationary in S', the position vectors of its ends being $\mathbf{r'_1}$ and $\mathbf{r'_2}$; then, the rod represents a fixed vector

$$\mathbf{x'} = \mathbf{r'_2} - \mathbf{r'_1} \qquad (1.42)$$

in the reference system S'.   An observer in S with respect to whom the rod is in motion, with velocity $\mathbf{v}$, would obtain the corresponding vector $\mathbf{x}$ by noting the position vectors of its ends at *a common time t* and taking their difference, as in (1.42).   In view of (1.38), one would have

$$\mathbf{x'} = \mathbf{x} + \frac{(\mathbf{x} \cdot \mathbf{v})\, \mathbf{v}}{v^2} (\gamma - 1). \qquad (1.43)$$

Decomposing $\mathbf{x'}$ and $\mathbf{x}$ into components, which are parallel and perpendicular to $\mathbf{v}$, (1.43) can be rewritten as

$$\left.\begin{array}{l} \mathbf{x'_\parallel} = \mathbf{x_\parallel}\, \gamma = \dfrac{\mathbf{x_\parallel}}{\sqrt{1 - v^2/c^2}} \\ \\ \mathbf{x'_\perp} = \mathbf{x_\perp}; \end{array}\right\} \qquad (1.44)$$

we thus find that only the component perpendicular to $\mathbf{v}$ has identical measures in the two systems; the parallel component has a larger measure in S', the rest system of the rod, than in S, with respect to which the rod is in motion.

Customarily, the measures obtained in the rest system of an object are referred to as *proper* measures while those obtained in an arbitrary system are referred to as *relative* or *relativistic* measures. The foregoing result thus means that the relativistic length l of a moving rod, aligned parallel to the direction of motion, is less than its proper length l_0 by a factor of $\sqrt{1-v^2/c^2}$ (usually referred to as the *Lorentz factor*) :

$$l = l_0 \sqrt{1 - v^2/c^2}. \tag{1.45}$$

In other words, the dimensions of a moving object would appear 'contracted' in the direction of motion of the object — the *length contraction*, as it is usually called. Of course, dimensions orthogonal to the direction of motion remain unaffected.

It is not difficult to see that the phenomenon of length contraction is a mutually reciprocal effect, that is, a rod at rest in S would appear contracted to an observer in S'. In this case, we would have to use transformation equations inverse to (1.38), viz. (1.39), for now the time coordinate to be eliminated would be t', and not t.

It must be pointed out here that the contraction hypothesis, put forward by FITZGERALD and LORENTZ (see Sec. 1.3), was of an entirely different character and must not be confused with the effect obtained here. That hypothesis did not refer to a mutually reciprocal effect ; it rather suggested a contraction in the *absolute* sense, arising from the motion of an object with respect to the aether or, so to say, from its 'absolute' motion. According to a relativistic standpoint, neither absolute motion nor any effect accruing therefrom has any physical meaning.

Another consequence of relations (1.44) is worth noting. Since the inclination of a given vector with the direction of relative motion would be given by the ratio of its perpendicular and parallel components and since relative motion affects these components differently, the vector would have different inclinations for the two observers—the only exceptional cases being those for which one of the two components vanishes, that is, when the vector in question is either parallel or perpendicular to **v**. By the same reason, the mutual inclination of two given vectors would be different for the two observers ; the only general exception in this case would be a pair of parallel (or antiparallel) vectors. In particular, two vectors orthogonal in S' would not, in general, be orthogonal in S; the only exception here would be when one of the vectors is parallel and the other perpendicular to **v**.

The next effect we propose to discuss here is concerned with the rates of clocks in relative motion. We consider a clock at rest with respect to S' and stationed at the point **r'**. The time interval between two instants of time,

t'_1 and t'_2 , is then given by

$$t' = t'_1 - t'_2 .$$ (1.46)

The corresponding recordings of an observer in S would be t_1 and t_2, as given by the second formula in (1.39); the corresponding measure of the time interval would then be

$$t = t_1 - t_2 = \gamma(t'_1 - t'_2) = \gamma t' .$$ (1.47)

Thus, the time interval t, as determined by the clocks of S, exceeds the value t', as determined by the clock in (relative) motion, which means that the clock in motion appears to be going slower in comparison with the ones at rest :

$$t' = t \sqrt{1 - v^2/c^2}.$$ (1.48)

This effect is usually referred to as *time dilation*.

Just like length contraction, this effect is also mutually reciprocal. One can show that if an observer in S' compares the time measurement of a clock at rest with respect to S (and stationed at the point $\mathbf{r}$, say) with the time measurement of his own (stationary) clocks,[1] then he would also arrive at the same conclusion, viz. that the clock in motion appears to be going slower in comparison with the ones at rest, the slowing factor being again $\sqrt{1 - v^2/c^2}$.

The *relativistic* time dilation effect may be compared with a hypothesis due to LARMOR[2], which was inspired by the contraction hypothesis of FITZGERALD, viz. a clock moving with velocity v *with respect to the aether* must run slower, in the ratio $\sqrt{1 - v^2/c^2} : 1$. Now, just as the null result of the MICHELSON-MORLEY experiment was regarded as explicable on the basis of the FITZGERALD hypothesis, the null result of an experiment performed later by KENNEDY and THORNDIKE,[3] in which they modified the MICHELSON-MORLEY arrangement by not only keeping the lengths of the two arms unequal but also making them oblique rather than orthogonal, could be considered as explicable on the basis of the LARMOR hypothesis and the FITZGERALD hypo-

---

[1] In this case one will have to employ the second formula in (1.38) because now it is $\mathbf{r}$, and not $\mathbf{r}'$, that has to be eliminated.

[2] J. LARMOR, *loc. cit.*

[3] R. J. KENNEDY and E. W. THORNDIKE, *Phys. Rev.* 42, 400 (1932).

thesis, taken together. However, these explanations suffer from an undesirable appeal to aether, and the effects referred to are not the mutually reciprocal effects predicted by relativistic kinematics; rather, they are hypothetical and, in view of their absolute character, unphysical.

A striking verification of the time-dilation formula (1.48) is provided by the observed increase in the apparent lifetimes of high energy μ-mesons.[1] It was found by Rossi and Hall[2] that the decay time of the mesons traversing Earth's atmosphere, as members of the penetrating component of cosmic rays, varied significantly, over a range of the order of 1—10, with their energies. They attributed this variation to the time dilation effect, according to which the meson-clock, being in fast motion relative to a terrestrial clock, would go slower by a factor of $\sqrt{1 - v^2/c^2}$, which in turn depends upon the energy of the meson; see Eq. (3.23). Making use of the relevant formulae, Rossi and Hall showed that the velocity dependence of the apparent lifetimes of the mesons was in reasonable agreement with formula (1.48). These investigations also enabled them to determine the 'proper' or 'true' lifetime of these mesons—the one that would be recorded by a clock moving along with the meson.[3,4]

The foregoing explanation may be put in slightly different, but physically equivalent, terms by looking at the observed effect from another angle, that is, to an observer moving along with the meson, distances in the atmosphere (which are at rest with respect to a terrestrial observer) would appear contracted by a factor of $\sqrt{1 - v^2/c^2}$. Accordingly, the distance traversed by a meson during its 'proper' life span would, in a terrestrial system, be more than the one normally expected, by a factor of $(1 - v^2/c^2)^{-1/2}$; this, in turn, would be interpreted as an increase in the lifetime of the meson. Thus, the observational fact studied by Rossi and Hall can be looked upon as providing indirect verification of the length contraction formula (1.45).

[1] See also Secs. 4.2 and 8.6.

[2] B. Rossi and D. B. Hall, *Phys. Rev.* **59**, 223 (1941). More recently this experiment has been repeated by D. H. Frisch and J. H Smith, *Am. J. Phys.* **31**, 342 (1963); a filmed version of the experiment has been discussed by A. P. French, *Special Relativity* (W. H. Norton and Co., New York, 1968).

[3] R. Durbin, H. H. Loar and W. W. Havens, *Phys. Rev.* **88**, 179 (1952), have measured the lifetime of artificially produced π-mesons of kinetic energy 73 MeV, corresponding to a time-dilation factor of about 1.5. The measured value is found to be in agreement with this factor to about 10 per cent.

[4] Farley *et al.*, *Nuovo Cimento* **45A**, 281 (1966), have observed time dilation effect with μ-mesons confined to circular orbits. With momentum $\approx$ 1.3 GeV/c, the time dilation factor for mesons in this experiment was $\approx$ 12.

A direct study of the phenomenon of length contraction is complicated by the fact that if we take a *visual* look at a moving object we must allow for retardation effects arising from the time taken by light in travelling to our receiving system (for instance, the eye) from different parts of the object. It is obvious that the appropriate instants at which light must start from various parts of the object so as to reach the observer at the same instant would be different, and hence would be associated with different positions of the object as a whole. The general result, when both length contraction and retardation effects are taken into account, is that the shape of the object appears distorted! There is, however, a special result of considerable interest, viz. that if the moving object is far enough to subtend a very small solid angle (so that the rays of light reaching the observer are essentially parallel), then the *projected* view of the object corresponds to the object being *at rest but rotated*. For instance, a moving metre stick would appear as foreshortened but only to an extent that corresponds to an apparent rotation of the stick (without involving contraction as such). This problem became prominent through the observations of TERRELL[1], and over the intervening years has generated considerable literature,[2] to which the interested reader may refer.

1.8. Transformation of velocities

Once again we consider two inertial systems of reference S and S', with relative velocity **v**; their space-time coordinates are connected through Eqs. (1.38) and (1.39). The motion of an arbitrary point (which may be a material particle or merely a geometrical position) will be described by a function $\mathbf{r}(t)$ in system S and a function $\mathbf{r}'(t')$ in system S'. The *instantaneous* velocities of the moving point, **u** (as measured in S) and **u**' (as measured in S'), will then be given by

$$\mathbf{u}\,(u_x, u_y, u_z) = \frac{d\mathbf{r}}{dt} \left[\frac{dx}{dt}, \frac{dy}{dt}, \frac{dz}{dt} \right] \tag{1.49}$$

and

$$\mathbf{u}'(u_x', u_y', u_z') = \frac{d\mathbf{r}'}{dt'} \left[\frac{dx'}{dt'}, \frac{dy'}{dt'}, \frac{dz'}{dt'} \right], \tag{1.50}$$

[1] J. TERRELL, *Phys. Rev.* **116**, 1041 (1959).

[2] V. F. WEISSKOPF, *Phys. Today* **13**, 24 (1960); R. PENROSE, *Proc. Camb. Phil. Soc.* **55**, 137 (1959); M. L. BOAS, *Am. J. Phys.* **29**, 283 (1961); G. D. SCOTT and M. R. VINER, *Am. J. Phys.* **33**, 534 (1965); N. C. McGILL, *Contemp. Phys.* **9**, 33 (1968); G. D. SCOTT and VAN DRIEL, *Am. J. Phys.* **38**, 971 (1970).

respectively. Carrying out the relevant differentiations in (1.38), we obtain

$$\mathbf{u}' = \left[\frac{d\mathbf{r}'}{dt}\right] \Big/ \left[\frac{dt'}{dt}\right]$$

$$= \frac{\sqrt{1 - v^2/c^2}\,\mathbf{u} + \dfrac{(\mathbf{u} \cdot \mathbf{v})\mathbf{v}}{v^2}[1 - \sqrt{1 - v^2/c^2}] - \mathbf{v}}{1 - \dfrac{(\mathbf{u} \cdot \mathbf{v})}{c^2}}. \tag{1.51}$$

The inverse relation follows similarly from formulae (1.39) or else by inter-changing the roles of $\mathbf{u}$ and $\mathbf{u}'$ and replacing $\mathbf{v}$ by $-\mathbf{v}$. Any way we obtain

$$\mathbf{u} = \frac{\sqrt{1 - v^2/c^2}\,\mathbf{u}' + \dfrac{(\mathbf{u}' \cdot \mathbf{v})\mathbf{v}}{v^2}[1 - \sqrt{1 - v^2/c^2}] + \mathbf{v}}{1 + \dfrac{(\mathbf{u}' \cdot \mathbf{v})}{c^2}}. \tag{1.52}$$

Equation (1.51), after some vector algebra, yields

$$\sqrt{1 - u'^2/c^2} = \sqrt{1 - u^2/c^2} \cdot \frac{\sqrt{1 - v^2/c^2}}{1 - \dfrac{(\mathbf{u} \cdot \mathbf{v})}{c^2}}. \tag{1.53}$$

The corresponding relation following from (1.52) is

$$\sqrt{1 - u^2/c^2} = \sqrt{1 - u'^2/c^2} \cdot \frac{\sqrt{1 - v^2/c^2}}{1 + \dfrac{(\mathbf{u}' \cdot \mathbf{v})}{c^2}}. \tag{1.54}$$

Combining (1.53) and (1.54), we obtain an interesting relation, viz.

$$\left[1 - \frac{(\mathbf{u} \cdot \mathbf{v})}{c^2}\right]\left[1 + \frac{(\mathbf{u}' \cdot \mathbf{v})}{c^2}\right] = \left[1 - \frac{v^2}{c^2}\right]. \tag{1.55}$$

Equations (1.51) — (1.55) lead to a number of significant results. First of all, we note that for velocities much smaller than c, we recover the well-known Newtonian formulae, viz.

$$\mathbf{u}' = \mathbf{u} - \mathbf{v} \quad \text{and} \quad \mathbf{u} = \mathbf{u}' + \mathbf{v}. \tag{1.56}$$

Next, we note that, according to formulae (1.53) and (1.54), if the speed of a moving point is c in one of the systems of reference, it is c in the other as well; it is important that this conclusion is independent of the *directions* of the vectors u and u'. This is consistent with the second postulate of the theory. It also follows from these relations that (i) for all $(u, v) < c$, $u' < c$, and (ii) for all $(u', v) < c$, $u < c$. This means that so long as we feed into our formulae velocities with magnitude less than c, the outcome is also of a magnitude less than c. We have already noticed that the validity of LORENTZ transformations requires that no velocity exceeding that of light be considered; we now see that our transformation formulae for velocities are consistent with this requirement.[1]

In the special case when u' is *perpendicular* to v, Eq. (1.52) becomes

$$\mathbf{u} = \sqrt{1 - v^2/c^2}\, \mathbf{u}' + \mathbf{v}, \tag{1.57}$$

and when u' is *parallel* to v, we have

$$\mathbf{u} = \frac{\mathbf{u}' + \mathbf{v}}{1 + \dfrac{u'v}{c^2}}. \tag{1.58}$$

In the latter case, u lies in the same direction as u' and v and its magnitude in terms of u and v is given by the same *law of addition* as encountered in the case of two successive special LORENTZ transformations; see Eq. (1.34). Relation inverse to (1.58) would be

$$\mathbf{u}' = \frac{\mathbf{u} - \mathbf{v}}{1 - \dfrac{uv}{c^2}}. \tag{1.59}$$

In the case of a special LORENTZ transformation without rotation, the

---

[1] Possibility of the existence of faster-than-light particles was first discussed by O. M. P. BILANIUK, V. K. DESHPANDE and E. C. G. SUDARSHAN, *Am. J. Phys.* 30, 718 (1962). For a detailed discussion of this question, see G. FEINBERG, *Phys. Rev.* 159, 1089 (1967); R. G. NEWTON, *Phys. Rev.* 162, 1274 (1967); M. ARONS and E. C. G. SUDARSHAN, *Phys. Rev.* 173, 1622 (1968); M. M. BROIDO and J. G. TAYLOR, *Phys. Rev.* 174, 1606 (1968); J. DHAR and E. C. G. SUDARSHAN, *Phys. Rev.* 174, 1808 (1968); E. C. G. SUDARSHAN, *Arkiv för Physik* 39, 585 (1969).

For a less technical account, see O. M P. BILANIUK and E. C. G. SUDARSHAN, *Phys. Today* 22(5), 43 (1969) and *Nature* 223, 386 (1969); E. C. G. SUDARSHAN, in *Symposia on Theoretical Physics and Mathematics* (Plenum Press, New York, 1970), Vol. 10, p. 129; G. FEINBERG, *Scientific American* 222 (2), 68 (1970).

components of the relative velocity $\mathbf{v}$ are $(v, 0, 0)$; relation (1.51) then gives

$$u'_x = \frac{u_x - v}{1 - \dfrac{u_x v}{c^2}}, \quad u'_y = \frac{u_y \sqrt{1 - v^2/c^2}}{1 - \dfrac{u_x v}{c^2}}, \quad u'_z = \frac{u_z \sqrt{1 - v^2/c^2}}{1 - \dfrac{u_x v}{c^2}}, \quad (1.60)$$

which could also be obtained directly from the transformation formulae (1.29). Now, if l, m and n denote the instantaneous direction cosines of the motion of the point in system S and l', m' and n' denote the corresponding quantities in system S', then

$$(l, m, n) = \frac{(u_x, u_y, u_z)}{\sqrt{u_x^2 + u_y^2 + u_z^2}}, \quad (1.61)$$

and

$$(l', m', n') = \frac{(u'_x, u'_y, u'_z)}{\sqrt{u_x'^2 + u_y'^2 + u_z'^2}}. \quad (1.62)$$

From (1.60), it then follows that

$$(l', m', n') = \frac{l - \dfrac{v}{u}, \; m\sqrt{1 - v^2/c^2}, \; n\sqrt{1 - v^2/c^2}}{D}, \quad (1.63)$$

where

$$D = \frac{u'}{u}\left(1 - \frac{u_x v}{c^2}\right) = \left[1 - 2l\frac{v}{u} + \frac{v^2}{u^2} - (1 - l^2)\frac{v^2}{c^2}\right]^{1/2}. \quad (1.64)$$

Relation (1.63) may be looked upon as the *relativistic aberration formula* for the motion of a point as referred to a pair of inertial systems whose relative velocity is confined to the direction of one of the axes.

A special case of this study has been of great practical interest. This is concerned with the *propagation of light in a moving medium*, the two motions being in the same direction (either parallel or antiparallel). The system S' may be taken to be at rest with respect to the medium (in which the velocity of light is c/n, n being the refractive index, and is directed along the x'-axis). And the system S may be taken to be the laboratory system with respect to which the medium, and hence the system S', is moving with velocity

v along the x-axis; the x- and x'-axes being parallel, the transformation will be of the 'special' type. We then have

$$(u'_x, u'_y, u'_z) = \left[\frac{c}{n}, 0, 0\right];$$ (1.65)

from (1.60), or from relations inverse to (1.60), we then get

$$(u_x, u_y, u_z) = \left[\frac{(c/n) \pm v}{1 \pm v/(nc)}, 0, 0\right].$$ (1.66)

Thus, the velocity of light with respect to the laboratory system would have a magnitude given by

$$u = \left[\frac{c}{n} \pm v\right]\left[1 \pm \frac{v}{nc}\right]^{-1}$$

$$\simeq \frac{c}{n} \pm v\left[1 - \frac{1}{n^2}\right];$$ (1.67)

the last approximation holds for $v \ll c$, as is normally true in the laboratory experiments.

A comparison of formula (1.67) with formula (1.2), which was obtained experimentally by FIZEAU in 1851, shows that the coefficient $(1 - 1/n^2)$, sometimes referred to as FRESNEL's *drag coefficient*, arises merely as a consequence of the relativistic law of addition of velocities. FIZEAU's result may, therefore, be looked upon as a *direct* verification of this law.

An extension of this problem to the case of a moving *dispersive* medium will be carried out in Sec. 4.4.

Problems

1.1. Consider a double-star system with two stars, A and B, moving in circular orbits, of period T, about their centre of mass. The Earth lies in the plane of these orbits, at a distance R (which is several light-years). If the speed of light emitted by A were modified by the motion of A (the RITZ hypothesis), so that it becomes $c + v_r$ where v_r ($= v \cos \theta$) is the *radial* velocity of the star, then show that the value of v_r, as inferred from spectroscopic observations on the Earth, would *appear* to be varying with time in accordance with the implicit relationship

$$v_r = v \sin \frac{2\pi}{T}\left[t - \frac{R}{c} + \frac{Rv_r}{c^2}\right]; \quad (v \ll c).$$

Examine the peculiarities of the function $v_r(t)$ over the interval $0 \leqslant t \leqslant T$.

1.2. Prove that the temporal sequence of two events is the same in *all* inertial systems of reference if, and only if, their time separation is more than it takes light to traverse the space between them. Hence show that the existence of signals propagating faster than light is incompatible with the concept of causality.

1.3. Consider two inertial systems of reference in standard configuration, as shown in Fig. 1.2. Prove that, at any instant of time, there is just one plane in S on which the clocks agree with those in S', and that this plane moves with velocity $v^* = (c^2/v) \{1 - (1 - v^2/c^2)^{1/2}\}$. Verify that $\tanh^{-1}(v^*/c) = \frac{1}{2} \tanh^{-1}(v/c)$, and interpret this result physically.

1.4. In the inertial system S' a straight rod parallel to the x'-axis moves in the y'-direction with velocity u. Show that in the inertial system S the rod appears inclined, at an angle $-\tan^{-1}\{(uv/c^2)(1 - v^2/c^2)^{-1/2}\}$, to the x-axis. Further show that a rod parallel to y'-axis, and moving arbitrarily, in S' remains parallel to the y-axis in S.

1.5. A 'rigid' rod of *proper* length 10 cm moves longitudinally over a flat table, the LORENTZ factor of the motion being 0.1. In the path of the rod lies a hole of *proper* width 20 cm. To an observer moving with the rod the hole would appear only 2 cm wide; so the rod should just slide over the hole. However, to an observer at rest with respect to the table it is the rod that would appear contracted (to an apparent length of 1 cm); so the rod must fall down into the hole. Resolve the paradox.

1.6. Two uniformly moving clocks pass each other with relative velocity v. An observer on one clock keeps the other in sight. By what factor will the *visually* observed rate of the other clock lag behind his own? What does he observe if the velocity of the receding clock is suddenly reversed to $-v$? [Compare with the phenomenon of *Doppler effect*; see Sec. 4.2.]

1.7. In the KENNEDY-THORNDIKE experiment, as outlined in Sec. 1.7, the difference between the arm lengths l_1 and l_2 of the interferometer was 16 cm. If the Sun is assumed to be at rest with respect to the aether, the velocity of the apparatus at any instant would be the vector sum of the Earth's orbital velocity $\mathbf{v}(\approx 30 \text{ km/sec})$ and the surface velocity $\mathbf{u}$ due to the rotation of the Earth (≈ 0.5 km/sec at the equator). Show that the maximum fringe shift that might be expected during the course of a day is given by

$$\delta n = \frac{4uv}{c^2} \frac{l_1 - l_2}{\lambda}.$$

Evaluate the magnitude of δn for $\lambda = 6 \times 10^{-7}$ m.

1.8. Derive LORENTZ transformation equations for the acceleration of a moving point, referred to the inertial systems S and S'. Show that if S' is the *instantaneous* rest system of the motion, then the magnitude of the instantaneous acceleration in it is given by

$$a' = \frac{\left[a^2 - \dfrac{(\mathbf{a} \times \mathbf{v})^2}{c^2}\right]^{1/2}}{\sqrt{(1 - v^2/c^2)^2}}.$$

Further show that, in the particular case when $\mathbf{v} = (v, 0, 0)$, the transformation equations for the components of the acceleration vector reduce to

$$(a'_x, a'_y, a'_z) = \left[\frac{a_x, \ a_y \sqrt{1 - v^2/c^2}, \ a_z\sqrt{1 - v^2/c^2}}{\sqrt{(1 - v^2/c^2)^3}}\right].$$

1.9. Discuss the motion of a particle, initially at rest, whose 'proper' acceleration is constant. Such a motion is generally referred to as *hyperbolic* motion.

1.10. Show that the result obtained by the relativistic addition of a velocity v_1 to a velocity v_2 is not, in general, the same as that obtained by adding velocity v_2 to velocity v_1. Determine the cases that form an exception. Also show that the magnitude of the resultant is, nevertheless, independent of the order in which the velocities are added, and is given by

$$V^2 = \frac{(\mathbf{v}_1 + \mathbf{v}_2)^2 - \dfrac{(\mathbf{v}_1 \times \mathbf{v}_2)^2}{c^2}}{\left[1 + \dfrac{(\mathbf{v}_1 \cdot \mathbf{v}_2)}{c^2}\right]^2}.$$

1.11. Consider three inertial systems S, S' and S'' such that the transformation $S \to S'$ is of the special LORENTZ type with relative velocity v_1 in the x-direction while the transformation $S' \to S''$ is of the same type but with relative velocity v_2 in the y'-direction. If the relative motion between S and S'' makes an angle θ with the x-axis of S and θ'' with the x''-axis of S'', then show that, for $(v_1, v_2) \ll c$,

$$(\theta'' - \theta) \simeq \frac{1}{2} \frac{v_1 v_2}{c^2}.$$

Hence deduce the THOMAS precession encountered in Sec. 1.6.

1.12. Derive an expression for the magnitude u of the *relative* velocity of two particles, one of which is moving with velocity $\mathbf{u}_1$ and the other with velocity $\mathbf{u}_2$ with respect to a given system of reference. Hence show that the expression for the 'line element' in velocity space, in terms of the coordinates (u, θ, φ), is

$$dl_u^2 = \frac{(du)^2}{(1 - u^2/c^2)^2} + \frac{u^2}{(1 - u^2/c^2)} [(d\theta)^2 + \sin^2\theta \, (d\varphi)^2].$$

Note that by introducing the 'rapidity' $\alpha \{\equiv \tanh^{-1}(u/c)\}$, the line element can be written as

$$dl_u^2 = c^2[(d\alpha)^2 + \sinh^2\alpha\{(d\theta)^2 + \sin^2\theta(d\varphi)^2\}];$$

geometrically, this corresponds to a line element in the three-dimensional space of LOBA-CHEVSKII—which is a space of constant negative curvature; *cf.* (9.13).

CHAPTER 2

FOUR-DIMENSIONAL FORMULATION
OF THE THEORY

2.1. The world of events

For the description of a physical event, one must specify four numbers, say x, y, z and t, of which the first three refer to the site where the event takes place while the fourth refers to the time of its occurrence. For another observer, the same event would be specified by a different set of numbers, say x', y', z' and t'. The mathematical connection between these two sets of numbers would depend upon the characteristic *features* of the transition from one system of reference to the other. Since there is a large variety of systems of reference, there is a large variety of transformations one has to deal with. Of these, the ones occurring in the special theory of relativity have already been considered in Secs. 1.5 and 1.6 ; they involve either one or more of the following features:

(i) a rotation of the coordinate axes in the (x, y, z)-space, with *three* degrees of freedom,

(ii) a displacement of the origin of the coordinate system, again with *three* degrees of freedom,

(iii) a shift in the zero of the time measurement, with a *single* degree of freedom, and finally

(iv) a uniform translational motion, with *three* degrees of freedom.

In this classification, we observe a clear-cut distinction between the spatial and temporal aspects of the transformation process. However, as revealed by the transformation equations derived in the preceding chapter, this distinction is not fundamental; in fact, it appears advisable to treat the fourth member of the set x, y, z and t on *formally* the same footing as the first three.   One may, therefore, regard the complete set of four numbers as the *coordinates* of a (world) point, representing the event in question in a four-dimensional 'world' of x, y, z and t, rather than regard x, y and z separately as the spatial coordinates of the event in the three-dimensional physical space and t separately as the time of its occurrence in a one dimensional continuum (of time).

The foregoing scheme was first suggested by MINKOWSKI in 1908 and has since been employed extensively.   According to this scheme, a physical event is represented by a *world point* in the four-dimensional world of events—usually referred to as the *Minkowski world* or the space-time continuum—which may be defined as the *totality of representative points of all possible physical events.* The 'position' of the representative point, corresponding to a particular event, in the MINKOWSKI world is determined by four coordinates x, y, z and t, as referred to a set of four axes—the so-called *coordinate axes.* The development of an event (for instance, the motion of a point) is depicted by a trajectory, usually referred to as the *world line*, along which the corresponding world point evolves.   The relative position of one event with respect to another is denoted by a vector—the *line element*—that joins the two events in the MINKOWSKI world, and so forth.

In this representation the study of relativistic kinematics reduces to a study of the geometry of the four-dimensional continuum.   In consequence, the formalism of the theory of relativity assumes so elegant a form that the importance and usefulness of MINKOWSKI's contribution can hardly be overestimated.   As regards transformations, their classification would now be based on the fact whether they involve one or both of the following features :

(i)* a rotation of the 'coordinate' axes in the (x, y, z, t)-continuum, with *six* degrees of freedom, and

(ii)* a displacement of the origin of the 'coordinate' system, with *four* degrees of freedom.

Class (i)* here contains classes (i) and (iv) of the previous classification, while (ii)* contains (ii) and (iii).   Further, in class (i)*, if the t-axis does not participate in the rotation of the axes, it would mean a reorientation of the space axes alone, which corresponds exactly to class (i) of the previous classification.   On the other hand, if the t-axis takes part in the rotation of the axes but there are no rotations in the (x, y), (y, z) and (z, x) planes, then we have a uniform translational motion *without rotation* ; this corresponds to class (iv) of the previous classification.

2.2. Geometrical representation of the special Lorentz transformation

We saw in Sec. 1.5 that the special LORENTZ transformation is characterized by the fact that the equality (1.21),

$$(ct)^2 - x^2 = (ct')^2 - x'^2, \tag{2.1}$$

must hold for all physical events. In terms of the four-dimensional geometry, this implies that in the (x, t)-plane of the MINKOWSKI world the sequence of events represented by the world line

$$(ct)^2 - x^2 = \text{const}, \tag{2.2}$$

when referred to the axes employed by the observer S, be represented by a similar equation

$$(ct')^2 - x'^2 = \text{const}, \tag{2.3}$$

when referred to the axes employed by the observer S': moreover, the constants appearing on the right-hand sides of Eqs. (2.2) and (2.3) must be identical. By a suitable choice of units, we can make these constants equal to unity; the two equations then become

$$(ct)^2 - x^2 = 1 \tag{2.4}$$

and

$$(ct')^2 - x'^2 = 1, \tag{2.5}$$

respectively.

For observer S, we adopt a set of rectangular axes, as shown in Fig. 2.1, and draw the world line (2.4), which is a rectangular hyperbola with apex at the point $A(0, 1)$.[1] The axes for observer S' must be chosen in such a way that an object at rest in S' moves with velocity v with respect to S. Now, the world line for an object at rest in S' would be a straight line parallel to the t'-axis (because, in the case of an object at rest, it is only the time-

---

[1] Throughout this section, we will be using the coordinates (x, ct) rather than (x, t); however, the time axis will usually be referred to as the t-axis.

coordinate of the representative point that changes) ; however, the same

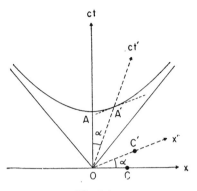

Fig. 2.1.

world line, when viewed from the coordinate axes of S, should correspond to a motion with velocity v, i.e. its slope with the t-axis should be

$$\frac{dx}{d\,(ct)} = \frac{v}{c}. \tag{2.6}$$

This means that the t'-axis of S' must be inclined to the t-axis of S at an angle α, such that

$$\tan \alpha = \frac{v}{c}. \tag{2.7}$$

Having chosen the t'-axis as desired, we have now to choose the x'-axis. This must be done in such a way that the hyperbola of Fig. 2.1, when referred to the new axes, satisfies Eq. (2.5). For this, we make use of a well known result in coordinate geometry, viz. that the equation of a hyperbola (actually, of any conic section) is formally the same when referred to *any* pair of 'conjugate' diameters; one may, however, have to modify the scale of the axes. In the case under discussion, the new unit to be adopted is the length OA', where A' is the point of intersection of the t'-axis with the hyperbola. Further, the x'-axis (which should be conjugate to the t'-axis) would be a straight line, passing through the origin, parallel to the tangent to the hyperbola at the point A'; the corresponding unit on the x'-axis is denoted by the length OC', which is the same as OA'. One can show that the x'-axis is inclined to the x-axis at the same angle α as we have between the t'- and t-axes. Thus, our new axes are *oblique*, though symmetrical about an asymptote of the hyperbola.

From the geometry of the figure and the equation of the hyperbola, it follows that

$$OA' = OC' = \frac{1}{\sqrt{\cos{(2\alpha)}}}, \tag{2.8}$$

which is the scale factor involved in the transformation. Now, if an arbitrary world point is referred to the two sets of axes *independently*, the equations connecting the two sets of coordinates would be

$$\left.\begin{aligned} ct &= \frac{ct' \cos\alpha + x' \sin\alpha}{\sqrt{\cos{(2\alpha)}}}, \\ \text{and} \\ x &= \frac{x' \cos\alpha + ct' \sin\alpha}{\sqrt{\cos{(2\alpha)}}}; \end{aligned}\right\} \tag{2.9}$$

the factor in the denominator takes account of the fact that the units adopted in S and S' differ by the scale factor (2.8). Eliminating α from Eqs. (2.7) and (2.9), we obtain

$$\left.\begin{aligned} t &= \frac{t' + \dfrac{v}{c^2} x'}{\sqrt{1 - v^2/c^2}}, \\ \text{and} \\ x &= \frac{x' + vt'}{\sqrt{1 - v^2/c^2}}, \end{aligned}\right\} \tag{2.10}$$

which are precisely the formulae for a special LORENTZ transformation; *cf.* (1.30).

We thus obtain a geometrical representation of LORENTZ transformation in the MINKOWSKI world. Now, as the physical parameter characterizing the transformation, viz. v/c, assumes various values in the range $(-1, +1)$, the corresponding geometrical parameter α takes values in the range $(-\pi/4, +\pi/4)$, i.e. the t'-axis sweeps out from one asymptote of the hyperbola to the other. Correspondingly, the inclination of the (oblique) axes, x'- and t'-, decreases from π to 0. The central position in the range is occupied by the initial system of reference S, corresponding to the case $v/c = 0$, i.e. $\alpha = 0$. However, this does not imply that system S is in some sort of a privileged position compared to other systems. Such an impression would be totally unreal because, in drawing the geometrical figure, one could equally well start with the reference system S' (and for that matter *any* inertial system of reference), allot rectangular axes to this one and oblique axes to others; the final transformation equations, and subsequent conclusions, would always exhibit the reciprocity which is basic in relativity.

Before we close this section, we make a geometrical study of the phenomenon of length contraction; see Sec. 1.7. Consider a rod, of length l, at rest in system S, placed parallel to the x-axis. Its representation in the (x, t)-plane of the MINKOWSKI world would consist of a 'world band', of width $P_1P_2 = l$, running parallel to the t-axis; see Fig. 2.2. From the point of view

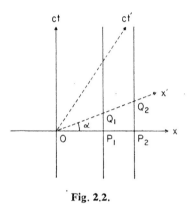

Fig. 2.2.

of S', the rod is moving uniformly with velocity $-v$, i.e. in the negative direction of the x'-axis. The length of the rod in S' would then be given by the difference between the *simultaneous* measures of the coordinates of its ends (of course, corrected by the relevant scale factor). Thus, one would have

$$l' = Q_1Q_2 \sqrt{\cos (2\alpha)}$$

$$= \frac{P_1P_2}{\cos \alpha} \sqrt{\cos (2\alpha)} = l \sqrt{1 - v^2/c^2}, \qquad (2.11)$$

which is the desired result.

It would be instructive to perform a similar calculation in the case of a rod at rest in S', placed parallel to the x'-axis. The representation of the rod would now be a 'world band' running parallel to the t'-axis; see Fig. 2.3. The length of the rod in the (rest) system S' would be

$$Q_1Q_2 \sqrt{\cos (2\alpha)} = l', \quad \text{say.}$$

In system S, with respect to which the rod is in motion, with velocity v in the

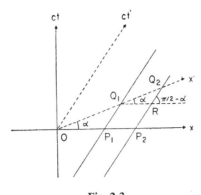

Fig. 2.3.

positive direction of the x-axis, the length would appear to be

$$l = P_1 P_2 = Q_1 R = Q_1 Q_2 \frac{\sin \angle Q_1 Q_2 R}{\sin \angle Q_1 R Q_2}$$

$$= \frac{l'}{\sqrt{\cos (2\alpha)}} \frac{\sin (\pi/2 - 2\alpha)}{\sin (\pi/2 + \alpha)} = \frac{l' \sqrt{\cos (2\alpha)}}{\cos \alpha}$$

$$= l' \sqrt{1 - v^2/c^2}. \tag{2.12}$$

A comparison of the formulae (2.11) and (2.12) shows that, although in both these calculations we employed the scheme of axes in which the observer S is *apparently* privileged, the conclusions arrived at conform to the fact that all inertial observers are on an equal footing.

2.3. World regions and the light cone

In this section we propose to examine the relationship various events bear to one another. For this, it would be sufficient to study the case of two events, which may be denoted as $E_1 (x_1, y_1, z_1, ct_1)$ and $E_2 (x_2, y_2, z_2, ct_2)$; the resulting considerations would apply to *any* pair of events.

From the coordinates of the events E_1 and E_2, we construct the quantity

$$(x_2 - x_1)^2 + (y_2 - y_1)^2 + (z_2 - z_1)^2 - c^2 (t_2 - t_1)^2, \tag{2.13}$$

which is *invariant* under LORENTZ transformation, i.e.

$$(x'_2 - x'_1)^2 + (y'_2 - y'_1)^2 + (z'_2 - z'_1)^2 - c^2(t'_2 - t'_1)^2$$

$$= (x_2 - x_1)^2 + (y_2 - y_1)^2 + (z_2 - z_1)^2 - c^2(t_2 - t_1)^2,$$

or

$$s'^2 = s^2, \quad \text{say.} \tag{2.14}$$

In analogy with a similar quantity in the three-dimensional space, the quantity (2.14) may be looked upon as a measure of the 'separation', between the events E_1 and E_2, in the four-dimensional MINKOWSKI world; however, because of the presence of a negative sign in this expression, neither its physical significance nor its basic properties correspond to those of its counterpart,

$$r^2 = (x_2 - x_1)^2 + (y_2 - y_1)^2 + (z_2 - z_1)^2, \tag{2.15}$$

in the physical space. Thus, whereas r^2 is always non-negative, s^2 can be negative as well as positive ; moreover, while r^2 vanishes only when the space points (x_1, y_1, z_1) and (x_2, y_2, z_2) coincide, s^2 can vanish even if the events E_1 and E_2 do not coincide. In fact, in the four-dimensional world of events we can demarcate distinct 'regions' in which the quantity s^2 is positive, negative or zero.

For carrying out the study of these regions, let us take one of the events as the origin and denote it by the symbol O (0, 0, 0, 0). At the same time, let us denote the other event by the symbol E (x, y, z, ct). Then, the square of the 'separation' (or the 'interval' as it is usually referred to) between the two events would be given by

$$s^2 = x^2 + y^2 + z^2 - c^2t^2 = x'^2 + y'^2 + z'^2 - c^2t'^2. \tag{2.16}$$

We now define the various 'world regions' as follows :

  (i)   region A, with s^2 positive,
  (ii) region B, with s^2 negative,
  (iii) sheet C, with s^2 vanishing;

the sheet C, which partitions the MINKOWSKI world into regions A and B, is defined by the equation

$$(ct)^2 = x^2 + y^2 + z^2, \quad \text{i.e. } ct = r. \tag{2.17}$$

In analogy with the three-dimensional case, where

$$z^2 = x^2 + y^2$$

represents the surface of a cone, with apex at the origin and axis in the direction of the z-axis (the semi-vertical angle being $\pi/4$), the sheet C is also referred to as a 'cone', with apex at the origin of the MINKOWSKI world and axis in the direction of the t-axis (the semi-vertical angle again being $\pi/4$). Further, since the value of s^2 is zero on this 'cone', it is appropriately called a 'null cone'. Next, since (2.17) represents the totality of all the world points which, sooner or later, are reached by a light signal that is once at O, the null cone provides a space-time representation of the propagation of light; hence, it is also referred to as the 'light cone'.

We shall now discuss the relationship of the various events, such as E, with the event O. For this, we refer to Fig. 2.4 where only one of the spatial

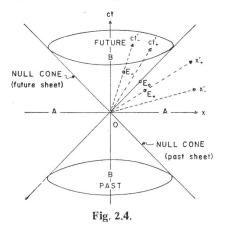

Fig. 2.4.

degrees of freedom is shown explicitly; another one is provided by a rotation about the t-axis, while the third one has got to be suppressed. The regions A and B, and the null cone, have been demarcated.

For an event in region A, such as E_+, s^2 is positive; hence, $r^2 > c^2 t^2$. It is, therefore, possible to introduce a new system of reference, S', such that the time coordinate, t', of the event vanishes[1]. In S', the event E_+ would appear to be *simultaneous* with the event O. However, the spatial separation between E_+ and O *cannot* be made to vanish; actually, this separation is *minimum* in

---

[1] A reference to Eqs. (1.38) shows that this would require a choice of $\mathbf{v}$ such that $(\mathbf{r} \cdot \mathbf{v}) = c^2 t$. With one condition and three unknowns (v_x, v_y and v_z), there are $(\infty)^2$ choices for the system S'.

the system S' (in which the two events are simultaneous), and is given by

$$r'^2 = r^2 - c^2t^2, \tag{2.18}$$

i.e. $r' = s$. Thus, in physical space, these events are *absolutely separated*. Further, since $t < r/c$, i.e. the time separation between the events is less than the time taken by light in traversing the spatial distance between them, the two events *cannot* be connected by any real physical process. Consequently, there *cannot* be a causal connection between such events.

For an event in region B, such as E_-, s^2 is negative; hence, $c^2t^2 > r^2$. It is, therefore, possible to introduce a new system of reference, S', such that the spatial coordinates, x', y' and z', of the event vanish.[1] In S', the event E_- would appear to be *spatially coincident* with the event O. However, the time separation between E_- and O *cannot* be made to vanish (nor can we change its sign); actually, the time separation is *minimum* in the system S' (in which the two events are spatially coincident), and is given by

$$c^2t'^2 = c^2t^2 - r^2, \tag{2.19}$$

i.e. $t' = \sqrt{(-s^2)}/c$. Thus, in time, the two events are *absolutely separated*; moreover, the events in the upper half of this region occur in *abolute future* with respect to O, while those in the lower half occur in *absolute past*. Further, since $t > r/c$, i.e. the time separation between the events is more than the time taken by light in traversing the spatial distance between them, the two events *can* be connected by real physical processes, such as the propagation of a signal or the motion of a particle. Consequently, there *can* be a causal connection between such events.

For events on sheet C, such as E_0, s^2 is equal to zero; hence, $r = ct$. Consequently, E_0 and O can be connected by a *light signal*, the world line joining them being a generator of the light cone.

By way of notation, the interval s between the event E and the event O is said to be *space-like*, *time-like* or *null* according as the event E lies in region A, in region B, or on sheet C, respectively.

It must be noted here that since the foregoing division of the world of events into various regions is based on the quantity s^2, which is *invariant*, the process of division is itself an invariant process. Thus, the relationship

---

[1] Here, the choice of S' is unique. We have: $\mathbf{v} = \mathbf{r}/t$, which means that the system S' is such as would follow the *uniform* motion of a particle corresponding to the world line OE_-. Obviously, the t'-axis would coincide with OE_-. Hence, t' determines the *proper time* of the motion concerned.

one event bears to another possesses an *absolute* character, irrespective of the observer!

2.4. The nature of a time track

The world line corresponding to the motion of a particle is usually referred to as the *time track* of the particle. Mathematically, it may be represented by the parametric equations

$$x_i = x_i(s), \quad i = 1, 2, 3, 4, \tag{2.20}$$

where x_i are the coordinates of a world point on the time track while s, the 'length' of the track from an arbitrary point on it, is taken as a parameter for the description of the track. Since the velocity of a material particle is always less than c, the quantity

$$\begin{aligned} ds^2 &= dx^2 + dy^2 + dz^2 - c^2 dt^2 \\ &= -c^2 dt^2 (1 - u^2/c^2), \end{aligned} \tag{2.21}$$

where u is the *instantaneous* velocity of the particle, is negative everywhere on the track[1]. We may, therefore, introduce a real parameter τ, defined by

$$\tau = \frac{s}{ic}, \tag{2.22}$$

so that

$$d\tau = \frac{ds}{ic} = dt\sqrt{1 - u^2/c^2}; \tag{2.23}$$

τ is thus identical with the *proper time* of the motion (as observed in the *instantaneous* rest system of the particle) and is a measure of the 'length' of the time track.

In the case of a *linear* track, which corresponds to a *uniform* motion (with velocity u_1, say), the length of a finite segment, AB, of the track is given by

$$s_{AB} = ic\, \tau_{AB} = ict_{AB}\sqrt{1 - u_1^2/c^2}, \tag{2.24}$$

---

[1] Consequently, if the time track passes through the origin O of the MINKOWSKI world, it *must* lie wholly in region B; this would also be true if we take any world point on the track as origin and construct regions A and B on the basis of that choice. In other words, the slope of the 'tangent' to the time track, drawn at any point, measured with respect to the time axis, *must* be less than the slope of a generator of the light cone; this should also be true for any 'chord' of the track.

where $t_{AB} (= t_B - t_A)$ is the corresponding time duration. On the other hand, if we connect the world points A and B by an *arbitrary* time track,[1] which would correspond to a *nonuniform* motion, the length of this segment would be

$$s^*_{AB} = ic\ \tau^*_{AB} = ic \int_A^B \sqrt{1 - u^2/c^2}\ dt, \qquad (2.25)$$

where u, the instantaneous velocity of motion, is no longer independent of t. To make a comparison of the *invariant* measures (2.24) and (2.25), we are at liberty to choose any inertial system that appears convenient for the purpose. Let us adopt a system whose time axis is parallel to the straight line AB, i.e. the one which follows a uniform motion connecting A and B. It is clear that in this system the right-hand sides of (2.24) and (2.25) would be (since $u'_1 = 0$)

$$ic\ t'_{AB} \qquad (2.24')$$

and

$$ic \int_A^B \sqrt{1-u'^2/c^2}\ dt', \qquad (2.25')$$

respectively. Now, since the factor $\sqrt{1 - u'^2/c^2}$ is everywhere less than (or, at certain points, equal to) unity, the integral in (2.25') is certainly less than t'_{AB}. Accordingly, the measure s_{AB}/ic for the linear track is greater than the corresponding measure s^*_{AB}/ic for an arbitrary track. Speaking of the square of the interval between the world points A and B, we conclude that it has a *minimum* value in the case of the linear track connecting the two events.

The foregoing result, which expresses an 'extremal' property of the world line corresponding to a free motion, is a particular case of a more general variational principle adopted in Sec. 6.5.

2.5.   Quasi-Euclidean geometry of the four-dimensional world

We have seen that the fundamental invariant of the homogeneous LORENTZ transformations is

$$s^2 \equiv x^2 + y^2 + z^2 - c^2 t^2$$
$$= x'^2 + y'^2 + z'^2 - c^2 t'^2 \equiv s'^2. \qquad (2.26)$$

---

[1] Of course, this track must also conform to the conditions outlined in the preceding footnote; otherwise, it is quite arbitrary.

If we employ the set of coordinates

$$x_1 = x, \quad x_2 = y, \quad x_3 = z \quad \text{and} \quad x_4 = ict \tag{2.27}$$

in S and the corresponding ones,

$$x_1' = x', \quad x_2' = y', \quad x_3' = z' \quad \text{and} \quad x_4' = ict', \tag{2.28}$$

in S', then (2.26) assumes the form

$$s^2 \equiv x_1^2 + x_2^2 + x_3^2 + x_4^2 = x_1'^2 + x_2'^2 + x_3'^2 + x_4'^2 \equiv s'^2. \tag{2.29}$$

Written in this form, the analogy with the three-dimensional world becomes complete, except for the fact that the fourth square in the sum here is really negative. Nevertheless, Eq. (2.29) enables us to apply rather freely the rules of Euclidean geometry to the four-dimensional continuum of x, y, z and ict. However, because of the imaginary character of the fourth-coordinate employed, the geometry of this continuum is only formally identical with Euclidean geometry; that is why it is usually referred to as *quasi-Euclidean*.

The immediate advantage of the transition from (2.26) to (2.29) is that now the rotation of the axes of an inertial observer would not lead to an oblique coordinate system, as met with in Sec. 2.2. On the other hand, these rotations would now involve only rectangular coordinate systems.[1] LORENTZ transformations would, therefore, become *linear orthogonal*.

Confining to the case of homogeneous transformations, viz. the ones that do not involve a displacement of the origin, we can write

$$x_i' = \sum_{k=1}^{4} \alpha_{ik} \, x_k, \qquad i = 1, 2, 3, 4, \tag{2.30}$$

where α_{ik}'s are the coefficients of transformation; geometrically, the coefficient α_{ik} may be understood as the cosine of the angle between the x_i'-axis and the x_k-axis. In view of the specific nature of the coordinates (2.27) and

---

[1] Referring to the case of a special LORENTZ transformation, see Sec. 2.2, the hyperbola would now become a circle, and the x'-axis would now have the *same* sense of rotation with respect to the x-axis as t'-axis would have with respect to t-axis. Thus, the (x', t')-system would also be rectangular.

(2.28), the coefficients α_{pq} (p, $q = 1$, 2, 3) and α_{44} would be real while the coefficients α_{p4} and α_{4q} would be purely imaginary.

From now onward, we adopt the so-called *summation convention*, i.e. if an index occurs *twice* in a term, a summation over all possible values of that index is *automatically* implied. The convention applies to all the equations that appear in the sequel, unless a statement is made to the contrary.

Our problem now consists in studying the invariance of the quantity

$$s^2 \equiv x_i\, x_i = x'_i\, x'_i \equiv s'^2 \tag{2.31}$$

under the transformations

$$x'_i = \alpha_{ik}\, x_k, \quad i = 1, 2, 3, 4. \tag{2.32}$$

Substituting (2.32) in (2.31), we get

$$x_i\, x_i = (\alpha_{ik}\, x_k)\, (\alpha_{il}\, x_l)$$
$$= (\alpha_{ik}\, \alpha_{il})\, x_k\, x_l;$$

a comparison of coefficients on the two sides gives

$$\alpha_{ik}\, \alpha_{il} = \delta_{kl}, \tag{2.33}$$

where δ_{kl} is the KRONECKER symbol:

$$\delta_{kl} = \begin{cases} 1 & \text{if } k = l, \\ 0 & \text{if } k \neq l. \end{cases} \tag{2.34}$$

Conditions (2.33) are the so-called *orthogonality conditions* which the coefficients of transformation must satisfy in order that s^2 be an invariant quantity; they also imply that the transformations under consideration are orthogonal, i.e. the ones that connect rectangular coordinates of one observer with similar coordinates of another. These conditions are, in all, ten in number (four for $k=l$, six for $k \neq l$); consequently, they leave, for the sixteen coefficients α_{ik}, precisely six degrees of freedom, as must be the case with homogeneous transformations (see Secs. 1.6 and 2.1).

From Eqs. (2.32)−(2.34), we obtain

$$\alpha_{ik}\, x'_i = \alpha_{ik}\, (\alpha_{il}\, x_l) = (\alpha_{ik}\, \alpha_{il})\, x_l$$
$$= \delta_{kl}\, x_l = x_k, \quad k = 1, 2, 3, 4, \tag{2.35}$$

which give transformations *inverse* to the ones we started with. The invariance (2.31), under the transformations (2.35), leads to the orthogonality conditions

$$\alpha_{ik}\, \alpha_{jk} = \delta_{ij}\,;\tag{2.36}$$

however, these conditions are essentially the same as the ones embodied in (2.33).

From the coefficients α_{ik} we construct the determinant

$$\alpha = |\alpha_{ik}| = \begin{vmatrix} \alpha_{11} & \alpha_{12} & \alpha_{13} & \alpha_{14} \\ \alpha_{21} & \alpha_{22} & \alpha_{23} & \alpha_{24} \\ \alpha_{31} & \alpha_{32} & \alpha_{33} & \alpha_{34} \\ \alpha_{41} & \alpha_{42} & \alpha_{43} & \alpha_{44} \end{vmatrix}.\tag{2.37}$$

For the evaluation of this determinant, we consider its square, viz.

$$\alpha^2 = |\,(\alpha_{il}\, \alpha_{lk})\,|\,.\tag{2.38}$$

Now, if we write one of the factor determinants with its rows and columns interchanged it would not make any difference to the result. Equation (2.38) would, however, become

$$\alpha^2 = |\,(\alpha_{li}\, \alpha_{lk})\,|\quad \text{or}\quad |\,(\alpha_{il}\, \alpha_{kl})\,|,\tag{2.38'}$$

which, on making use of the orthogonality conditions (2.33) or (2.36), gives

$$\alpha^2 = |\,\delta_{ik}\,| = 1.\tag{2.39}$$

Thus, the value of the determinant α is either $+1$ or -1.

For a *continuous* transformation — the one obtainable from the *identity* transformation $(S \to S)$ by a *continuous* rotation of the axes — the value of the determinant must be $+1$. [For, a *continuous* rotation of the axes implies a *continuous* variation of the coefficients α_{ik} (starting from the initial values, $\alpha_{ik} = \delta_{ik}$, pertaining to the identity transformation), which in turn implies a *continuous* variation of α (starting from the value $+1$ for the identity transformation); clearly, α must remain $+1$]. Such transformations are usually referred to as *proper* transformations.

On the other hand, we can have transformations which involve 'reflections' of the coordinate axes[1] and may, therefore, be *discontinuous* (in the sense that they may not be obtainable from the identity transformation by a *continuous* rotation of the axes). Of these, the ones that involve an *odd* number of reflections would have : $\alpha = -1$. Such transformations are necessarily discontinuous; they are usually referred to as *improper* transformations. Of course, transformations that involve an *even* number of reflections must again have: $\alpha = +1$.

2.6. Elements of Cartesian tensors

In this section we propose to outline the elements of *tensor analysis* which has proved exceedingly useful in the analytical development of the theory of relativity. Tensors are mathematical entities defined in respect of their behavior under coordinate transformations ; among them there is a class, called *Cartesian tensors*, which are concerned exclusively with linear orthogonal transformations. Since, in the special theory of relativity, we are concerned only with linear orthogonal transformations, our interest in tensors would be confined, for the time being, to Cartesian tensors alone. The study of general tensors will be taken up in Chapter 6.

Let us consider two neighboring points $P(x_i)$ and $Q(x_i + dx_i)$ in the four-dimensional continuum. The infinitesimal line element connecting P and Q may be looked upon as a 'displacement vector' with components dx_i whose relative magnitudes determine the direction of the vector while the sum of their squares determines its magnitude. Under a coordinate transformation, these components would change over to

$$dx_i' = \alpha_{ik}\, dx_k\,, \qquad (2.40)$$

where $\alpha_{ik}\,(=\partial x_i'/\partial x_k)$ are the coefficients of transformation; see (2.32).

Now, a *four-vector* is defined as an entity which, in every coordinate system, has four components a_i such that they transform *in the same manner* as the components of a line element. Thus, we have

$$a_i' = \alpha_{ik}\, a_k, \qquad (2.41)$$

along with their inverse

$$a_k = \alpha_{ik}\, a_i'\,; \qquad (2.42)$$

[1] These include 'time reversal' as well.

see (2.35). By virtue of the orthogonality conditions, satisfied by the coefficients α_{ik}, the quantity $(a_i a_i)$ is an invariant:

$$a'_i\, a'_i = a_i a_i\; ; \tag{2.43}$$

this is equal to the square of the magnitude of the four-vector (a_i) and is called its *norm*. In analogy with the invariant $x_i x_i\ (\equiv s^2)$, we adopt the following nomenclature:

The four-vector (a_i) would be called *space-like*, *time-like* or *null* according as its norm is positive, negative or zero, respectively.

From two four-vectors (a_i) and (b_i) we can form a linear combination $(\lambda a_i + \mu b_i)$, where λ and μ are certain constants ; this combination is again a four-vector. Its norm,

$$\lambda^2(a_i a_i) + 2\lambda\mu(a_i b_i) + \mu^2(b_i b_i)$$

would naturally be invariant. However, $(a_i a_i)$ and $(b_i b_i)$ are themselves invariant; hence, the quantity $(a_i b_i)$ would also be invariant. This quantity is called the *scalar product* (sometimes, the *inner product*) of the given vectors.

We speak of two vectors as being *orthogonal* to each other if their scalar product vanishes.

In any system of coordinates, a given four-vector may be split into a *spatial* part and a *temporal* part:

$$(a_i) = (\mathbf{a}, a_4); \tag{2.44}$$

however, this splitting is not an invariant process.[1] If a four-vector is space-like we can always transform to a new system S' in which the temporal part of the vector vanishes; this requires that the t'-axis of S' be orthogonal to the four-vector concerned. On the other hand, if a four-vector is time-like we can transform to a new system S' in which the spatial part of the vector vanishes; this requires that the t'-axis of S' be parallel to the four vector.

Next, we define a *four-tensor* of rank 2 as an entity which, in every coordinate system, has sixteen components t_{ik} such that they transform accord-

---

[1] For instance, in the case of a general LORENTZ transformation, as in (1.38), the spatial and temporal parts of a four-vector would transform according to the formulae

$$\mathbf{a}' = \mathbf{a} + \frac{(\mathbf{a} \cdot \mathbf{v})\, \mathbf{v}}{v^2}\ (\gamma - 1) + i\gamma\ \frac{\mathbf{v}}{c}\ a_4, \qquad a'_4 = \gamma\left[a_4 - i\ \frac{(\mathbf{a} \cdot \mathbf{v})}{c} \right].$$

ing to the formulae

$$t'_{ik} = \alpha_{il}\, \alpha_{km}\, t_{lm},\qquad(2.45)$$

along with their inverse

$$t_{lm} = \alpha_{il}\, \alpha_{km}\, t'_{ik},\qquad(2.46)$$

where the coefficients α_{ik} are the same as the ones appearing in the transformation of the components of a four-vector.

Now, just as a four-vector can be split into a spatial part and a temporal part, a four-tensor of rank 2 can be split into a 'purely spatial' part,

$$t_{pq}\ (p,\, q = 1,\, 2,\, 3),$$

a 'purely temporal' part t_{44} and two 'mixed' parts t_{p4} and t_{4q}. These parts behave very differently if the tensor is subjected to certain specialized transformations; for instance, under an ordinary rotation in the three-dimensional space (for which $\alpha_{p4} = \alpha_{4q} = 0$ and $\alpha_{44} = 1$), formulae (2.45) become

(i) $t'_{pq} = \alpha_{p\lambda}\, \alpha_{q\mu}\, t_{\lambda\mu}$, $\lambda,\, \mu$ going from 1 to 3,

which means that the purely spatial part behaves like a *three*-tensor of rank 2,

(ii) $t'_{p4} = \alpha_{p\lambda}\, t_{\lambda 4}$ and $t'_{4q} = \alpha_{q\mu}\, t_{4\mu}$,

which means that the mixed parts behave like *three*-vectors, and

(iii) $t'_{44} = t_{44}$,

which means that the purely temporal part behaves like an invariant.[1]

The Kronecker symbol δ_{ik}, defined by (2.34), is a second-rank tensor of an especially simple character. On coordinate transformation, its components become

$$\delta'_{ik} = \alpha_{il}\, \alpha_{km}\, \delta_{lm} = \alpha_{il}\, \alpha_{kl} = \delta_{ik},\qquad(2.47)$$

i.e. their numerical values remain unchanged; consequently, the definition

---

[1] Under the transformation considered, the spatial and temporal parts of a four-vector would also behave like a *three*-vector and an invariant, respectively.

(2.34) holds in *all* coordinate systems.   Actually, among tensors of rank 2, this is the only one having this particular character.

We shall now mention some of the basic properties of second-rank four-tensors (hereafter referred to simply as tensors):

(i) A linear combination of two tensors is again a tensor.

(ii) If (t_{ik}) is a tensor, then (t_{ki}) is also a tensor; this is known as the *transpose* of the given tensor.   Symbolically, one may write

$$(t_{ki}) = (\widetilde{t_{ik}}). \tag{2.48}$$

(iii) If (a_i) and (b_i) are two vectors, then the quantities

$$(t_{ik}) = (a_i b_k) \tag{2.49}$$

constitute a tensor ; this is referred to as the *direct product* (or the *outer product*) of the given vectors.

(iv) A tensor (t_{ik}) is said to be *symmetric* if, for all i and k,

$$t_{ki} = t_{ik}. \tag{2.50}$$

For instance, the tensor

$$(a_i b_k + a_k b_i)$$

is symmetric.   Because of the conditions (2.50), only ten of the sixteen components of a symmetric tensor are truly independent.

(v) A tensor (t_{ik}) is said to be *antisymmetric* if, for all i and k,

$$t_{ki} = -t_{ik}. \tag{2.51}$$

For instance, the tensor

$$(a_i b_k - a_k b_i)$$

is antisymmetric.   Because of the conditions (2.51), the diagonal components of an antisymmetric tensor are identically zero ; accordingly, the number of independent components in this case is only six.

(vi) The property of a tensor being symmetric or antisymmetric is an invariant property.   Thus, a tensor which is symmetric (or antisymmetric) in one system of reference is symmetric (or antisymmetric) in all systems of reference.

(vii) Two tensors (a_{ik}) and (b_{ik}) are said to be *equal* if, for all i and k,

$$a_{ik} = b_{ik}. \tag{2.52}$$

The property of two tensors being equal is obviously an invariant property; consequently, *a tensor equation is an invariant equation.* This suggests that the most direct way of putting a physical law into a *covariant* form would be to formulate it as a tensor equation!

We shall now introduce tensors of *arbitrary rank*. By definition, a tensor of rank n is an entity which, in every coordinate system, has 4^n components $t_{ikl} \ldots$ such that they transform according to the formulae

$$t'_{ikl} \ldots = \alpha_{im}\, \alpha_{kn}\, \alpha_{lp} \ldots t_{mnp} \ldots, \tag{2.53}$$

along with their inverse

$$t_{mnp} \ldots = \alpha_{im}\, \alpha_{kn}\, \alpha_{lp} \ldots t'_{ikl} \ldots, \tag{2.54}$$

the coefficients α_{ik} being the same as before. Formally speaking, each of the n indices acts in accordance with the rule laid down for a vector. Accordingly, a vector may be looked upon as a tensor of rank 1 and an invariant as a tensor of rank 0.

We now introduce an important operation which can be carried out with tensors of rank $n \geqslant 2$, viz. equating any two of the indices, say i and k, and summing over all possible values of an index (here, from 1 to 4); thus, we obtain

$$t_{ikl} \ldots \rightarrow t_{iil} \ldots (= T_l \ldots, \text{ say}). \tag{2.55}$$

It can be verified that the resulting set of quantities $T_l \ldots$ is again a tensor (of course, of rank $n - 2$). This process is referred to as the *contraction* of the given tensor and the resulting tensor is called the *contracted tensor.* Clearly, a tensor of rank 2, (t_{ik}), would give, on contraction, an invariant t_{ii} (which, by definition, is the sum of the diagonal components of the given tensor).

A tensor of rank m, on *direct* multiplication with a tensor of rank n, leads to a tensor of rank $m + n$:

$$(t_{ik} \ldots {}_{lm} \ldots) = (a_{ik} \ldots b_{lm} \ldots); \tag{2.56}$$

this is called the *direct* or the *outer* product of the given tensors. Applying contraction to this product, one obtains the so-called *inner* product of the given tensors.

Example 1. Two vectors, on direct multiplication, lead to a tensor of rank 2 which, on contraction, yields an invariant:

$$(t_{ik}) = (a_i b_k) \rightarrow (a_i b_i). \tag{2.57}$$

This will be identified as the scalar product of the given vectors.

Example 2. A tensor of rank 2 and a vector may enter into an inner product, yielding a new vector:

$$(t_{ik} a_m) \rightarrow t_{ik} a_k (= b_i, \quad \text{say}). \tag{2.58}$$

Another inner product obtainable in this case is

$$(t_{ik} a_m) \rightarrow t_{ik} a_i (= c_k, \quad \text{say}). \tag{2.59}$$

Example 3. A tensor of rank 2, on direct multiplication with itself, leads to a tensor of rank 4 which, on two contractions, yields an invariant:

$$(t_{iklm}) = (t_{ik} t_{lm}) \rightarrow t_{ik} t_{ik}. \tag{2.60}$$

Sometimes one would wish to ascertain whether a given set of quantities constitutes a tensor. The direct method for doing this would be to check whether the given quantities conform to the transformation equations appropriate to the components of a tensor. However, there is a useful alternative which may be stated as follows:

A given set of 4^n quantities, where n is an integer, would constitute a tensor, of rank n, provided that an inner product of this set with an *arbitrary* tensor is itself a tensor.

This rule is commonly known as the *quotient law*.

It appears worthwhile to introduce at this stage the concept of *pseudo-tensors*. A pseudo-tensor of rank n differs from a tensor of rank n in that its transformation formulae contain an extra factor $\alpha (= |\alpha_{ik}|$; see (2.37)):

$$t'_{ikl} \ldots = \alpha \cdot \alpha_{im} \alpha_{kn} \alpha_{lp} \ldots t_{mnp} \ldots , \tag{2.53'}$$

and

$$t_{mnp} \ldots = \alpha \cdot \alpha_{im} \alpha_{kn} \alpha_{lp} \ldots t'_{ikl} \ldots . \tag{2.54'}$$

It is obvious that the product of an even number of pseudo-tensors is a tensor (since $\alpha^2 = 1$), while the product of an odd number of pseudo-tensors is again a pseudo-tensor. In general, all operations that can be carried out with tensors are applicable to pseudo-tensors as well.

For proper transformations, $\alpha = +1$; the distinction between a tensor and a pseudo-tensor, then, disappears.

A striking example of a pseudo-tensor of rank 0, which may be called a pseudo-scalar, is provided by the 'volume element' in the four-dimensional continuum; this follows from the fact that the only factor appearing in its transformation is the Jacobian

$$\left| \frac{\partial x'_i}{\partial x_k} \right| = |\alpha_{ik}| = \alpha. \tag{2.61}$$

Finally, we consider *tensor fields*. One speaks of a tensor field when with each point (x_i) of the continuum (or of a 'region' therein) is associated a tensor, whose components are continuously varying functions of the coordinates x_i. In particular, we have a *scalar field* if with each point (x_i) is associated an *invariant* quantity, which is a continuously varying function of the x_i:

$$\varphi'(x'_i) = \varphi(x_i). \tag{2.62}$$

In general, φ' will not be the same function of the coordinates x'_i as φ is of x_i; thus, a scalar function is generally not a form-invariant function. Of course, if φ is a function of the quantity (2.31) alone, which itself is an invariant, then it will also be form-invariant.

Similarly, we have a *vector field*, with components $a_i(x)$ in S and $a'_i(x')$ in S', such that

$$a'_i(x') = \alpha_{ik} a_k(x), \tag{2.63}$$

and so on for tensor fields of higher rank.

We can perform operations of differentiation on a given tensor field and obtain new tensor fields. For instance, from the scalar field (2.62) we can obtain a vector field, whose components are $\partial\varphi/\partial x_i$ in S and $\partial\varphi'/\partial x'_i$ in S':

$$\frac{\partial\varphi'}{\partial x'_i} = \frac{\partial\varphi'}{\partial\varphi} \frac{\partial\varphi}{\partial x_k} \frac{\partial x_k}{\partial x'_i} = \alpha_{ik} \frac{\partial\varphi}{\partial x_k}; \tag{2.64}$$

here, we have made use of the result

$$\frac{\partial x_k}{\partial x_i'} = \alpha_{ik},$$ (2.65)

which follows from (2.35). The vector field (2.64) is the *Gradient*[1] of the scalar field (2.62).

Next, we obtain from the vector field (2.63) a tensor field, whose components in S are $\partial a_i/\partial x_k$; this field, on combination with its transpose, leads to another tensor field, viz.

$$\left[\frac{\partial a_k}{\partial x_i} - \frac{\partial a_i}{\partial x_k} \right],$$ (2.66)

which is antisymmetric in character. The tensor field (2.66) is the *Curl* of the vector field (2.63). Contracting the tensor field ($\partial a_i/\partial x_k$), we obtain a scalar field ($\partial a_i/\partial x_i$), which is the *Divergence* of the vector field (a_i).

From a scalar field φ, we can obtain another scalar field by first taking the Gradient of the original field and then the Divergence of the resulting vector field, viz. $\partial^2\varphi/(\partial x_i\,\partial x_i)$. The scalar operator $\partial^2/(\partial x_i\,\partial x_i)$ is called the D'-Alembertian operator and is usually denoted by the symbol $\square$; this operator is the four-dimensional analogue of the Laplacian operator.

The foregoing operations can be carried out with tensor fields of arbitrary rank as well. Thus, from a tensor field of rank n we can obtain its Gradient (of rank $n + 1$), its Curl (of rank $n + 1$), its Divergence (of rank $n - 1$) and finally, by the application of the D'Alembertian operator, another tensor field (of rank n).

Problems

2.1. Using a MINKOWSKI diagram, demonstrate the phenomenon of time dilation. Also verify the reciprocal character of this phenomenon.

2.2. Illustrate Problems 1.2 and 1.6 with the help of MINKOWSKI diagrams.

2.3. Write down the coefficients α_{ik} for (i) a special and (ii) a general LORENTZ transformation (without rotation), and verify that they satisfy the conditions of orthogonality.

2.4. Discuss the properties of an *infinitesimal* LORENTZ transformation

$$x_i' = \alpha_{ik}\, x_k,$$

where

$$\alpha_{ik} = \delta_{ik} + \varepsilon_{ik} \qquad (\varepsilon_{ik} \ll 1).$$

---

[1] It is customary to capitalize these terms so as to distinguish them from their counterparts in the ordinary *three*-dimensional space.

Show that the orthogonality conditions now reduce to the requirement that the quantities ε_{ik} be antisymmetric in i and k.

2.5. Prove that the four-vector joining any two events in an inertial observer's *instantaneous* space ($t=$const.) must be orthogonal to his time axis.

2.6. (i) Prove that any four-vector orthogonal to a time-like vector must be space-like, but that two space-like vectors can be mutually orthogonal.

(ii) Prove that any four-vector orthogonal to a null vector (other than a null vector itself) must be space-like.

2.7. Prove that the sign of the fourth component of a time-like or a null four-vector is invariant under a general LORENTZ transformation. Illustrate this point with the help of a MINKOWSKI diagram.

CHAPTER 3

MECHANICS

3.1. Mass, momentum and their conservation

The most fundamental concept we come across on passage from kinematics to mechanics is that of *mass*, which is commonly defined as the *quantity of matter contained in a body*. This definition has been useful in the elaboration of basic physical theories, particularly in the interpretation of the law of *conservation of mass*. However, when examined closely, it does not appear to be self-supporting, for it meets serious conceptual difficulties when applied to the very entities of which matter in bulk is made up, viz. elementary particles. In fact, it would be far more satisfactory to base our definition of mass on some physically realizable *property* of matter, rather than on its *quantity*.

In this connection one recalls a fundamental property of matter, viz. the resistance it offers to an external agency which tends to change its state of motion, i.e. its *inertia* towards any change that is intended to be brought about in its state of motion. The mass of a material body may, then, be defined as a *measure of its inertia*; consequently, this may be referred to as the 'inertial' mass of the body.[1]

Now, one of the basic assumptions in Newtonian mechanics has been that the mass of a body, being one of its inherent characteristics, is *independent of its*

---

[1] There is yet another way of introducing the concept of mass, that is, through the gravitational attraction between two given bodies ; the resulting concept is referred to as the 'gravitational' mass. Further details, especially the relationship between the 'inertial' mass and the 'gravitational' mass, are discussed in Sec. 7.1.

state of motion with respect to the observer.[1]   Thus, equal forces impressed upon a body would produce equal accelerations, whatever the instantaneous velocity of the body.   Hence, if we continue to apply a force indefinitely, the velocity of the given body would go on increasing at a *constant* rate and finally it could exceed *any* pre-assigned value.   This, however, negates our earlier result that there exists an upper limit, given by the velocity of light in free space, on the velocities that material objects can have.   The conclusion is, therefore, inescapable that *the inertia of a body must increase with velocity, tending to infinity as the velocity approaches c.*   Thus, a given force acting on a body should, in the initial stages, produce an effectively *constant* acceleration, but its influence should gradually decrease as the velocity of the body increases until finally, as the velocity approaches its limiting value *c*, the influence of the force should tend to vanish.   In view of this, we may write

$$m_u = m_0 f(u), \tag{3.1}$$

where m_u denotes the mass of the body when it is moving with velocity u with respect to the observer while $f(u)$ is a (yet unknown) function of u, such that

$$f(u) \rightarrow \begin{cases} 1 & \text{as } u \rightarrow 0, \tag{3.2} \\ \infty & \text{as } u \rightarrow c. \tag{3.3} \end{cases}$$

The quantity m_0 denotes the 'rest' mass of the body (or, since it is measured by an observer moving with the body, its 'proper' mass) ; m_u, on the other hand, is referred to as the 'relativistic' mass (or simply the mass) of the body.

Our problem then is to determine the form of the function $f(u)$.   However, before we do this we must introduce the concept of *momentum*, which is commonly defined as the *quantity of motion possessed by a body* (as given by the product of the quantity of matter contained in it and its velocity).   And since the quantity of matter contained in a body was recognized as its mass, the momentum was taken to be the product of mass and velocity.   We, however, noticed that mass is better defined as a measure of inertia of the body, i.e. the 'inertial' mass.   It is, therefore, appropriate if we define momentum as the product of *inertial* mass and velocity :

$$\mathbf{p} = m_u \mathbf{u} = m_0 f(u) \mathbf{u}. \tag{3.4}$$

---

[1] This seemed to be fully borne out by experiment until the beginning of the present century when observations on β-particles emitted by radioactive materials provided evidence to the contrary ; see the relevant account on p. 81.

The nonrelativistic counterpart of this definition was

$$\mathbf{p}_{N.R.} = m_0 \mathbf{u}. \tag{3.5}$$

We shall now consider the law of conservation of momentum, which states that *for a system of mutually interacting bodies the vector sum[1] of their momenta remains unchanged, provided that all the forces acting on the system can be ascribed to the members of the system themselves.* Now, by the principle of relativity, the validity of this law cannot be restricted to any privileged subgroup of the inertial systems of reference ; in other words, its validity in any one inertial system of reference should imply its validity in all inertial systems.[2] However, it can be seen that, with momentum defined as in (3.5), the foregoing requirement of relativity can be satisfied only if the transformation of velocity vectors conforms to the Newtonian formulae (1.56) and not if it conforms to the relativistic formulae (1.51) and (1.52). Since only relativistic formulae are the correct ones to use, the law of conservation of momentum would violate the principle of relativity if we stick to definition (3.5).

On the other hand, the definition (3.4), because of the presence of an extra factor $f(u)$, is relatively *more involved* than definition (3.5), but this is the case with transformation formulae (1.51) and (1.52) in comparison with formulae (1.56). One, therefore, hopes that, with a suitable choice of the function $f(u)$, the afore-mentioned requirement of relativity may be satisfied when, both for the definition of momentum and for the transformation of velocities, use is made of relativistic expressions. As will be seen in the following section, this hope is indeed fulfilled.

3.2. Variation of mass with velocity

It was LORENTZ who gave, in his monumental work of 1904, the formula

$$m_u = \frac{m_0}{\sqrt{1 - u^2/c^2}} \tag{3.6}$$

for the mass of an electron as a function of its velocity; this formula was

---

[1] In Secs. 3.1—3.4, only three-vectors appear.

[2] This becomes all the more obligatory when one notes that the law of conservation of momentum is directly related to the 'homogeneity' of space : see L. D. LANDAU and E. M. LIFSHITZ, *Mechanics* (Pergamon Press, London, 1960), Sec. 7. Since the homogeneity of space is common to all inertial observers, so must be the validity of the law of momentum conservation.

obtained on the assumption that electrons in motion underwent FITZGERALD contraction.   In his paper of 1905, EINSTEIN also touched upon this question but did not derive the formula (3.6); of course, he did obtain correct expressions for the so-called 'longitudinal' mass of the electron and its kinetic energy; see the following section.   It was PLANCK[1] who, in 1906, carried out a systematic study of relativistic dynamics and discovered equations which were to replace the Newtonian equations of motion of a material particle. PLANCK thereby obtained correct relativistic expressions for momentum and kinetic energy, and, hence, for the function $f(u)$ of Eqs. (3.1)—(3.4).   However, the arguments on which PLANCK based his derivations were not regarded as fully convincing.   This led LEWIS and TOLMAN[2] to follow an approach of a very different character; they obtained the relevant formulae by considering a collision between two particles and stipulating that the laws of conservation of mass and momentum hold *equally well* for the inertial observers S and S' who watch and analyze the details of the collision process.

Clearly, the LEWIS-TOLMAN approach is in the same spirit as advocated towards the end of the preceding section.   However, we consider here a more generalized case, viz. that of a system of n particles P_i ($i=1, 2, \ldots, n$), and proceed in a manner which does not require any detailed information regarding the mode of interaction of these particles.[3]   At any given instant of time, let $\mathbf{u}_i$ and $\mathbf{u}'_i$ be the velocities of the i-th particle as measured in systems S and S', respectively.   Defining γ_i and γ'_i by the relations

$$\gamma_i = \frac{1}{\sqrt{1 - u_i^2/c^2}} \quad \text{and} \quad \gamma'_i = \frac{1}{\sqrt{1 - u_i'^2/c^2}},$$

we obtain from (1.54)

$$\gamma_i = \gamma'_i \, \gamma \left[1 + \frac{(\mathbf{u}'_i \cdot \mathbf{v})}{c^2} \right], \tag{3.7}$$

where

$$\gamma = \frac{1}{\sqrt{1 - v^2/c^2}}. \tag{3.8}$$

[1] M. PLANCK, *Verh. d. Deutschen Phys. Ges.* **4**, 136 (1906).

[2] G. N. LEWIS and R. C. TOLMAN, *Phil. Mag.* **18**, 510 (1909).

[3] This derivation is an adaptation from W. H. McCREA, *Relativity Physics* (Methuen, London, 1960).

v being the velocity of system S' with respect to system S. Further, dividing (1.52) by (1.54) we obtain

$$\gamma_i \, \mathbf{u}_i = \gamma_i' \left[\mathbf{u}_i' + (\gamma - 1) \frac{(\mathbf{u}_i' \cdot \mathbf{v}) \, \mathbf{v}}{v^2} + \gamma \mathbf{v} \right]. \qquad (3.9)$$

Now, the total mass and total momentum of the system under consideration, as measured in S, would be

$$\sum_i m_i \quad \text{and} \quad \sum_i (m_i \mathbf{u}_i), \qquad (3.10)$$

respectively; here, m_i denotes the inertial mass of the i-th particle. Making use of (3.9), the second quantity in (3.10) can be written as

$$\sum_i (m_i \mathbf{u}_i) = \sum_i \left[m_i \frac{\gamma_i'}{\gamma_i} \left\{ \mathbf{u}_i' + (\gamma - 1) \frac{(\mathbf{u}_i' \cdot \mathbf{v}) \, \mathbf{v}}{v^2} + \gamma \mathbf{v} \right\} \right]$$

$$= \sum_i (m_i' \, \mathbf{u}_i') + (\gamma - 1) \frac{\left\{ \sum_i (m_i' \, \mathbf{u}_i') \cdot \mathbf{v} \right\} \mathbf{v}}{v^2} + \gamma \mathbf{v} \sum_i (m_i'),$$

$$(3.11)$$

where

$$m_i' = m_i \frac{\gamma_i'}{\gamma}. \qquad (3.12)$$

Next, making use of (3.7), the first quantity in (3.10) can be written as

$$\sum_i m_i = \sum_i \left[m_i \frac{\gamma_i'}{\gamma_i} \gamma \left\{ 1 + \frac{(\mathbf{u}_i' \cdot \mathbf{v})}{c^2} \right\} \right]$$

$$= \gamma \left[\sum_i m_i' + \frac{\left\{ \sum_i (m_i' \, \mathbf{u}_i') \cdot \mathbf{v} \right\}}{c^2} \right], \qquad (3.13)$$

where m_i' is the same as before.

We have thus expressed quantities (3.10) as linear functions of the quantities

$$\sum_i m_i' \quad \text{and} \quad \sum_i (m_i' \, \mathbf{u}_i'), \qquad (3.14)$$

with coefficients depending *solely* on the relative velocity **v** of the two systems of reference.   Moreover, one can solve Eqs. (3.11) and (3.13) for the quantities (3.14) and obtain relations expressing them as linear functions of the quantities (3.10), with coefficients similar to the ones we now have.[1]   It then follows that a *necessary* and *sufficient* condition for the quantities (3.10) to be conserved throughout the motion of the particles is that the quantities (3.14) be conserved, and *vice versa*.   But we are already asserting that the validity of the laws of conservation of mass and momentum in any given inertial system, say S, must imply their validity in any other such system, say S'.   It is, therefore, essential that the observer S' recognize quantities (3.14) as the total mass and total momentum, respectively, of the system under consideration.   However, for this to have an unambiguous meaning the quantities m'_i must be independent of anything relating to the system S, such as the particle velocities u_i or the relative velocity **v** of the systems.   For the same reason, the quantities m_i must be independent of anything relating to the system S', such as u'_i and **v**.   Thus, we must have, see (3.12),

$$\frac{m'_i}{\gamma'_i} = \frac{m_i}{\gamma_i} = m_{i0},\tag{3.15}$$

where m_{i0} should be independent of anything relating to the system S or S' (or, for that matter, to *any* system of reference). It must, then, be a characteristic of the particle itself; actually, it is just the 'proper' mass of the particle.

Dropping the index i from (3.15), we can write, for *any* material particle,

$$m = \frac{m_0}{\sqrt{1 - u^2/c^2}},\tag{3.16}$$

where u is the velocity of the particle with respect to the observer.   Correspondingly, the momentum of the particle would be

$$\mathbf{p} = \frac{m_0\mathbf{u}}{\sqrt{1 - u^2/c^2}}.\tag{3.17}$$

Formulae (3.16) and (3.17) can be put to test through experiments on the

---

[1] Comparing (3.11) and (3.13) with (1.39), one finds that the total momentum and total mass of the given system transform in the same way as the quantities **r** and t; see, in this connection, Sec. 3.5.

deflection of fast-moving cathode rays or β-rays under the influence of electric and magnetic fields. In the earliest of such experiments, carried out by KAUFMANN,[1] a qualitative evidence for the increase of mass with velocity was indeed obtained; however, the precise validity of the relativistic expressions could not be established. Later experiments, carried out by BUCHERER and others,[2] showed that the relativistic expressions were accurate to a high degree of precision for (u/c) ranging from 0.38 to 0.69.

Nowadays, relativistic formulae are taken for granted in all considerations, theoretical or experimental, on high-energy particles, whether in cosmic rays or in man-made machines. In the latter case, the synchronization of the acceleration process is liable to be upset as a result of increase in the particle mass as its speed becomes a significant fraction of the speed of light. The design of the machine must, then, be such that the relativistic effect does not impair the mechanism of further acceleration. Relativistic formulae of mass and momentum are employed towards this end, and the resulting design is found to work exceedingly well. This may, therefore, be regarded as an indirect, but convincing, evidence for the relativistic expressions.

In passing, mention may be made of an experiment, carried out by ROGERS, McREYNOLDS and ROGERS,[3] which verified the relativistic mass formula for electrons in a range of velocities up to $0.8c$.

3.3. Mass-energy equivalence

We shall now derive an appropriate expression for the kinetic energy of a moving particle. For this, we must first of all introduce the concept of *force*. This may be done in terms of NEWTON's second law of motion, according to which the rate of change of momentum of a body, with a suitable choice of units, is equal to the net force experienced by the body and is directed in the line of this force. There is, however, one fundamental difference between the nonrelativistic and relativistic cases, i.e. while in the former case our defini-

---

[1] W. KAUFMANN, *Nachr. Ges. Wiss. Göttingen*, 143 (1901), 291 (1902), 90 (1903); *Berlin Sitzungsberichte* 45, 949 (1905); *Ann. der Phys.* 19, 487 (1906) and 20, 639 (1906).

[2] A. H. BUCHERER, *Verh. d. Deutschen Phys. Ges.* 6, 688 (1908); *Phys. ZS.* 9, 755 (1908). See also E. HUPKA, *Ann. der Phys.* 31, 169 (1910); G. NEUMANN, *ibid.* 45, 529 (1914); C. SCHÄFER, *ibid.* 49, 934 (1916); CH. E. GUYE and CH LAVANCHY, *Arch. des Sci. Phys. natur.*, Geneva 41, 286, 353, 441 (1916). For a comprehensive account of these findings, see W. GERLACH in *Handbuch der Physik* 22, 61 (1926).

[3] M. M. ROGERS, A. W. McREYNOLDS and F. T. ROGERS, JR., *Phys. Rev.* 57, 379 (1940).

tion of force reduces to the product of the mass of the body and its acceleration, it is not so in the latter case. In fact, in the general case, force and
acceleration may not even be parallel:

$$\mathbf{F} = \frac{d\mathbf{p}}{dt} = \frac{d}{dt} \left[\frac{m_0 \mathbf{u}}{\sqrt{1 - u^2/c^2}} \right]$$

$$= \frac{m_0}{\sqrt{1 - u^2/c^2}} \dot{\mathbf{u}} + \frac{m_0 (\mathbf{u} \cdot \dot{\mathbf{u}})/c^2}{\sqrt{(1 - u^2/c^2)^3}} \mathbf{u}, \tag{3.18}$$

which shows that, in general, $\mathbf{F}$ is given by the sum of two vectors, one of
which is parallel to $\dot{\mathbf{u}}$ and arises because of the change in the velocity of the
body (at a fixed mass), while the other is parallel to $\mathbf{u}$ and arises because of
the change in the mass of the body as a result of a change in the magnitude
of its velocity. In two specific cases, $\mathbf{F}$ and $\dot{\mathbf{u}}$ are found to be parallel:

(i)   When the acceleration of the body is *perpendicular* to its velocity, in
      which case (3.18) reduces to

$$\mathbf{F} = \frac{m_0}{\sqrt{1 - u^2/c^2}} \dot{\mathbf{u}} \; ; \tag{3.19}$$

the coefficient of $\dot{\mathbf{u}}$ in this equation may be referred to as the 'transverse' mass of the body.

(ii) When the acceleration of the body is either *parallel* or *antiparallel* to
      its velocity, in which case (3.18) becomes

$$\mathbf{F} = \frac{m_0}{\sqrt{(1 - u^2/c^2)^3}} \dot{\mathbf{u}} \; ; \tag{3.20}$$

the coefficient of $\dot{\mathbf{u}}$ in this equation may be referred to as the 'longitudinal' mass of the body.

The 'transverse' mass and the 'longitudinal' mass of a body arise in the
nature of an 'effective' mass when the body is subjected to a transverse or a
longitudinal acceleration, respectively. It must, however, be noted that only
(3.16) can be regarded as *the* expression for the mass of a body, for it is
this quantity which, on multiplication with velocity, gives the momentum of
the body and also obeys the relevant law of conservation. An 'effective'

mass, in the sense of (3.19) or (3.20), cannot even be defined for cases other than these two.

We shall now calculate the *kinetic energy* of a body, of 'proper' mass m_0, moving with velocity u. Obviously, this will be equal to the 'total work done by the forces acting on the body in bringing it from a state of rest with respect to the observer to its actual state of motion'. Now, the rate at which work is done by a force F, acting on a body moving with (instantaneous) velocity u, is given by the formula

$$\frac{dW}{dt} = \mathbf{F} \cdot \mathbf{u}.$$ (3.21)

Substituting from (3.18), we get

$$\frac{dW}{dt} = \frac{m_0 (\mathbf{u} \cdot \dot{\mathbf{u}})}{\sqrt{(1 - u^2/c^2)^3}}$$

$$= \frac{d}{dt}\left[\frac{m_0 c^2}{\sqrt{1 - u^2/c^2}}\right].$$ (3.22)

Integrating from the initial to the final state of motion, we obtain

$$W = \frac{m_0 c^2}{\sqrt{1 - u^2/c^2}}\Big|_0^u$$

$$= \left[\frac{m_0}{\sqrt{1 - u^2/c^2}} - m_0\right] c^2.$$ (3.23)[1]

Now, $m_0/\sqrt{1 - u^2/c^2}$ is nothing but the relativistic mass of the body; we can, therefore, write

$$W = (m - m_0) c^2,$$ (3.24)

which means that the kinetic energy acquired by the body is directly proportional to the mass change brought about by its motion. Now, the total energy of a body is arbitrary to the extent of an additive constant; formula (3.24), therefore, suggests that we may regard mc^2 as the (*total*) *energy* of the body when moving with velocity u and $m_0 c^2$ as the corresponding quantity

---

[1] We note that, for $(u/c) \ll 1$, the relativistic expression for W reduces to the Newtonian expression, viz.

$$W \simeq \frac{1}{2} m_0 u^2.$$

when the body is at rest; the latter may be called the *rest energy* of the body. The kinetic energy is, then, given by the difference between the total energy and the rest energy :

$$W = E - E_0, \tag{3.25}$$

where

$$E = mc^2. \tag{3.26}$$

Formula (3.26) constitutes one of most remarkable conclusions of the special theory of relativity ; in some sense, this is *the* most outstanding result of this theory.

We have thus established a linear relationship between changes in the kinetic energy (or in the total energy) of a body and changes in its mass :

$$\Delta W = \Delta E = (\Delta m)c^2. \tag{3.27}$$

It is, however, well known that energy may be converted from any one of its various forms into any other; hence, the foregoing relationship should apply to energy in any form whatsoever. We, therefore, conclude that energy of amount ΔE, *of any form whatsoever*, is equivalent to a mass Δm, given by

$$\Delta m = \frac{\Delta E}{c^2} ; \tag{3.28}$$

this is the famous principle of *mass-energy equivalence*. Further, since inertia is associated with energy, energy in motion must be accompanied by a momentum flux and, hence, exert pressure.

A case of special interest is provided by a quantum of radiation—the photon—whose energy is given by hv, v being the frequency of the radiation. According to the foregoing results, the photon would be regarded as endowed with a mass hv/c^2 and a momentum hv/c. Further, since its speed of motion is exactly c, a finite value of its energy implies that its rest mass be taken as zero.

One of the most significant consequences of mass-energy equivalence is that the law of conservation of mass, considered in Secs. 3.1 and 3.2, automatically implies (rather includes) the law of conservation of energy.[1] These

---

[1] One can show that energy conservation is directly related to the 'homogeneity' of time. Since this property of time is common to all inertial observers, so must be the validity of the law of energy conservation (*cf.* footnote on p. 77). See also L. D. LANDAU and E. M. LIFSHITZ, *Mechanics, loc. cit.*, Sec. 6.

fundamental laws of mechanics are, thus, no longer distinct from one another; they actually blend into a single law of conservation, which may be referred to as the law of conservation of mass or of energy, as it suits the context.

3.4. Mass-energy equivalence (contd.)

Having established mass-energy equivalence, we shall now trace, at some length, the gradual emergence and subsequent verification of this concept. It was as early as 1881 that J. J. THOMSON arrived at the result that a charged spherical conductor in rectilinear motion behaves as if it had an additional mass given by $(4/3c^2)$ times the energy of its electrostatic field.[1] The first to give the correct relationship was POINCARE[2] who, in 1900, suggested that if electromagnetic momentum, in free aether, is given by $(1/c^2)$ times the energy flux—a result following from MAXWELL's theory—then electromagnetic energy should possess a mass density given by $(1/c^2)$ times the energy density; moreover, a Hertzian oscillator which sends out electromagnetic energy preponderantly in one direction should recoil as a gun does on firing.

In 1905, EINSTEIN[3] arrived at the same result as did POINCARE; however, whereas POINCARE had made his suggestion without proof, EINSTEIN did give a proof (though only an approximate one). EINSTEIN also remarked that the energy involved need not be electromagnetic in character, and asserted the general principle : *'the mass of a body is a measure of its energy-content; if the energy changes by L, the mass changes (in the same sense) by L/c^2'*. A year later, he claimed that this principle is a *necessary and sufficient* condition that the law of conservation of momentum be valid for systems in which electromagnetic as well as mechanical processes take place.[4]

In 1908, LEWIS[5] proved EINSTEIN's result on the basis of the theory of

[1] See E. T. WHITTAKER, *loc. cit.*, Vol. I, pp. 306-310. It was shown by FERMI in 1922 that the transport of the stress system set up in the material of the sphere must be taken into account and that, when this is done, THOMSON's result gets multiplied by a factor of 3/4.

[2] H. POINCARE, *Archives Neerland* **5**, 252 (1900).

[3] A. EINSTEIN, *Ann. der Phys.* **18**, 639 (1905); English translation available in *The Principle of Relativity, loc. cit.*

[4] A. EINSTEIN, *Ann. der Phys.* **20**, 627 (1906).

[5] G. N. LEWIS, *Phil. Mag.* **16**, 705 (1908); in this work, LEWIS also derived, with the help of the 'postulate' $E = mc^2$ and the equation $dE/dt = \mathbf{u} \cdot d(m\mathbf{u})/dt$, the formula $m = m_0 / \sqrt{1 - u^2/c^2}$.

For an account of the development of mechanics on the basis of the 'postulate' $E = mc^2$, see J. MANDELKAR, *Matter Energy Mechanics* (Philosophical Library, New York, 1954).

radiation pressure and, making a reference to PLANCK's work[1] of 1906, put his results in the form in which we have obtained them here. In 1911, LORENTZ[2] showed quite generally that in formulae (3.26) and (3.28) all forms of energy must be included, e.g. potential energy, thermal energy, etc.

As for the experimental verification of the various features of relativistic mechanics, especially (i) the inertia of energy and (ii) the association of momentum flux with energy flow, evidence has come from different directions. We have, first of all, the COMPTON effect, in which an x-ray photon is scattered by an electron and, transferring a part of its energy to the latter, leaves with an increased wavelength. As was shown by COMPTON,[3] and independently by DEBYE,[4] the process of scattering could be visualized as an elastic collision between the photon and the electron and, employing relativistic expressions for the energy and momentum of both the particles and assuming the validity of the relevant *conservation laws*, complete details of the collision process could be worked out. Subsequent observations[5] fully confirmed the theoretical results and, thus, verified the correctness of the relativistic expressions. About a decade later, CHAMPION[6] investigated collisions between rapidly moving electrons (β-particles) and electrons 'at rest' in a WILSON cloud chamber. He also found good agreement between experimental results and theoretical predictions (based on the relativistic formulae and the conservation laws).

Next, we discuss evidence for the relation (3.28) which comes from the considerations of *mass defect* of atomic nuclei.[7] This quantity represents the deficit in the mass of a nucleus (compared with the sum of the masses of its constituent nucleons), and corresponds to the *binding energy* of the nucleus concerned. Obviously, this much energy must be supplied to the nucleus before it can break up into individual nucleons; conversely, when a

[1] M. PLANCK, *Verh. d. Deutschen Phys. Ges.* **4**, 136 (1906).

[2] H. A. LORENTZ, *Versl. gewone Vergad. Akad. Amsterdam* **20**, 87 (1911).

[3] A. H. COMPTON, *Phys. Rev.* **21**, 207, 483 (1923); *Phil. Mag.* **46**, 897 (1923).

[4] P. DEBYE, *Phys. ZS.* **24**, 161 (1923).

[5] A. H. COMPTON, *Phys. Rev.* **22**, 409 (1923); C. T. R. WILSON, *Proc. Roy. Soc. London* A, **104**, 1 (1923); A. H. COMPTON and J. C. HUBBARD, *Phys. Rev.* **23**, 439 (1924).

[6] F. C. CHAMPION, *Proc. Roy. Soc. London* A, **136**, 630 (1932*)*.

[7] P. LANGEVIN. in 1913 (*J. Phys. theor. appl.* (5), **3**, 553), attempted to explain the deviations of atomic weights, from integral values, in terms of the 'mass equivalents' of internal energies of nuclei. However, it was stressed by R. SWINNE, *Phys. ZS.* **14**, 145 (1913), that one would have to consider possible isotopes as well, discovered later by ASTON (in 1919-20). It must, however, be emphasized that it is *not* the deviations *from integral values* that are of real fundamental significance; *cf.* text.

nucleus is formed from the individual nucleons an equivalent amount of energy must be released.[1] For instance, the mass of a helium nucleus is 4.0028 U, whereas the sum of the masses of its constituents (two protons and two neutrons), in the free state is 4.0331 U. The mass defect of 0.0303 U corresponds, by virtue of the relation (3.28), to a binding energy of about 4.5×10^{-5} erg or 28 MeV. This much energy would be released in the *fusion* of the constituent particles into a single helium nucleus. Considerations such as these form the basis of the explanation[2] for the source of stellar energy that has kept, and is continuing to keep, the Sun and the stars radiant for billions of years; thus, stellar objects are shining at the expense of their mass-content.[3]

On the other hand, if nuclei with a given mass defect were transformed into nuclei with a different mass defect, formula (3.28) could be readily tested. In a nuclear reaction of this type, the net energy released would be equivalent to the difference between the total mass defect of the reactants and the products. This quantity could be negative as well as positive; correspondingly, energy could be absorbed or released, depending upon the nature of the reaction.

The first quantitative verification of mass-energy equivalence in a nuclear reaction was made by COCKCROFT and WALTON[4] who investigated the reaction

$$_3\text{Li}^7 + _1\text{H}^1 \rightarrow 2\,_2\text{He}^4,$$

which involved a mass difference of

$$(7.0166 + 1.0076 - 2 \times 4.0028) = 0.0186 \text{ U,}$$

with an equivalent energy difference of 27.7×10^{-6} erg. Measurements of the difference between the total kinetic energy of the α-particles produced and that of the incident proton gave the value

$$17.28 \pm 0.03 \text{ MeV, i.e. } (27.6 \pm 0.05) \times 10^{-6} \text{ erg,}$$

---

[1] For details, see R. D. EVANS, *The Atomic Nucleus* (McGraw-Hill Book Company, Inc., New York, 1955), Chap. 9.

[2] See, e.g., M. SCHWARZSCHILD, *Structure and Evolution of the Stars* (The University Press, Princeton, 1958), Sec. 10.

[3] As early as 1908, D. F. COMSTOCK (*Phil. Mag.* **15**, 1) had remarked that 'assuming the loss of mass accompanying the dissipation of energy, the solar mass must have decreased steadily through millions of years'.

[4] J. D. COCKCROFT and E. T. S. WALTON, *Proc. Roy. Soc. London* A, **137**, 229 (1932).

in excellent agreement with the theoretical result.[1]

Numerous evidences of the type given above are now available; reference may be made to any standard treatise dealing with nuclear reactions.[2]

The *most direct* evidence of mass-energy equivalence is provided by reactions in which material particles are completely *annihilated* and give rise to radiation or by reactions in which material particles are *created* out of the energy-content of radiation. Several reactions, of either type, have been observed in the realm of elementary particles and *in each case* the principle of mass-energy balance is found to be satisfied. A specific mention, in this connection, may be made of the following reactions :

(i) *mutual annihilation* of an electron and a positron, with the emission of radiation quanta whose energy is equal to the sum of the rest energies of the two particles and their kinetic energies,[3]

(ii) *pair production*, of an electron and a positron, by the materialization of the energy of radiation quanta (which provides for the rest masses of the two particles, the balance going in towards providing the kinetic energies),[4] and

(iii) *pair production*, of a proton and an antiproton, by the materialization of high-energy radiation (along with a subsequent *annihilation* of the products).[5]

We may also consider, in this connection, the decay processes of elementary particles through weak interactions. Here, too, the principle of mass-energy

---

[1] N. M. SMITH, JR., *Phys. Rev.* **56**, 548(1939).

[2] See, for instance, E. SEGRÉ (Ed.) *Experimental Nuclear Physics* (Chapman and Hall, London, 1953), Vol. I, Part V—Sec. 4E of K. T. BAINBRIDGE's article, especially Table 15 which gives extensive data on nuclear reactions; Vol. II, Part VI—PHILLIP MORRISON's article entitled 'A Survey of Nuclear Reactions'. See also R. D. EVANS, *loc. cit.*, Chap. 12.

[3] C. D. ANDERSON and S. H. NEDDERMEYER, *Phys. Rev.* **43**, 1034 (1933); F. RASETTI, L. MEITNER and K. PHILLIP, *Naturw.* **21**, 286 (1933). For a quantitative study, see J. DuMOND, D. A. LIND and B. B. WATSON, *Phys. Rev.* **75**, 1226 (1949).

[4] C. D. ANDERSON, *Phys. Rev.* **41**, 405 (1932); P. M. S. BLACKETT and G. P. S. OCCHIALINI, *Proc. Roy. Soc. London* A, **139**, 699 (1933). For quantitative investigations, refer to R. WALKER and B. MCDANIEL, *Phys. Rev.* **74**, 315 (1948).

[5] O. CHAMBERLAIN, E. G. SEGRÉ, C. E. WIEGAND and T. YPSILANTIS, *Phys. Rev.* **100**, 947 (1955). See also E. SEGRÉ, *Am. J. Phys.* **25**, 363 (1957).

balance is found to be in perfect order ; for details, reference may be made to other sources.[1]

In fact, the inertial aspect of energy forms one of the most certain foundations of present-day nuclear physics and elementary particle physics. It has also given rise to the ambitious programme of interpreting the masses of the various elementary particles as energy eigenvalues of a certain Hamiltonian.[2]

3.5. Basic four-vectors of mechanics

We shall now consider a four-dimensional formulation of relativistic mechanics ; this will enable us to express our results in the form of four-vector and four-tensor equations, i.e. in a form manifestly *covariant* with respect to LORENTZ transformations.

We start with the consideration of the time track of a moving particle ; see Sec. 2.4. It has already been noted that the coordinate increments dx_i along the track are the components of a four-vector (dx_i), which is the differential of the four-vector (x_i) that represents the position of a world point on the track. Let the corresponding increment in the *proper* time of the motion be $d\tau$, which is an *invariant* measure related to the observer's time interval dt through the relation (2.23), viz.

$$d\tau = \frac{ds}{ic} = dt\sqrt{1 - u^2/c^2};$$ (3.29)

here, $\mathbf{u}$ ($= d\mathbf{r}/dt$) denotes the instantaneous velocity of the particle while ds represents the differential length of the track:

$$(ds)^2 = dx_i\, dx_i.$$ (3.30)

From the four-vector character of the quantities dx_i and the invariant character of $d\tau$, it follows that the quantities

$$U_i = \frac{dx_i}{d\tau}$$ (3.31)

[1] See, for instance, J. D. JACKSON, *The Physics of Elementary Particles* (Oxford University Press, London, 1958), Part III; earlier references are also available there.

[2] H. P. DUERR, W. HEISENBERG, H. MITTER, S. SCHLIEDER and K. YAMAZAKI, *Z. Naturforsch.* 14, 441 (1959); W. HEISENBERG, *Proceedings of the 1960 Annual International Conference on High-Energy Physics at Rochester* (Interscience Publishers, Inc., New York, 1960), p. 851. See also Y. NAMBU, *ibid.* p. 858; Y. NAMBU and G. JONA-LASINIO, *Phys. Rev.* 122, 345(1961) and 124, 246 (1961).

would also be the components of a four-vector (U_i); we call it the 'velocity four-vector' or the 'four-velocity' of the particle. Its components, in conventional terms, are given by

$$(U_i) = \left[\frac{d\mathbf{r}, \, ic \, dt}{dt\sqrt{1 - u^2/c^2}}\right] = \left[\frac{\mathbf{u}}{\sqrt{1 - u^2/c^2}}, \, \frac{ic}{\sqrt{1 - u^2/c^2}}\right]; \qquad (3.32)$$

the temporal part here represents the LORENTZ factor of the motion, while the spatial part represents the three-dimensional velocity vector (along with the LORENTZ factor).

By definition, the four-vector (U_i) is 'tangential' to the time track at the point (x_i). Further, we have for its norm

$$U_i U_i = \frac{dx_i \, dx_i}{(d\tau)^2} = \frac{(ds)^2}{(d\tau)^2} = -c^2; \qquad (3.33)$$

hence, quite generally, (U_i) is a *time-like* four-vector.[1]

Differentiating (3.33) with respect to τ, we obtain

$$U_i \frac{dU_i}{d\tau} = 0. \qquad (3.34)$$

The four-vector $(dU_i/d\tau)$ is known as the 'acceleration four-vector' or the 'four-acceleration' of the particle and is usually denoted by the symbol (A_i). Thus

$$(A_i) = \left[\frac{dU_i}{d\tau}\right] = \frac{\left[\dfrac{dU_i}{dt}\right]}{\sqrt{1 - u^2/c^2}}. \qquad (3.35)$$

In view of (3.34), the four-vectors (U_i) and (A_i) are mutually orthogonal; the latter, therefore, lies in the direction of the 'normal' to the time track at the point (x_i), i.e. towards the 'centre of curvature' of the track. In the *instantaneous* rest system of the particle (where $\mathbf{u} = 0$), the components of the four-velocity $(U_i)^0$ are $(0, 0, 0, ic)$. In that system of reference, because of orthogonality, the temporal part of $(A_i)^0$ must vanish; hence, its norm must be positive. Because of the invariance of the norm, this would be true in *all* systems of reference; consequently, (A_i) must be a *space-like* vector.

From the four-vector (U_i) and the proper mass m_0 of the particle (which is

---

[1] In literature, one also comes across a 'unit' tangent four-vector $(u_i) = (U_i)/c$, for which $u_i u_i = -1$.

invariant), we may construct another four-vector in the direction of the 'tangent' to the time track, viz.

$$(p_i) = m_0 (U_i) ; \qquad (3.36)$$

this represents the 'momentum four-vector', or the 'four-momentum', of the particle. We have for its norm

$$p_i p_i = - m_0^2 c^2; \qquad (3.37)$$

this is again a *time-like* four-vector. Its spatial and temporal parts are

$$\left[\frac{m_0 \mathbf{u}}{\sqrt{1 - u^2/c^2}}, \frac{i m_0 c}{\sqrt{1 - u^2/c^2}} \right] = \left[\mathbf{p}, i \frac{E}{c} \right], \qquad (3.38)$$

where $\mathbf{p}$ is the three-dimensional momentum of the particle and E its total energy; see Eqs. (3.16), (3.17) and (3.26). The four-momentum is thus more comprehensive than the three-momentum, in that it includes, as its temporal part, the total energy as well. From (3.37) and (3.38), we get

$$p^2 - E^2/c^2 = - m_0^2 c^2. \qquad (3.39)$$

A particular case of this result may be noted. Keeping p and E finite, let $m_0 \to 0$ (and $u \to c$). Formula (3.39) then gives

$$p = E/c, \qquad (3.40)$$

while the four-momentum takes the form

$$(p_i) = (\mathbf{p}, ip); \quad p_i p_i = 0. \qquad (3.41)$$

This is a *null* vector, 'tangential' to the trajectory of the (nonmaterial) particle, that lies wholly on the light cone. An example of this is provided by a photon — the quantum of the electromagnetic radiation; for more details, see Sec. 4.1.

The four-vector character of the quantities (3.38) is very significant from an analytical point of view. Firstly, it enables us to write down immediately the transformation equations for the momentum and the energy of a particle, because they have to be similar to the ones for the coordinates of an event; one has merely to replace $\mathbf{r}$ by $\mathbf{p}$ and t by E/c^2. Thus, we obtain from

(1.38) and (1.39)[1],

$$\mathbf{p}' = \mathbf{p} + \frac{(\mathbf{p} \cdot \mathbf{v})\,\mathbf{v}}{v^2}\,(\gamma-1) - \gamma\mathbf{v}\,\frac{E}{c^2}, \quad \gamma = \left[\frac{1}{\sqrt{1 - v^2/c^2}}\right] \left.\rule{0pt}{40pt}\right\} \quad (3.42)$$

and

$$E' = \gamma\,[E - (\mathbf{p} \cdot \mathbf{v})],$$

along with their inverse

$$\mathbf{p} = \mathbf{p}' + \frac{(\mathbf{p}' \cdot \mathbf{v})\,\mathbf{v}}{v^2}\,(\gamma - 1) + \gamma\mathbf{v}\,\frac{E'}{c^2}, \left.\rule{0pt}{30pt}\right\} \quad (3.43)$$

and

$$E = \gamma\,[E' + (\mathbf{p}' \cdot \mathbf{v})];$$

cf. Eqs. (3.11) and (3.13).

Secondly, the laws of conservation of momentum and energy can now be coalesced to give a composite law of conservation (of four-momentum).[2] Since the conservation of a vector implies the conservation of all its components *individually*, the foregoing coalescence is only a matter of mathematical elegance. From a physical point of view, it represents nothing new; thus, it contrasts sharply with the case of the laws of conservation of mass and energy where one found that the two laws were no longer distinct and *had to be replaced by a single law of conservation* in which the mass-content and the energy-content of a system were to be considered together ; see the end of Sec. 3.3. In that case, an inter-conversion of the two hitherto distinct entities—the mass and the energy—became an allowed physical process. There is no such thing in the case of momentum and energy; physically, they continue to be distinct.

One comment on the relative roles of $\mathbf{p}$ and E, in the structure of the four-momentum (3.38), may be made. That is, there appears to be a deep-seated connection between the space coordinates of a particle and its momentum on one hand, and between the time coordinate and energy on the other.

---

[1] See also the footnote on p. 67.

[2] One can show that the law of conservation of four-momentum is directly related to the 'homogeneity' of the space-time continuum. Since this property of the continuum is common to all inertial observers, so must be the validity of this law (*cf.* footnotes on pp. 77 and 84). The invariance of this law is also evident from the fact that its mathematical expression would be in the form of a *tensor equation* (depicting the equality of the total four-momentum of a given closed system *before* and *after* a collision process).

These connections show up, rather conspicuously, in the quantum-mechanical formulation of particle dynamics.[1]

Finally, we differentiate the momentum four-vector (p_i) with respect to the proper time τ and obtain another four-vector (F_i) :

$$(F_i) = \left(\frac{dp_i}{d\tau}\right) = \left[\frac{d(m_0 U_i)}{d\tau}\right],$$ (3.44)

which is known as the 'force four-vector' or the 'four-force'. Assuming m_0 to be a constant in time, (3.44) can be written as

$$(F_i) = m_0 \left(\frac{dU_i}{d\tau}\right) = m_0 (A_i),$$ (3.45)

i.e. the four-force and four-acceleration vectors are parallel to each other. Equation (3.34) then leads to the result

$$U_i F_i = 0,$$ (3.46)

i.e. the four-force and four-velocity vectors are mutually orthogonal.

The foregoing property of (F_i) is closely connected with the constancy (in time) of the proper mass, which was assumed above. In general, we have

$$U_i F_i = U_i \frac{d(m_0 U_i)}{d\tau}$$

$$= m_0 U_i \frac{dU_i}{d\tau} + U_i U_i \frac{dm_0}{d\tau}$$

$$= - c^2 \frac{dm_0}{d\tau},$$ (3.47)

instead of (3.46). Thus, the scalar product of the four-vectors (U_i) and (F_i) is a direct measure of the rate at which the proper mass of the object changes with (proper) time. If $dm_0/d\tau \neq 0$, then (3.45) and (3.46) do not hold; however, (3.34) is quite generally true.

From (3.38) and (3.44), we get (making use of the relations $dp/dt = \mathbf{F}$ and $dE/dt = \mathbf{F} \cdot \mathbf{u}$)

$$(F_i) = \left(\frac{\mathbf{F}}{\sqrt{1 - u^2/c^2}}, \frac{i(\mathbf{F} \cdot \mathbf{u})/c}{\sqrt{1 - u^2/c^2}}\right),$$ (3.48)

---

[1] See, for instance, P. A. M. Dirac, *The Principles of Quantum Mechanics* (Clarendon Press, Oxford, 1947; 3rd edition), p. 110.

which may also be written as

$$(F_i) = \left[\mathbf{F}_m, \, i \, \frac{(\mathbf{F}_m \cdot \mathbf{u})}{c} \right], \tag{3.49}$$

where $\mathbf{F}_m$ stands for $\mathbf{F}/\sqrt{1-u^2/c^2}$. The vector $\mathbf{F}_m$ is referred to as the *Minkowski force*.

We note that the four-vector equation (3.44) includes, as its spatial part, the three-dimensional equations of motion of the particle and, as its temporal part, the conservation of the particle energy (which is secured by equating the power expended on the particle with the rate of increase of its energy).

3.6. Mechanics of incoherent matter. The kinetic energy-momentum tensor

We shall now consider the motion of continuously distributed matter under the influence of external forces, neglecting elastic forces, if any, between the neighboring parts of the distribution.[1] In an arbitrary inertial system of reference S, the given mass distribution may be specified by a continuously varying function $\mu(x_i)$, of the space-time coordinates x_i, which denotes the *relativistic mass density* (or, simply, the *mass density*) at the world point (x_i). Then, $\mu \delta V$ determines the total mass contained in the volume element δV, around the space point $\mathbf{r}$, at time t.   Let $\mathbf{u}(\mathbf{r}, t)$ denote the local-cum-instantaneous velocity of the matter under consideration; then, $\mu \mathbf{u}$ would determine the local-cum-instantaneous *mass current density* or the *momentum density*.

Next, we denote by μ^0 the mass density, as determined by an observer in the *instantaneous* rest system S^0 of the matter around that point.   Then, by virtue of the transformation property of mass,

$$\mu \delta V = \frac{\mu^0 \delta V^0}{\sqrt{1 - u^2/c^2}}. \tag{3.50}$$

Combining (3.50) with the transformation equation

$$\delta V = \delta V^0 \sqrt{1 - u^2/c^2} \tag{3.51}$$

for the volume element, we obtain for the transformation of mass density

$$\mu = \frac{\mu^0}{(1 - u^2/c^2)}. \tag{3.52}$$

We now introduce the *proper mass density* μ_0 which is defined as the *proper*

---

[1] This explains the term 'incoherent' appearing in the title of this section.

mass ('computed' by an observer in system S) in a given element of space, divided by the volume of that element (determined by the same observer). Thus,

$$\mu_0 = (\mu \delta V) \sqrt{1 - u^2/c^2} / \delta V$$

$$= \mu \sqrt{1 - u^2/c^2} = \frac{\mu^0}{\sqrt{1 - u^2 c/^2}}. \tag{3.53}$$

Let us now consider a portion of matter confined to a finite region, of volume V, bounded by a closed surface σ. The change in V during the time interval $(t, t + dt)$ is given by

$$dV = dt \int_\sigma (d\boldsymbol{\sigma} \cdot \mathbf{u}), \tag{3.54}$$

which, by GAUSS's theorem, becomes

$$dV = dt \int_V (\operatorname{div} \mathbf{u}) \, dV. \tag{3.55}$$

Hence, for a portion of matter within an *infinitesimal* volume δV,

$$\frac{d}{dt} (\delta V) = (\operatorname{div} \mathbf{u}) \, \delta V. \tag{3.56}$$

If the proper mass of the system is conserved, then

$$\frac{d}{dt} (\mu_0 \, \delta V) = 0. \tag{3.57}$$

Equations (3.56) and (3.57) give

$$\frac{d\mu_0}{dt} + \mu_0 \operatorname{div} \mathbf{u} = 0. \tag{3.58}$$

Now, the 'substantial' differential operator[1] d/dt is connected with the partial differential operators $\partial/\partial t$ and $\partial/\partial \mathbf{r}$ by the relation

$$\frac{d}{dt} = \frac{\partial}{\partial t} + \mathbf{u} \cdot \operatorname{grad}; \tag{3.59}$$

---

[1] Also called the 'total' differential operator or the 'mobile' operator.

hence, (3.58) can be written as

$$\frac{\partial \mu_0}{\partial t} + \mathbf{u} \cdot \text{grad } \mu_0 + \mu_0 \text{ div } \mathbf{u} = 0,$$

or

$$\frac{\partial \mu_0}{\partial t} + \text{div } (\mu_0 \, \mathbf{u}) = 0, \tag{3.60}$$

which is an *equation of continuity*, expressing *the conservation of the proper mass* in the system.

We now define a four-vector (c_i), parallel to the four-velocity (U_i), by the relation

$$(c_i) = \mu^0(U_i)/c, \tag{3.61}$$

μ^0/c being an invariant.   The spatial and temporal parts of this four-vector are, see (3.32) and (3.53),

$$(c_i) = \left[\frac{\mu^0 \mathbf{u}/c}{\sqrt{1 - u^2/c^2}} , \frac{i\mu^0}{\sqrt{1 - u^2/c^2}} \right] = (\mu_0 \mathbf{u}/c, \, i\mu_0). \tag{3.62}$$

In terms of the four-vector (c_i), Eq. (3.60) can be written as

$$\frac{\partial c_i}{\partial x_i} \equiv \text{Div } (c_i) = 0 \, ; \tag{3.63}$$

see Sec. 2.6.   In this form, the equation of continuity is manifestly *covariant* under LORENTZ transformations.

Now, the external forces impressed upon the system may be represented by a *force density* $\mathbf{f}$, so that $\mathbf{f}\delta V$ determines the force experienced by matter in the volume δV. The four-momentum (p_i) of this element of matter is given by

$$(p_i) = (\mu^0 \delta V^0) \, (U_i), \tag{3.64}$$

and the corresponding four-force by

$$(F_i) = \left[\frac{\mathbf{f}\delta V}{\sqrt{1 - u^2/c^2}} , \frac{i(\mathbf{f} \cdot \mathbf{u}) \, \delta V/c}{\sqrt{1 - u^2/c^2}} \right]$$

$$= \{\mathbf{f}, \, i(\mathbf{f} \cdot \mathbf{u})/c\} \, \delta V^0 \, . \tag{3.65}$$

Since δV^0 is an invariant, the set of quantities

$$(f_i) \equiv \{\mathbf{f}, \, i(\mathbf{f} \cdot \mathbf{u})/c\} \tag{3.66}$$

would be a four-vector; we call it the *four-force density*. The spatial part of this vector is identical with the three-dimensional force density while the temporal part is a measure of the *mechanical* work done on the system per unit volume per unit time (that is, the *power density*). The covariant form of the equations of motion, then, turns out to be

$$\mu^0 \frac{d(U_i)}{d\tau} = (f_i) ; \qquad (3.67)$$

see Eqs. (3.64)−(3.66).

It may be emphasized here that the foregoing simple form of the equations of motion holds only if the proper mass of the system is conserved. However, if this assumption is dropped it becomes essential to introduce certain modifications into the formalism. First of all, we must add to the temporal part of (f_i) a term, q say, which represents the *nonmechanical* energy developed in the system per unit volume per unit time; this term will finally account for the change in the proper mass. Thus, our four-force density would become

$$(f_i) = \{\mathbf{f}, \, i(\mathbf{f} \cdot \mathbf{u} + q)/c\} ; \qquad (3.68)$$

cf. (3.66). For the equations of motion, we shall now have

$$(F_i) = \frac{d}{d\tau} (p_i) = \frac{d}{d\tau} (\mu^0 U_i \, \delta V^0)$$

$$= \left[\frac{d}{d\tau} (\mu^0 U_i) + \mu^0 U_i \, (\mathrm{div}^0 \, \mathbf{u}^0) \right] \delta V^0, \qquad (3.69)$$

where use has been made of the formula (3.56), as applicable in S^0. The symbol div^0 here means that the divergence has to be taken with respect to the (space) coordinates x_i^0 of S^0. No doubt, $\mathbf{u}^0 = 0$; however, this does not mean that $\mathrm{div}^0 \, \mathbf{u}^0$ is also equal to zero. Actually, we note that

$$\mathrm{Div}^0 (U_i)^0 \equiv \sum_{i=1}^{3} \frac{\partial}{\partial x_i} \left[\frac{u_i}{\sqrt{1 - u^2/c^2}} \right] + \frac{\partial}{\partial x_4} \left[\frac{ic}{\sqrt{1 - u^2/c^2}} \right] \Bigg|_{\mathbf{u}=0}$$

$$= \sum_{i=1}^{3} \frac{\partial}{\partial x_i} (u_i) \Bigg|_{\mathbf{u}=0} = \mathrm{div}^0 \, \mathbf{u}^0. \qquad (3.70)$$

Now, the left-hand side of (3.70) is an invariant quantity; it may, therefore, be replaced by a corresponding quantity in an arbitrarily chosen system S, i.e. by $\partial U_k/\partial x_k$. Further, the operator $d/d\tau$ in (3.69) is equivalent to the operator

$$\frac{\partial}{\partial x_k} \cdot \frac{dx_k}{d\tau} = \frac{\partial}{\partial x_k} \cdot U_k. \tag{3.71}$$

We thus obtain for the four-force density

$$(f_i) = \frac{\partial}{\partial x_k} (\mu^0 U_i) \cdot U_k + (\mu^0 U_i) \frac{\partial U_k}{\partial x_k}$$

$$= \frac{\partial}{\partial x_k} (\mu^0 U_i U_k), \tag{3.72}$$

which constitutes the fundamental equations of motion, for the given system, in the general case when the proper mass of the system may not be conserved. Clearly, the four-vector (f_i) is equal to the Divergence of a *symmetric* tensor

$$\theta_{ik} = \mu^0 U_i U_k. \tag{3.73}$$

The covariance of Eqs. (3.72) is evident.

Contracting (3.72) with the four-vector (U_i), we get

$$(f_i U_i) = \frac{\partial}{\partial x_k} (\mu^0 U_k)(U_i U_i) + \mu^0 U_k \left[\frac{\partial U_i}{\partial x_k} U_i \right]$$

$$= \frac{\partial}{\partial x_k} (\mu^0 U_k)(-c^2) + \mu^0 \left[\frac{dU_i}{d\tau} U_i \right]. \tag{3.74}$$

The second part on the right-hand side of this equation vanishes identically, see (3.34), while the left-hand side is equal to, see (3.32) and (3.68),

$$\frac{\mathbf{f} \cdot \mathbf{u}}{\sqrt{-u^2/c^2}} - \frac{(\mathbf{f} \cdot \mathbf{u}) + q}{\sqrt{1 - u^2/c^2}} = \frac{-q}{\sqrt{1 - u^2/c^2}} = -q^0 ; \tag{3.75}$$

the quantity q^0 represents the *nonmechanical* effect, as measured in the rest system S^0. We then have from (3.74)

$$\frac{\partial}{\partial x_k} (\mu^0 U_k) = \frac{q^0}{c^2}. \tag{3.76}$$

This represents a generalization of the equation of continuity (3.63) to situations in which there *are* sources (or sinks) of the proper mass; the quantity q^0/c^2 represents the relevant *source density*—in accordance with the principle of mass-energy equivalence.

Finally, we discuss the physical significance of the tensor θ_{ik}, defined by (3.73), viz.

$$\theta_{ik} = \mu^0 U_i U_k,$$

where

$$(U_i) = \left[\frac{\mathbf{u}}{\sqrt{1 - u^2/c^2}}, \frac{ic}{\sqrt{1 - u^2/c^2}} \right]. \tag{3.32}$$

First of all, we note that the purely temporal part of this tensor is

$$\theta_{44} = \mu^0 U_4 U_4 = \frac{-\mu^0 c^2}{(1 - u^2/c^2)} = -\mu c^2 = -h; \tag{3.77}$$

in the last step we have made use of the formula (3.52). The purely temporal part thus represents the *energy density*[1] of the system (around the point **r**, at time t). Next, we have for the mixed parts of the tensor

$$(\theta_{p4}) = \mu^0 \frac{\mathbf{u}}{\sqrt{1 - u^2/c^2}} \frac{ic}{\sqrt{1 - u^2/c^2}} = ic \, (\mu \mathbf{u}), \tag{3.78}$$

which represents the three-dimensional vector corresponding to the *mass current density* or *momentum density*, and

$$(\theta_{4q}) = \mu^0 \frac{ic}{\sqrt{1 - u^2/c^2}} \frac{\mathbf{u}}{\sqrt{1 - u^2/c^2}} = \frac{i}{c} \, h\mathbf{u}, \tag{3.79}$$

which represents the three-dimensional vector corresponding to the *energy current density*. The symmetry of the tensor θ_{ik} ensures the equality of the vectors (3.78) and (3.79), which in turn expresses the fact that *energy in motion is accompanied by a proportionate amount of momentum flux*; see Sec. 3.3. Finally, the purely spatial part of the tensor θ_{ik} is

$$\theta_{pq} = \mu^0 \frac{u_p}{\sqrt{1 - u^2/c^2}} \frac{u_q}{\sqrt{1 - u^2/c^2}} = \mu \, u_p u_q; \tag{3.80}$$

---

[1] This would be equal to the sum of the (instantaneous) kinetic energy and proper energy per unit volume of the system,

this represents the three-dimensional tensor corresponding to the *momentum current density*—the so-called *momentum current tensor*.

In view of these results, the tensor θ_{ik} is generally referred to as the *kinetic energy-momentum tensor*.[1]

In passing, we note that the invariant θ_{ii}, obtained from the tensor θ_{ik}, is given by

$$\theta_{ii} = \mu^0 U_i U_i = -\mu^0 c^2 = -h^0, \tag{3.81}$$

accordingly, it represents the *proper* or *rest energy density*. Moreover, contracting (3.73) with (3.32), we obtain

$$\theta_{ik} U_k = (-\mu^0 c^2) U_i = (-h^0) U_i ; \tag{3.82}$$

thus, (U_i) may be looked upon as an eigenvector of the matrix θ_{ik}, the quantity $(-h^0)$ being the corresponding eigenvalue.

Problems

3.1. EDDINGTON once remarked that if 1 g of electrons were confined to a sphere of 10 cm radius, the mass associated with their electric potential energy would be of the order of 10^7 tons. Check this statement, assuming that the electrons form a ball of uniform charge density.

3.2. If the light corpuscles of NEWTON possessed mass (by virtue of their energy-content), they would be susceptible to the force of gravitation just like normal material particles. Assuming this, examine the Newtonian orbit of such a corpuscle, coming from infinity and going to infinity, past a massive object such as the Sun. Show that the path of the corpuscle undergoes a bending given by the expression $2\,GM/c^2$, where M is the mass of the massive object; *cf.* the corresponding problem in Sec. 8 4.

3.3. Show that the relativistic motion of a particle, under inverse square law of force, takes place in a *precessing* ellipse. Derive an expression for the rate of precession of the orbit and verify that, if applied to the case of Mercury going round the Sun, it gives a precession of about 7 seconds of arc per century; *cf.* the corresponding problem in Sec. 8.3.

3.4. Two neutrons, A and B, are approaching each other along a common straight line. Each has a constant speed βc, as measured in the laboratory frame of reference. Show that the total energy of neutron B, as observed in the rest frame of neutron A, is $m_0 c^2 (1+\beta^2)/(1-\beta^2)$, where m_0 is the neutron rest mass.

3 5. A uniform rod of mass M and length $2L$ spins, with angular frequency $\omega(\ll c/L)$. about a perpendicular axis passing through its centre. Its angular momentum and kinetic

---

[1] H. MINKOWSKI, (II), Nachr. Ges Wiss. Göttingen (1908), p. 53 ; also *Math. Ann.* **68**, 472 (1910).

energy of rotation may be written as

$$\frac{1}{3} ML^2 \omega \left(1 + A \ \frac{\omega^2 L^2}{c^2} + \ldots \right) \quad \text{and} \quad \frac{1}{6} ML^2 \omega^2 \left(1 + B \ \frac{\omega^2 L^2}{c^2} + \ldots \right),$$

respectively. Determine the values of A and B.

3.6. A particle, of constant proper mass m_0, moves along the x-axis of an inertial frame of reference under a *relativistic* force $-kx$ directed towards the origin. If the amplitude of the resulting oscillation is a, prove that the time-period T is given by

$$T = \frac{4}{c} \int_0^a \gamma (\gamma^2 - 1)^{-1/2} \, dx,$$

where $\gamma = 1 + \frac{1}{2} k (a^2 - x^2)/(m_0 c^2)$.

3.7. Prove that a particle of finite rest mass *cannot* distintegrate into a single photon.

3.8. A particle, of proper mass m_0, decays from rest into a particle, of proper mass m_0', and a photon. Show that the energy of the outgoing photon is given by $c^2 (m_0 - \gamma m_0')$, where

$$\gamma = (1 - u^2/c^2)^{-1/2} \quad \text{and} \quad u = c \, (m_0^2 - m_0'^2)/(m_0^2 + m_0'^2).$$

Note that the special case $m_0' = m_0$ corresponds to Problem 3.7.

3.9. Prove that the outcome of a collision between two distinct particles (of finite or zero rest mass) *cannot* be a single photon.

3.10. An excited atom, of proper mass m_0, is at rest in a certain system of reference. It emits a photon, losing thereby an internal energy ΔE. Allowing for the recoil of the atom, show that the frequency of the emitted photon would be

$$\nu = \frac{\Delta E}{h} \left(1 - \frac{1}{2} \frac{\Delta E}{m_0 c^2} \right).$$

3.11. An atom, of proper mass m_0, is in an excited state of energy Q_0 above the ground state, and is moving with speed v towards a scintillation counter. It decays to its ground state by emitting a photon of energy Q (as recorded by the counter), and itself comes to rest. Show that

$$Q = Q_0 \left(1 + \frac{Q_0}{2 m_0 c^2} \right).$$

3.12. A photon, of frequency ν, collides with an electron, of mass m_0, at rest. The scattered photon goes in a direction making an angle θ with its original direction, while the electron recoils at an angle φ. Show that

$$\tan \left(\frac{1}{2} \theta \right) \tan \varphi = \frac{1}{1 + \alpha},$$

where $\alpha = (h\nu/m_0c^2)$ If ν' denotes the frequency of the scattered photon, then

$$\frac{\nu'}{\nu} = \frac{1}{1 + 2\alpha \sin^2 \left(\frac{1}{2}\theta\right)},$$

so that the increase in the photon wavelength is given by

$$\delta\lambda = c\left(\frac{1}{\nu'} - \frac{1}{\nu}\right) = 2\lambda_c \sin^2\left(\frac{1}{2}\theta\right);$$

here, $\lambda_c \left(= \frac{c}{\nu} \; \alpha = \frac{h}{m_0 c}\right)$ denotes the *Compton wavelength* of the electron.

3.13. A particle moving with velocity u collides with an identical particle at rest. After collision the two particles leave in directions making angles θ and φ with the initial direction of motion of the colliding particle. Prove that

$$\tan\theta \tan\varphi = \frac{2}{\gamma + 1},$$

where $\gamma = 1/\sqrt{1 - u^2/c^2}$. Study, quantitatively, the departure of the angle $(\theta+\varphi)$ from its nonrelativistic value, viz. $\pi/2$.

Further show that the kinetic energy W of the particle going at angle θ is given by

$$W = \frac{W_0 \cos^2\theta}{1 + \dfrac{W_0 \sin^2\theta}{2E_0}},$$

where W_0 and E_0 are the kinetic energy and the rest energy, respectively, of the colliding particle.

3.14. A meson, of mass π, comes to rest and disintegrates into a lighter meson, of mass μ, and a neutrino, of mass ν. Show that the total energies of the product particles are

$$\frac{\pi^2 + \mu^2 - \nu^2}{2\pi} c^2 \quad \text{and} \quad \frac{\pi^2 - \mu^2 + \nu^2}{2\pi} c^2,$$

respectively.

3.15. A particle moving with velocity u disintegrates 'in flight' into two particles. Determine the relationship between the angles of emergence of these particles and their energies.

Discuss, in some detail, the special case when the decay products are photons.

3.16. A particle, of proper mass m_1, strikes a particle, of proper mass m_2, at rest. They undergo an inelastic collision leading to a new set of particles with mass excess ΔM. Show that the *threshold* kinetic energy, of the first particle, for this reaction is given by

$$c^2 \Delta M \left(1 + \frac{m_1}{m_2} + \frac{\Delta M}{2m_2}\right).$$

[Note that the criterion for the threshold is that there be enough energy available in the centre-of-mass system so that the new particles *can* be created, though with zero kinetic energy.]

3.17. A system, of proper mass M, decays at rest into a number of particles with mass deficit ΔM. Show that the maximum kinetic energy acquired by the i-th particle (of proper mass m_i) would be

$$c^2 \, \Delta M \left(1 - \frac{m_i}{M} - \frac{\Delta M}{2M} \right).$$

3.18. Using the fact that the quantities $(U_i) = \gamma(\mathbf{u}, ic)$ constitute a four-vector, derive the transformation equations $(1.51) - (1.54)$.

3.19. An inertial observer 0 has a four-velocity (U_0) while a particle P has a (variable) four-acceleration (A). If $(U_0 \cdot A) \equiv 0$, what can be concluded about the motion of P in 0's rest frame?

3.20. Show that, in the rest frame of a particle, its four-acceleration takes the form $(\mathbf{a_0}, 0)$. Making use of this result, determine the four-acceleration of a particle undergoing uniform circular motion.

3.21. A particle with four-momentum (p) is observed by an observer with four-velocity (U_0). Show that the energy of the particle, relative to this observer, would be $(U_0 \cdot p)$.

3.22. A photon with four-momentum (p) is observed by two observers having four-velocities (U_1) and (U_2), respectively. Show that the observed frequencies would be in the ratio $(U_1 \cdot p)$ to $(U_2 \cdot p)$; see also Sec. 4.2.

CHAPTER 4

OPTICS

4.1. Wave motion and its characteristics

In this chapter we propose to discuss certain phenomena in the domain of optics. Of vital importance in this connection is the analysis of wave motion, with a view to determining the transformation equations for various physical quantities characterizing the process involved. This analysis has been instrumental in bringing out the close association between the phenomenon of wave motion on the one hand and of corpuscular motion on the other.

We start by considering a plane wave whose wave normal, with respect to an observer in system S, points in the direction (l, m, n), with the velocity of propagation given by $\mathbf{w}$ $(= wl, wm, wn)$, the frequency of vibration by ν, the time period by T $(= 1/\nu)$ and the wavelength λ $(= w/\nu = wT)$; the corresponding quantities in system S' will be denoted by the same symbols but *primed*. The wave motion is then described by a wave function of the type

$$\psi = A \exp\left[2\pi i \left(\frac{lx + my + nz}{\lambda} - \frac{t}{T} \right) \right]; \tag{4.1}$$

here, A is the amplitude while (x, y, z, t) are the coordinates of the event corresponding to the 'wave condition' at the point (x, y, z) at time t; it is assumed for simplicity, though without any loss of generality, that for $x_i = 0$, $\psi = A$, which is another 'wave condition' of interest.

Let us take a close look at the quantity

$$\frac{lx + my + nz}{\lambda} - \frac{t}{T}, \tag{4.2}$$

which determines the *phase angle* of the wave at the point (x, y, z) at time t. Since $(lx + my + nz)$ is the perpendicular distance between the wave fronts passing through the points $(0, 0, 0)$ and (x, y, z), the first part in (4.2) gives the number of waves between the two wave fronts. Next, t/T gives the number of waves that pass through any point P during the time interval $(0, t)$. Hence, the combination (4.2) gives us the number of waves lying between the two labelled conditions : one at the point $(0, 0, 0)$ at time 0 and the other at the point (x, y, z) at time t. With this picture in mind, we conclude that, since the number of waves in between two labelled conditions must be the *same* for all observers (although, the coordinates of one, or generally both, of the labels would be different for different observers), *the phase of a wave must be an invariant quantity.*

We now introduce a set of four quantities

$$(k_i) = (\mathbf{k}, i\omega/c), \tag{4.3}$$

where

$$\mathbf{k} = \frac{2\pi}{\lambda} (l, m, n), \qquad k = \frac{2\pi}{\lambda}, \qquad \omega = \frac{2\pi}{T} = 2\pi\nu; \tag{4.4}$$

here, $\mathbf{k}$ is a vector in the direction of wave propagation (whose magnitude is 2π times the wave number $1/\lambda$), while ω is the angular frequency of vibration. In terms of the set (k_i) and the four-vector $(x_i) = (\mathbf{r}, ict)$, the phase of the wave can be written as

$$2\pi \left[\frac{lx + my + nz}{\lambda} - \frac{t}{T} \right] \equiv (k_i x_i). \tag{4.5}$$

Thus, the set (k_i), on entering into an *inner* product with the arbitrary four-vector (x_i), leads to an invariant (which is a tensor of rank zero). Hence, by the quotient law (see Sec. 2.6), the set (k_i) must itself be a four-vector; we call it the *four-wave vector.*

Denoting (4.5) as a scalar field φ, we can write

$$(k_i) = \text{Grad } \varphi; \tag{4.6}$$

hence, (k_i) would be 'normal' to the hyper-surface $\varphi = $ const. passing through the world point (x_i).

As for the transformations, the quantities (4.3) must behave in the same manner as the components of a four-vector. Actually, on comparing the four-vector (4.3) with the four-momentum (3.38), we notice a striking resemblance, i.e. the characteristics $\mathbf{k}$ and ω of the wave motion transform in *exactly* the same manner as the characteristics $\mathbf{p}$ and E of a corpuscular motion. One could, therefore, *in a manner that is Lorentz invariant*, associate with a given motion of one type a corresponding motion of the other type. Thus, with the motion of a particle of momentum $\mathbf{p}$ and energy E we may associate a wave motion whose characteristics are given by

$$(k_i) = \text{const } (p_i), \tag{4.7}$$

where the constant appearing here must be an invariant; in this form, the relation (4.7) is obviously *covariant* and, hence, the foregoing association would hold in *all* inertial systems of reference.

In order to determine the constant in (4.7), let us consider the particular case of a photon in free space. Here, we already have sufficient evidence for the relationship between the corpuscular characteristics of the photon motion and the wave characteristics of the corresponding radiation ; these are

$$\mathbf{p} = \frac{h\nu}{c} (l, m, n) = \hbar\mathbf{k}, \tag{4.8}$$

and

$$E = h\nu = \hbar\omega, \tag{4.9}$$

so that

$$(p_i) = \hbar(k_i) \quad \text{or} \quad (k_i) = (p_i)/\hbar ; \tag{4.10}$$

here, $\hbar = h/2\pi$, h being PLANCK's constant. This case is, thus, already in conformity with the scheme (4.7), with the (unknown) constant given by $1/\hbar$.

The inner consistency of the process of associating a wave motion with a corpuscular motion, in accordance with the four-vector relation (4.7), and the well established evidence for the special case considered above made LOUIS DE BROGLIE put forward the bold suggestion that *to every particle motion be associated a corresponding wave motion*, the respective characteristics of the two motions being connected by the relation

$$(k_i) = (p_i)/\hbar. \tag{4.11}$$

It was emphasized that this association may not be regarded as a mere formality; it should be taken as a concrete physical reality, as in the case of the photon and the light wave. This introduces into our considerations an entirely new concept—the *duality* of physical objects as regards the representation of their motion and the manner of their behavior in respect of mutual interactions.[1] This radical concept, in the hands of ERWIN SCHRÖDINGER,[2] led to the development of a new type of mechanics—the so-called *wave mechanics*—which continues to govern the analysis of the most fundamental processes in physics. The emergence of this concept may, therefore, be regarded as one of the most significant contributions the special theory of relativity has made to physical science.

Separating spatial and temporal parts of (4.11), we get

$$\mathbf{k} = \mathbf{p}/\hbar \quad \text{and} \quad \omega = E/\hbar, \tag{4.12}$$

that is

$$\lambda = h/p \quad \text{and} \quad \nu = E/h. \tag{4.13}[3]$$

Combining these results, we obtain for the *phase velocity* of the associated wave

$$w = \lambda\nu = E/p, \tag{4.14}$$

[1] The ideas on duality were developed by L. DE BROGLIE in a series of notes published in the *Comptes Rendus*: 177, 507, 548, 630(1923); 179, 39, 676, 1039(1924), next in his doctorate thesis: *Theses*, Paris (1924), and finally in *Phil. Mag.* 47, 446(1924) and *Annales de Phys.* (10), 3, 22(1925).

[2] E. SCHRÖDINGER, *Ann. der Phys.* 79, 361, 489, 734(1926); 80, 437(1926); 81, 109 (1926). See also SCHRÖDINGER's *Collected Papers on Wave Mechanics* (Blackie and Son, London, 1928).

[3] The first experimental confirmation of the relation $\lambda = h/p$ came from the study of *diffraction* of electron beams passing through crystals: C. J. DAVISSON and L. H. GERMER, *Phys. Rev.* 30, 707(1927); *Nature* 119, 558(1927). E. RUPP, *Ann. der. Phys.* 1, 801(1929). Similar confirmation came from the study of *scattering* of cathode rays from thin metallic films: G. P. THOMSON and A. REID, *Nature* 119, 890(1927); G. P. THOMSON, *Proc. Roy. Soc. London* A, 117, 600 (1928) and 128, 641(1930); A. REID, *ibid.* 119, 663(1928); S. KIKUCHI, *Proc. Tokyo Acad.* 4, 271, 354, 471(1928).

The DE BROGLIE relation has also been verified in the case of protons by A.J. DEMPSTER, *Phys. Rev.* 34, 1493(1929); 35, 1405(1930), and in the case of lighter atoms and molecules by I. ESTERMANN and O. STERN, *Zeits. f. Phys.* 61, 95(1930); I. ESTERMANN, R. FRISCH and O. STERN, *Phys. ZS.* 32, 670(1931); *Zeits. f. Phys.* 73, 348(1931); T. H. JOHNSON, *Phys. Rev.* 35, 1299(1930); 37, 847(1931); R. M. ZABEL, *Phys. Rev.* 42, 218(1932).

For details, refer to J. D. STRANATHAN, *The Particles of Modern Physics* (The Blakiston Company, Philadelphia, 1942), Chap. 14.

which, in view of the relations $E = mc^2$ and $p = mu$, becomes

$$w = c^2/u. \tag{4.15}$$

Thus, the phase velocity of the associated wave, though in the same direction as that of the particle, has a magnitude inversely proportional to the particle velocity u. Since $u \leqslant c$, $w \geqslant c$.

At first sight the foregoing result seems unacceptable because we cannot admit velocities exceeding that of light.[1] However, we must recall that the relativity restriction applies only to those velocities that carry energy with them and thus can serve as signals for communicating information. There is no such restriction on velocities which merely indicate the rate of advance of a certain 'condition' rather than that of an 'object'. We need not, therefore, worry about the *phase velocity* of the wave exceeding c; we must, however, look at the group velocity instead. We have

$$v_g = \frac{d\nu}{d(1/\lambda)} = \frac{dE}{dp}; \tag{4.16}$$

cf. (4.14). The right-hand side of (4.16), by virtue of the formulae (3.16), (3.17) and (3.26), gives: $dE/dp = u$; hence

$$v_g = u, \tag{4.17}$$

which not only conquers the foregoing objection but also brings out the interesting fact that the speed at which the wave motion transmits energy is identical with the speed of the particle. Thus, the particle moves in coincidence with the *wave packet* associated therewith.[2]

The case of the photon is, obviously, of a limiting nature. It corresponds to $u = c$; consequently, both w and v_g are equal to the speed of light. Moreover, since λ in this case is equal to c/ν, the norm of the vector (k_i) is

---

[1] A similar situation arises in connection with the propagation of light in a medium in which there is anomalous dispersion. One can have in such a case a refractive index less than unity, whence it would appear that light propagates in such a medium faster than in free space. This difficulty was removed, in the same manner as has been done here, by BRILLOUIN and by SOMMERFELD: L. BRILLOUIN, *Comp. Rend.* 157, 914(1913); A. SOMMERFELD, *Ann. der. Phys.* 44, 177(1914).

[2] This incidentally means that we must associate with a particle a *group of waves* rather than a *monochromatic wave*, the *average* values of the various characteristics of the group being related to those of the particle through the equations given in the text.

identically zero; *cf.* (3.41).   The four-wave vector of a photon is, therefore, a *null* vector whereas that of a material particle is *time-like*.

We now proceed to the discussion of various results that follow from the transformation equations for the set (k_i).

4.2.   Doppler effect

In this section we study measures of the frequency of a wave, as obtained in two systems of reference, S and S', in relative motion.   In view of the four-vector character of the set (k_i), as defined by Eqs. (4.3) and (4.4), we have

$$\mathbf{k}' = \mathbf{k} + \frac{(\mathbf{k} \cdot \mathbf{v}) \mathbf{v}}{v^2} (\gamma - 1) - \gamma \mathbf{v} \frac{\omega}{c^2}, \quad \left[\gamma = \frac{1}{\sqrt{1 - v^2/c^2}} \right]$$

and

$$\omega' = \gamma[\omega - (\mathbf{k} \cdot \mathbf{v})] ;$$

$$(4.18)$$

see Eqs. (3.42).   The first three equations determine the relation between the directions of propagation of the wave as observed in the two systems, while the fourth one connects the respective frequency measures.   It is only the latter that concerns us in the present context; the former ones will be needed in the next section.

Since $\omega = 2\pi\nu$ and $\mathbf{k} = (2\pi/\lambda) \, (\mathbf{w}/w) = (2\pi\nu/w^2)\mathbf{w}$, we obtain from (4.18)

$$\nu' = \gamma\nu \left[1 - \frac{(\mathbf{w} \cdot \mathbf{v})}{w^2} \right], \tag{4.19}$$

which becomes the relativistic formula for the *Doppler effect* in the case of an arbitrary plane wave.   Let us consider, in particular, the case of a *light wave* propagating in free space, its frequency in the *rest* system S' of the source being $\nu_0 \, (\equiv \nu')$.   Equation (4.19) then becomes

$$\nu_0 = \gamma\nu \left[1 - \frac{(\mathbf{c} \cdot \mathbf{v})}{c^2} \right], \tag{4.20}$$

whence it follows that

$$\nu = \frac{\nu_0 \sqrt{1 - v^2/c^2}}{1 - \frac{(\mathbf{c} \cdot \mathbf{v})}{c^2}} . \tag{4.21}$$

The following special cases of this formula may be noted:

(i) when the source of light is receding *radially* (away from the observer), $(\mathbf{c} \cdot \mathbf{v}) = -cv$ and hence

$$\nu_1 = \nu_0 \frac{\sqrt{1 - v/c}}{\sqrt{1 + v/c}}; \tag{4.22a}$$

(ii) when the source of light is approaching *radially* (towards the observer), $(\mathbf{c} \cdot \mathbf{v}) = + cv$ and hence

$$\nu_2 = \nu_0 \frac{\sqrt{1 + v/c}}{\sqrt{1 - v/c}}; \tag{4.22b}$$

(iii) when the source of light is moving *transversely* with respect to the observer, $(\mathbf{c} \cdot \mathbf{v}) = 0$ and hence

$$\nu_3 = \nu_0 \sqrt{1 - v^2/c^2}. \tag{4.22c}$$

One finds that the relativistic results are strikingly different from the corresponding nonrelativistic ones, where we have the general result

$$\nu = \nu_0 \frac{1 - \dfrac{\mathbf{c} \cdot \mathbf{v}_0}{c^2}}{1 - \dfrac{\mathbf{c} \cdot \mathbf{v}_s}{c^2}} = \nu_0 \left[1 + \frac{\mathbf{c} \cdot (\mathbf{v}_s - \mathbf{v}_0)}{c^2} + \cdots \right] \tag{4.23}$$

$$= \nu_0 \left[1 + \frac{\mathbf{c} \cdot \mathbf{v}}{c^2} \right], \tag{4.24}$$

where $\mathbf{v}_s$ and $\mathbf{v}_0$ are the velocities of the source and the observer *with respect to the medium of propagation* while $\mathbf{v}$ is the velocity of the source with respect to the observer. In particular, if the observer is at rest and the source is in motion, then

$$\nu = \nu_0 \frac{1}{1 - \dfrac{\mathbf{c} \cdot \mathbf{v}}{c^2}}, \tag{4.25}$$

and

$$\nu_1 = \nu_0 \frac{1}{1 + \dfrac{v}{c}}, \qquad \nu_2 = \nu_0 \frac{1}{1 - \dfrac{v}{c}}, \qquad \nu_3 = \nu_0. \tag{4.26}$$

On the other hand, if the source is at rest and the observer is in motion, then

$$\mathbf{\nu} = \nu_0 \left[1 + \frac{\mathbf{c} \cdot \mathbf{v}}{c^2} \right], \qquad (4.27)$$

and

$$\nu_1 = \nu_0 \left[1 - \frac{v}{c} \right], \quad \nu_2 = \nu_0 \left[1 + \frac{v}{c} \right], \quad \nu_3 = \nu_0. \qquad (4.28)$$

To the *first* order in (v/c), the relativistic and nonrelativistic formulae are the same. To higher orders, they differ significantly. The most serious drawback in the nonrelativistic formula (4.23) is that it makes an undue appeal to the *medium of propagation* (with respect to which the velocities $\mathbf{v}$ and $\mathbf{v}_0$ of the source and the observer are supposed to be measured); the relative velocity $\mathbf{v}$ plays only a secondary role. The immediate consequence of this approach is that we are led to markedly different (and annoyingly asymmetric) formulae, namely (4.25–26) or (4.27–28), depending upon whether the source is in motion or the observer. This is quite contrary to the spirit of relativity, according to which the final effect should depend on $\mathbf{v}$ alone; this is indeed true for the relativistic formulae (4.21–22).

Deviations from nonrelativistic formulae become significant in the second order itself, a decisive contribution being made by the time dilation factor $\sqrt{1 - v^2/c^2}$; see (4.21). In the *transverse* case, this factor alone is responsible for the Doppler effect. We then obtain formula (4.22c), viz. $\nu_3 = \nu_0 \sqrt{1 - v^2/c^2}$, whereas in the nonrelativistic theory, the corresponding result is : $\nu_3 = \nu_0$; see (4.26) and (4.28).

The foregoing formulae provide another means of testing the validity of the relativistic results, especially time dilation ; see also Secs. 8.6 and 8.7.

Soon after J. STARK[1] had observed Doppler effect in the light emitted by rapidly moving H_2-ions in a positive-ray tube, EINSTEIN[2] suggested that the *transverse Doppler effect* might be verified by observations on positive rays. The accuracy of STARK's work, however, was not sufficient to decide the issue so far as terms of *second* order in (v/c) were concerned. It was more than three decades later that IVES and STILWELL[3] succeeded in carrying out

[1] J. STARK, *Ann. der Phys.* **21**, 401(1906); J. STARK and K. SIEGEL, *ibid.* **21**, 457(1906); J. STARK, W. HERMANN and S. KINOSHITA, *ibid.* **21**, 462 (1906).

[2] A. EINSTEIN, *Ann. der Phys.* **23**, 197(1907).

[3] H. E. IVES and G. R. STILWELL, *J. Opt. Soc. America* **28**, 215 (1938); **31**, 369 (1941); for a summary of their final results, see G. STEPHENSON and C. W. KILMISTER, *Special Relativity for Physicists* (Longmans, Green and Co., London, 1958), Chap. 3. See also G. OTTING, *Phys. ZS.* **40**, 681(1939).

the desired measurement with a reasonable degree of success. The major difficulty in such experiments lay in setting the spectroscope *exactly* transverse to the direction of the ion beam, for any deviation from exactitude would introduce an effect of *first* order in (v/c), which could easily mask the second-order effect being looked for. IVES and STILWELL circumvented this difficulty by comparing the arithmetic mean v_m of the frequencies corresponding to the *longitudinal* cases (i) and (ii) with the normal frequency v_0 corresponding to the case of a source (practically) at rest. One can show, using (4.22a) and (4.22b), that

$$\frac{v_m - v_0}{v_m} = 1 - \sqrt{1 - v^2/c^2} \simeq \frac{1}{2}\,(v^2/c^2)\;; \qquad (4.29)$$

according to the nonrelativistic formulae (4.26),

$$\frac{v_m - v_0}{v_m} = 1 - (1 - v^2/c^2) = (v^2/c^2). \qquad (4.30)$$

The experimental results were found to agree fairly well with formula (4.29), but not at all with (4.30).

4.3.   Aberration

Our next problem consists in comparing the directions of propagation of a plane wave, as observed in systems S and S'. The relevant formulae for this purpose are

with

$$\left.\begin{aligned}
\mathbf{k}' &= \mathbf{k} + \frac{(\mathbf{k} \cdot \mathbf{v})\,\mathbf{v}}{v^2}\,(\gamma - 1) - \gamma\mathbf{v}\,\frac{\omega}{c^2}, \\[2ex]
\gamma &= \frac{1}{\sqrt{1 - v^2/c^2}}.
\end{aligned}\right\} \qquad (4.18)$$

Assuming, for the sake of simplicity, that the components of the relative velocity $\mathbf{v}$, between systems S and S', are $(v, 0, 0)$, formulae (4.18) reduce to

$$\left.\begin{aligned}
k'_x &= \gamma\,(k_x - v\omega/c^2), \\[2ex]
k'_y &= k_y, \qquad k'_z = k_z.
\end{aligned}\right\} \qquad (4.18a)$$

In terms of the respective direction cosines,

$$(l, m, n) = \frac{(k_x, k_y, k_z)}{\sqrt{k_x^2 + k_y^2 + k_z^2}} \tag{4.31}$$

and

$$(l', m', n') = \frac{(k_x', k_y', k_z')}{\sqrt{k_x'^2 + k_y'^2 + k_z'^2}}, \tag{4.32}$$

formulae (4.18a) become (since $\omega = wk$)

$$(l', m', n') = \frac{\left[1 - \frac{vw}{c^2},\; m\sqrt{1 - v^2/c^2},\; n\sqrt{1 - v^2/c^2}\right]}{D}, \tag{4.33}$$

where

$$D = \frac{k'}{\gamma k} = \left[1 - 2l\frac{vw}{c^2} + \frac{v^2 w^2}{c^4} - (1 - l^2)\frac{v^2}{c^2}\right]^{1/2}. \tag{4.34}$$

A comparison of Eqs. (4.33) and (4.34) with the corresponding ones for particle motion, viz. (1.63) and (1.64), shows that the direction cosines in both cases transform *alike*, provided that the wave velocity w and the particle velocity u are connected by the relation (4.15), i.e. $w = c^2/u$.

For subsequent discussion we assume that the wave normal is perpendicular to the z-axis, i.e. $n = n' = 0$. In that case, the study of wave motion is confined to a plane perpendicular to the z-axis, say the x-y plane. Moreover, we shall be concerned specifically with the propagation of a *light wave*: $w = c$. Formulae (4.33) and (4.34) then reduce to

$$\{l', m'\} = \frac{\{l - (v/c),\; m\sqrt{1 - v^2/c^2}\}}{1 - l(v/c)}. \tag{4.35}$$

Let α be the angle between the wave normal and the positive direction of the x-axis in system S and α' the corresponding angle in system S'; then, from (4.35), we get

$$\cos \alpha' = \frac{\cos \alpha - (v/c)}{1 - \cos \alpha(v/c)} \tag{4.36}$$

and

$$\sin \alpha' = \frac{\sin \alpha\sqrt{1 - v^2/c^2}}{1 - \cos \alpha(v/c)}, \tag{4.37}$$

whence it follows that

$$\tan \alpha' = \frac{\sin \alpha \sqrt{1 - v^2/c^2}}{\cos \alpha - (v/c)}. \tag{4.38}$$

These formulae determine the *aberration* of light.   It will be noted that the formula (4.38) differs from its nonrelativistic counterpart through the factor $\sqrt{1 - v^2/c^2}$ alone ; thus, relativistic results are the same as nonrelativistic ones insofar as terms of first order in (v/c) are concerned.

Certain variations of these formulae are worth noting, viz.

$$\tan \frac{\alpha'}{2} = \frac{\sqrt{1 + v/c}}{\sqrt{1 - v/c}} \tan \frac{\alpha}{2}, \tag{4.39}$$

and

$$\sin \left(\frac{\alpha' - \alpha}{2} \right) = \tanh \left(\frac{1}{2} \tanh^{-1} \left(\frac{v}{c} \right) \right) . \sin \left(\frac{\alpha' + \alpha}{2} \right) ; \tag{4.40}$$

Of these the second one is particularly useful, in that it directly deals with the 'amount' of aberration $(\alpha' - \alpha)$.   For $v/c \ll 1$, $(\alpha' - \alpha)$ would itself be very small; one can then write

$$\delta\alpha \simeq \frac{v}{c} \sin \alpha, \tag{4.41}$$

which is the classical formula for the aberration of light.

In the present case, experimental results have remained unable to establish the validity of relativistic formulae as opposed to nonrelativistic ones, because the level of observational accuracy for this phenomenon has not been able to go beyond terms of the *first* order in v/c, to which order the two formulae are, of course, identical.

A problem closely related to the one discussed above is that of reflection of light from a moving mirror; it arises, particularly, in the study of the thermodynamic behavior of electromagnetic or thermal radiation in a perfectly reflecting enclosure *whose volume is slowly changing*. Its solution involves the application of aberration formulae twice, first to the incident ray and then to the reflected ray. For illustration, it would suffice to consider the simplest case, i.e. that of a plane mirror moving in a direction normal to itself.[1]

Let the reflecting surface of the mirror M constitute the (y', z')-plane of

---

[1] For more details, see W. PAULI, *loc. cit.*, Sec. 32 (δ).

system S' and, hence, be in motion with respect to system S in the direction of the x-axis; see Fig. 4.1. Let a ray of light be incident on the mirror, striking it at the point O' with the angle of incidence α', and be subsequently reflected from it with the angle of reflection, again, α'. Let the corresponding angles be α_1 and α_2, as observed in S. To be consistent in notation, all angles will

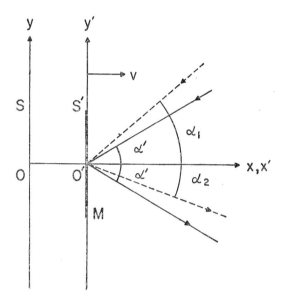

Fig. 4.1.

be measured with respect to the *positive* direction of the x'- or x-axis, keeping in view the directions of the rays ; only then will formulae (4.36)—(4.38) be applicable *as such*. Referring to the accompanying figure, we observe that the angle of incidence is actually $(\pi + \alpha')$ in S' and $(\pi + \alpha_1)$ in S, while the corresponding measures for the angle of reflection are $(2\pi - \alpha')$ and $(2\pi - \alpha_2)$, respectively. Substituting these in formula (4.38), we obtain for the incident ray

$$\tan \alpha' = \frac{\sin \alpha_1 \sqrt{1 - v^2/c^2}}{\cos \alpha_1 + (v/c)}, \qquad (4.38a)$$

and for the reflected ray

$$\tan \alpha' = \frac{\sin \alpha_2 \sqrt{1 - v^2/c^2}}{\cos \alpha_2 - (v/c)}. \qquad (4.38b)$$

Equating (4.38a) and (4.38b), we obtain the desired relationship between the angles of incidence and reflection, as observed in S:

$$\frac{\sin \alpha_1}{\cos \alpha_1 + (v/c)} = \frac{\sin \alpha_2}{\cos \alpha_2 - (v/c)}, \tag{4.42}$$

whence it follows that

$$\alpha_1 - \alpha_2 = \sin^{-1}\left[\frac{v}{c}(\sin \alpha_1 + \sin \alpha_2)\right] \tag{4.43}$$

$$= 2 \sin^{-1}\left[\frac{v}{c}\sin\left(\frac{\alpha_1 + \alpha_2}{2}\right)\right]; \tag{4.44}$$

cf. the corresponding formula (4.40) for aberration. Further, for $v/c \ll 1$, we have

$$\delta\alpha \simeq 2\frac{v}{c}\sin \alpha ; \tag{4.45}$$

cf. the corresponding result (4.41) for aberration.

In passing, we may also derive formulae for the change of wavelength on reflection. Making use of Eq. (4.20), we obtain for the incident ray

$$v' = \gamma v_1\left[1 - \frac{v}{c}\cos(\pi + \alpha_1)\right], \tag{4.46}$$

and for the reflected ray

$$v' = \gamma v_2\left[1 - \frac{v}{c}\cos(2\pi - \alpha_2)\right], \tag{4.47}$$

which give for the ratio of the incident and reflected wavelengths

$$\frac{\lambda_1}{\lambda_2} = \frac{v_2}{v_1} = \frac{1 + \frac{v}{c}\cos \alpha_1}{1 - \frac{v}{c}\cos \alpha_2}, \tag{4.48}$$

where α_1 and α_2 are related through (4.42). Again, for $v/c \ll 1$, we obtain the nonrelativistic result

$$\frac{\lambda_1}{\lambda_2} \simeq 1 + 2\frac{v}{c}\cos \alpha , \tag{4.49}$$

which was employed by WIEN in his investigations, of 1893, leading to the discovery of the *displacement law* of black-body radiation.

4.4.   Propagation of light in a moving dispersive medium

In Sec. 1.8 we considered the propagation of light in a moving medium, but neglected dispersion effects altogether ; of course, the result derived there was found to be in perfect agreement with the empirical one obtained in 1851 by FIZEAU.   In the case of a dispersive medium, however, a correction to the right-hand side of (1.67) becomes necessary.[1]   This arises from the fact that the refractive index of light is now a function of its wavelength and is no longer a constant of the medium.   It is straightforward to see that the parameter n in (1.65) should correspond to wavelength λ', as observed in the rest system S' of the medium, whereas n in the final result should correspond to wavelength λ, as observed in the laboratory system S.   We must, therefore, express the quantity $c/n(\lambda')$, appearing in (1.67), in terms of quantities appropriate to system S.

To the first approximation, we have

$$\frac{c}{n(\lambda')} \simeq \frac{c}{n(\lambda)} - \frac{c}{n^2(\lambda)} \frac{dn(\lambda)}{d\lambda} (\lambda' - \lambda). \tag{4.50}$$

Now, $(\lambda' - \lambda)$ can be obtained directly from formulae for the Doppler effect; of course, for a first-order calculation (as is being done in this problem), one may even employ nonrelativistic formulae.   In any case,

$$\lambda' - \lambda \simeq \lambda \frac{v}{(c/n)}. \tag{4.51}$$

In the last equation, we have written n without specifying whether it corresponds to wavelength λ or λ' because this distinction would affect only second-order terms of the final result (which are being neglected altogether).

Substituting (4.50) and (4.51) into (1.67), and writing n for $n(\lambda)$, we obtain the desired formula

$$u \simeq \frac{c}{n} \pm v \left[1 - \frac{1}{n^2} - \frac{\lambda}{n} \frac{dn}{d\lambda} \right]. \tag{4.52}$$

This formula was subjected to extensive verification by ZEEMAN[2] and others,[3]

---

[1] This was first pointed out by LORENTZ ; see p. 101 of the reference quoted in footnote 2 on p. 5 of the text.

[2] P. ZEEMAN, *Versl. gewone Vergad. Akad. Amsterdam* 23, 245 (1914); 24, 18 (1915); 28, 1451 (1919).

[3] See, in particular, G. SAGNAC, *C. R. Acad. Sci.* Paris 157, 708, 1410 (1913) ; *Journ. de Phys.* (5), 4, 177 (1914). Also P. ZEEMAN and A. SNETHLAGE, *Versl. gewone Vergad. Akad. Amsterdam* 28, 1462 (1919); *Proc. Akad. Sci. Amsterdam* 22, 462, 512 (1920).   The theory of these experiments was developed in detail by M. V. LAUE, *Ann. der Phys.* 62, 448 (1920).

using both liquid and solid media; the results were found to be in reasonable agreement with the predictions of the theory.[1]

Problems

4.1. (a) Derive the nonrelativistic formula (4.23) for the optical Doppler effect.

(b) Examine situations in which the relativistic and nonrelativistic formulae predict Doppler shifts in opposite directions, i.e. $(v/v_0) < 1$ in one case and $(v/v_0) > 1$ in the other.

4.2. There is a spaceship shuttle service between Earth and Mars. Each of the spaceships involved is equipped with two identical lights, one at the front and one at the rear. The spaceships normally travel at a speed v_0, relative to the Earth, with the result that the head-light of a spaceship approaching the Earth appears green ($\lambda_2 = 5000$ Å) while the tail light of a departing spaceship appears red ($\lambda_1 = 6000$ Å).

(a) Determine the values of v_0 and λ_0.

(b) If a Mars-bound spaceship accelerates to overtake another Mars-bound spaceship ahead of it, at what speed must the overtaking spaceship travel, relative to the Earth, so that the tail light of the other spaceship appears to it as a headlight?

4.3. (a) Consider two inertial frames, S and S', such that their relative velocity **v** is confined to the x-direction. A light signal of frequency v_0 in S is emitted from the point $x = -l$ at time $t = -l/c$. What is the frequency of that signal in S'? Simplify your answer for the case $v \ll c$.

(b) Note that, at time $t = 0$, the frame S' has the same velocity as a *noninertial* frame which began accelerating from rest at time $t = -l/c$ with acceleration $a = vc/l$. By the *principle of equivalence*, see Sec. 7.1, the result in (a) implies that the frequency shift of light in a uniform gravitational field g should be given by

$$| \Delta v |/v_0 \simeq v/c = gl/c^2 = | \Delta\varphi |/c^2,$$

where l is the distance between the emitter and the receiver while $\Delta\varphi$ is the corresponding difference in the gravitational potential. Obtain the foregoing result on the basis of *mass-energy equivalence*, as applied to a photon ; for a rigorous treatment of this problem, see Sec. 8.5.

---

[1] In the case of a moving solid medium, such as glass or quartz, the theory has to be slightly modified and a formula different from (4.52) results. See J. L. SYNGE, *Relativity: The Special Theory* (North-Holland Publishing Company, Amsterdam, 1956), Chap. V, Sec. 8; See also A. A. EVETT and D. C. FRIED, *Am. J. Phys.* **28**, 733 (1960); P. T. LANDSBERG, *Nature* **189**, 654 (1961).

CHAPTER 5

ELECTROMAGNETISM

5.1 Electromagnetic field equations in free space

In this chapter we propose to undertake the relativistic study of electromagnetic phenomena *in free space*, i.e. in the absence of conductors or dielectric and magnetic substances; thus, the only type of (electric) currents appearing in this study would be the 'convection' currents arising from the motion of charges, if any. As is well known, the basic equations governing these phenomena are the MAXWELL-LORENTZ field equations:[1]

$$\text{div } \mathbf{B} = 0, \tag{5.1a}$$

$$\text{curl } \mathbf{E} + \frac{1}{c} \frac{\partial \mathbf{B}}{\partial t} = 0, \tag{5.1b}$$

$$\text{div } \mathbf{E} = 4\pi\rho, \tag{5.2a}$$

and

$$\text{curl } \mathbf{B} - \frac{1}{c} \frac{\partial \mathbf{E}}{\partial t} = 4\pi \frac{\rho\mathbf{u}}{c}. \tag{5.2b}$$

Here, $\mathbf{E}$ and $\mathbf{B}$ are the field vectors representing electric field intensity and magnetic induction, respectively, ρ is the charge density and, $\mathbf{u}$ being the flow velocity of the charge, $\rho\mathbf{u}$ gives the corresponding current density;[2] all

---

[1] For a systematic study of the subject-matter leading to Equations (5.1) and (5.2), refer to J. D. JACKSON, *Classical Electrodynamics* (John Wiley & Sons, New York, 1962).

[2] We use here the Gaussian system of units.

these quantities are supposed to be continuously varying functions of the space and time coordinates.

Eliminating the field vectors $\mathbf{E}$ and $\mathbf{B}$, by taking the divergence of (5.2b) and making use of (5.2a), we get

$$\text{div}(\rho\mathbf{u}) + \frac{\partial\rho}{\partial t} = 0, \tag{5.3}$$

which is the *equation of continuity*, expressing one of the most fundamental experiences of electrodynamics, viz. the *conservation of electric charge*.

In the case of free space, with $\rho = 0$, we obtain the following set of equations:[1]

$$\nabla^2\mathbf{E} - \frac{1}{c^2}\frac{\partial^2\mathbf{E}}{\partial t^2} = 0, \tag{5.4a}$$

and

$$\nabla^2\mathbf{B} - \frac{1}{c^2}\frac{\partial^2\mathbf{B}}{\partial t^2} = 0, \tag{5.4b}$$

where the differential operator ∇^2, the familiar Laplacian operator, stands for $(\partial^2/\partial x^2 + \partial^2/\partial y^2 + \partial^2/\partial z^2)$. Equations (5.4) are wave equations for the propagation of electromagnetic disturbances in free space, with velocity c.

In view of (5.1), we may introduce a vector potential $\mathbf{A}$ and a scalar potential φ, such that $\mathbf{E}$ and $\mathbf{B}$ can be written as

$$\mathbf{B} = \text{curl } \mathbf{A}, \tag{5.5a}$$

and

$$\mathbf{E} = -\text{ grad } \varphi - \frac{1}{c}\frac{\partial\mathbf{A}}{\partial t}; \tag{5.5b}$$

with $\mathbf{E}$ and $\mathbf{B}$ related to $\mathbf{A}$ and φ as in (5.5), field equations (5.1) are *identically* satisfied. However, there does remain an element of arbitrariness in the choice of the potentials, for the fields $\mathbf{E}$ and $\mathbf{B}$ (which are the quantities of direct physical interest) remain unaffected if we add to the vector potential $\mathbf{A}$ the quantity grad f, f being an arbitrary scalar function, and at the same time subtract from the scalar potential φ the quantity $\frac{1}{c}\frac{\partial f}{\partial t}$. Such a

---

[1] For deriving these equations, we take the curl of (5.1b) and (5.2b), make use of the vector relation

$$\text{curl curl } \mathbf{F} = \text{grad div } \mathbf{F} - \Delta^2\mathbf{F},$$

and substitute from (5.2a) and (5.1a), respectively.

transformation (of the potentials) is called a *gauge transformation*:

$$\mathbf{A}^* = \mathbf{A} + \text{grad } f, \qquad \varphi^* = \varphi - \frac{1}{c}\frac{\partial f}{\partial t}; \qquad (5.6)$$

this constitutes a mathematical contrivance which can prove to be of great analytical importance, though it *does not* modify the physical situation in any manner whatsoever.

Different choices of the scalar function f lead to a variety of gauges. Most important of these is the so-called *Lorentz gauge,* in which the potentials $\mathbf{A}$ and φ (we omit the asterisk here) satisfy the condition

$$\text{div } \mathbf{A} + \frac{1}{c}\frac{\partial \varphi}{\partial t} = 0; \qquad (5.7)$$

this is usually referred to as the *Lorentz condition.* In this gauge,[1] field equations (5.2) become

$$\nabla^2 \varphi - \frac{1}{c^2}\frac{\partial^2 \varphi}{\partial t^2} = -4\pi\rho, \qquad (5.8a)$$

and

$$\nabla^2 \mathbf{A} - \frac{1}{c^2}\frac{\partial^2 \mathbf{A}}{\partial t^2} = -4\pi \frac{\rho \mathbf{u}}{c}. \qquad (5.8b)$$

These equations have well known solutions:[2]

$$\varphi(\mathbf{r}, t) = \int \frac{[\rho]_{t-R/c}}{R} dV' + \varphi_0, \qquad (5.9)$$

and

$$\mathbf{A}(\mathbf{r}, t) = \int \frac{[\rho \mathbf{u}]_{t-R/c}}{cR} dV' + \mathbf{A}_0, \qquad (5.9a)$$

where the subscript $(t-R/c)$ means that the charge density and the current density appearing in the integrands are to be taken at time $(t-R/c)$, R being

---

[1] It may be pointed out that, within the LORENTZ gauge, a restricted set of (gauge) transformations is still possible. For instance, the LORENTZ condition (5.7) continues to be satisfied under transformation (5.6), provided that the function f is so chosen as to satisfy the D'Alembert equation

$$\square f \equiv \left(\nabla^2 - \frac{1}{c^2}\frac{\partial^2}{\partial t^2} \right) f = 0.$$

Such a transformation will not modify, in any manner, the field equations (5.8).

[2] See, for instance, L. D. LANDAU and E. M. LIFSHITZ, *The Classical Theory of Fields* (Pergamon Press, London, 1962), Sec. 62.

the distance between the volume element dV' (from where an electromagnetic influence started at time $t' = t - R/c$) and the point of observation $\mathbf{r}$, (where the net effect was observed at time t). Moreover, φ_0 and $\mathbf{A}_0$ are the solutions of (5.8a) and (5.8b), respectively, with the right-hand sides equated to zero. Potentials (5.9), without φ_0 and $\mathbf{A}_0$, are referred to as the *retarded potentials*.

With these preliminaries, we now proceed to consider the fundamental question of *transformation equations* for the measures of various electromagnetic quantities, such as the charge density, the current density, the field intensities and the field potentials, as obtained in two inertial systems of reference S and S'. Closely associated with this question is the question of *covariance* of the basic equations of electromagnetism under LORENTZ transformation. We shall find that the formalism of the electromagnetic theory turns out to be *already* LORENTZ-covariant; any changes that the new theory introduces are of a purely mathematical character, in that one prefers to employ the four-tensor language instead of the ordinary three-vector language. However, this formal switch-over not only enables one to write down field equations in a form that is *manifestly* covariant but also to derive transformation equations for various physical quantities in a rather straightforward manner.

5.2   Charge conservation and the four-current density

We start with the equation of continuity,

$$\text{div}\,(\rho\mathbf{u}) + \frac{\partial\rho}{\partial t} = 0, \tag{5.3}$$

which, as indicated in the preceding section, expresses the fact that *electric charge is conserved*. We now postulate, with POINCARE, that the law of conservation of electric charge holds equally well in all inertial systems of reference; in other words, the equation of continuity is satisfied by the primed variables (appropriate to the system S') as well as by the unprimed variables (appropriate to the system S). Thus, in S', we must have

$$\text{div}'\,(\rho'\mathbf{u}') + \frac{\partial\rho'}{\partial t'} = 0, \tag{5.10}$$

the symbol div' means that the divergence in (5.10) is taken with respect to the space coordinates x'_i of S'. Since the LORENTZ transformation equations for the space-time coordinates and for the velocity components have already been obtained in Chapter 1, we must now determine transformation equations for the charge density such that, for arbitrary charge and current distributions, Eqs. (5.3) and (5.10) are mutually compatible!

Introducing a set of four quantities

$$(s_i) = \left(\rho \frac{\mathbf{u}}{c}, i\rho \right) \tag{5.11}$$

in S and the analogous set (s'_i) in S', Eqs. (5.3) and (5.10) become

$$\frac{\partial s_i}{\partial x_i} = 0 \tag{5.12}$$

and

$$\frac{\partial s'_i}{\partial x'_i} = 0, \tag{5.13}$$

respectively. Our problem, then, consists in determining the form of the functional relationships

$$s'_i = s'_i (s_1, s_2, s_3, s_4), \qquad i = 1, 2, 3, 4, \tag{5.14}$$

such that the quantities s'_i satisfy the continuity equation

$$0 = \frac{\partial s'_i}{\partial x'_i} = \frac{\partial s'_i}{\partial s_k} \frac{\partial s_k}{\partial x_l} \frac{\partial x_l}{\partial x'_i} = \left(\frac{\partial s'_i}{\partial s_k} \alpha_{il} \right) \frac{\partial s_k}{\partial x_l} \tag{5.15}$$

for all possible variations of the variables s_k and $\partial s_k/\partial x_l$, subject to the restrictive condition $(\partial s_i/\partial x_i) = 0$; note that we have written here: $\alpha_{il} = \partial x_l/\partial x'_i$ [see (2.42)]. Multiplying our restrictive condition by the LAGRANGE undetermined multiplier λ (which may be a function of the s_k) and subtracting it from (5.15), we get

$$\left(\frac{\partial s'_i}{\partial s_k} \alpha_{il} - \lambda \delta_{kl} \right) \frac{\partial s_k}{\partial x_l} = 0; \tag{5.16}$$

this equation must now hold for *arbitrary* variations of the variables s_k and $\partial s_k/\partial x_l$. Because of the variations of the latter, we must have

$$\frac{\partial s'_i}{\partial s_k}\,\alpha_{il} = \lambda\,\delta_{kl}$$

$$= \lambda\,\alpha_{ik}\,\alpha_{il}, \tag{5.17}$$

by (2.33).   Hence

$$\left(\frac{\partial s'_i}{\partial s_k} - \lambda\,\alpha_{ik}\right)\alpha_{il} = 0. \tag{5.18}$$

Now, in order that Eqs. (5.18) be true for all transformations in the LORENTZ group, i.e. for all values of the (transformation) coefficients α_{il}, we must have, quite generally,

$$\frac{\partial s'_i}{\partial s_k} = \lambda\,\alpha_{ik}\,. \tag{5.19}$$

Next we have to decide about the dependence of λ on s_k.   For this, we differentiate (5.17) with respect to s_m, with the result

$$\frac{\partial^2 s'_i}{\partial s_m\,\partial s_k}\,\alpha_{il} = \frac{\partial \lambda}{\partial s_m}\,\delta_{kl}\,. \tag{5.20}$$

Since the quantity on the left-hand side here is symmetrical with respect to the suffices m and k, so must be the quantity on the right-hand side.   Thus, we must have

$$\frac{\partial \lambda}{\partial s_m}\,\delta_{kl} = \frac{\partial \lambda}{\partial s_k}\,\delta_{ml}. \tag{5.21}$$

For $m = k$, this is merely an identity; for $m \neq k \neq l$, it reduces to the trivial statement: $0 = 0$; for $m \neq k = l$, it gives

$$\frac{\partial \lambda}{\partial s_m} = 0. \tag{5.22}$$

Hence, λ is independent of s_m and, for that matter, of all s_k; it is, therefore, a constant multiplier.   We can now integrate (5.19), with the result

$$s'_i = \lambda\,\alpha_{ik}\,s_k + \beta_i^{(k)}, \tag{5.23}$$

$\beta_i^{(k)}$ being the relevant constant of integration.

Physically, it is quite obvious that if the quantities s_k vanish, the quantities s'_k must also vanish. The constant(s) of integration must, therefore, be zero. Equations (5.23) then become

$$s'_i = \lambda \, \alpha_{ik} \, s_k \, . \tag{5.24}$$

Next, in order that for an identity transformation $(\alpha_{ik} = \delta_{ik})$, s'_i be equal to s_i, λ must be equal to unity. Hence, we finally obtain

$$s_i = \alpha_{ik} \, s_k, \tag{5.25}$$

which shows that the set (s_i) is a four-vector;[1] we call it the *four-current density*. The equation of continuity (5.12) may then be written as

$$\mathrm{Div}(s_i) = 0 \, ; \tag{5.26}$$

its covariance under LORENTZ transformations is self-evident. Actually, these conclusions are inevitably suggested by the very form of Eqs. (5.12) and (5.13).

From the invariance of the norm of the four vector (5.11), we get

$$\rho^2 \left(\frac{u^2}{c^2} - 1 \right) = \rho'^2 \left(\frac{u'^2}{c^2} - 1 \right) = (\rho^0)^2 \, (-1), \tag{5.27}$$

that is,

$$\rho\sqrt{1 - u^2/c^2} = \rho' \sqrt{1 - u'^2/c^2} = \rho^0 \, ; \tag{5.28}$$

this shows the manner in which charge density ρ transforms. Here, ρ^0 denotes the 'proper' charge density, i.e. the one measured in the local-cum-instantaneous rest system S^0 of the charge. Now, the transformation equations for the volume element are already known to be

$$\frac{\delta V}{\sqrt{1 - u^2/c^2}} = \delta V^0 = \frac{\delta V'}{\sqrt{1 - u'^2/c^2}} \, ; \tag{5.29}$$

---

[1] Four-vector character of the set (5.11) was first recognized by POINCARE in 1905; though not exactly in this language (because the four-dimensional formulation of the theory had yet to come), the results of this section, in all essentials, can be found in his papers, *loc. cit.*

hence, for an element of charge we obtain

$$\rho \, \delta V = \rho^0 \, \delta V^0 = \rho' \, \delta V'. \tag{5.30}$$

Thus, *the measure of the electric charge in each volume element is Lorentz-invariant*; obviously, the total charge would also be so.

We thus arrive at the invariance of electric charge after having /established the four-vector character of (s_i). SOMMERFELD,[1] on the other hand, showed that one could even reverse the argument. For instance, one may postulate that the charge element $\rho \delta V$ is invariant. Then, since the four-dimensional volume element $\delta V dx_4$ is already known to be invariant,[2] ρ/dx_4 must also be invariant; consequently, $\rho(dx_i/dx_4)$ must be a four-vector. This is nothing but $\rho \, [(\mathbf{u}/ic), 1] = (1/i) \, (s_i)$; the four-vector character of the set (s_i) is thereby established.

5.3. Four-potential and the field tensor

We shall now derive transformation equations for the electromagnetic potentials and field intensities. We take as our starting point the LORENTZ condition (5.7), which must be satisfied if the potentials A and φ are to conform to the LORENTZ gauge. Introducing another set of four quantities

$$(A_i) = (\mathbf{A}, \, i\varphi), \tag{5.31}$$

condition (5.7) can be written as

$$\frac{\partial A_i}{\partial x_i} = 0. \tag{5.32}$$

We now require that the LORENTZ condition (on the electromagnetic potentials) be itself LORENTZ-invariant; in other words, the adoption of the LORENTZ gauge be an absolute process. This clearly demands that (5.32) be a covariant equation. It then follows, in the same way as in the case of (5.12), that the set (5.31) be a four-vector;[3] we call this the electromagnetic *four-potential*. The LORENTZ condition then becomes

$$\mathrm{Div}(A_i) = 0. \tag{5.33}$$

---

[1] A. SOMMERFELD, *Ann. der Phys.* 32, 749 (1910).

[2] Actually, a pseudo-invariant; see Sec. 2.6.

[3] This was first noticed by MINKOWSKI in 1907; published posthumously in *Ann. der Phys.* 47, 927 (1915).

At the same time, Eqs. (5.8) take the form

$$\Box A_i = -4\pi \, s_i, \qquad (5.34)$$

(s_i) being the four-current density, as defined in (5.11). The covariance of Eqs. (5.33) and (5.34) is again self-evident.

It may be mentioned here that Eqs. (5.34), which are in the same form as the usual potential equations, can be solved by a method analogqus to the one employed in three dimensions.[1]

In view of Eqs. (5.5a) and (5.5b), we construct the (four-dimensional) Curl of the four-potential (5.31). Denoting the resulting tensor by (F_{ik}), we have, see (2.66),

$$(F_{ik}) = \mathrm{Curl}\,(A_i) = \left[\frac{\partial A_k}{\partial x_i} - \frac{\partial A_i}{\partial x_k}\right]. \qquad (5.35)$$

Being an antisymmetric tensor of rank 2, (F_{ik}) has only six independent components (those for which $i \neq k$). Comparing with (5.5), we obtain the following remarkable correspondence:

$$(F_{23}, F_{31}, F_{12}) = \mathbf{B}, \qquad (5.36)$$

$$(F_{41}, F_{42}, F_{43}) = i\mathbf{E}\,; \qquad (5.37)$$

of course, we also have $F_{ki} = -F_{ik}$. Consequently, the tensor (F_{ik}) may be depicted as

$$F_{ik} = \begin{bmatrix} 0 & B_z & -B_y & -iE_x \\ -B_z & 0 & B_x & -iE_y \\ B_y & -B_x & 0 & -iE_z \\ iE_x & iE_y & iE_z & 0 \end{bmatrix}. \qquad (5.38)$$

Thus, the field vectors $\mathbf{E}$ and $\mathbf{B}$ constitute an antisymmetric tensor of rank 2, viz. (F_{ik}), which is called the *electromagnetic field tensor*.

We again notice that, for a given tensor (F_{ik}), the choice of the vector (A_i) in (5.35) is not unique; we are at liberty to carry out a gauge transformation:

$$A_i^* = A_i + \frac{\partial \psi}{\partial x_i}, \qquad (5.39)$$

where ψ is a scalar function: *cf.* similar discussion in Sec. 5.1, especially Eqs. (5.6). One can readily verify that the replacement of (A_i), on the right-hand

---

[1] A. SOMMERFELD, *Ann. der Phys.* 33, 649 (1910) ; for details, refer to C. MOLLER, *The Theory of Relativity* (Clarendon Press, Oxford, 1952), Secs. 55 and 56.

side of (5.35), by (A_i^*) does not make any difference to the expression there. However, in order that the transformed four-potential (A_i^*) again conform to the LORENTZ gauge, as characterized by conditions (5.32), the function ψ must be such that it satisfies the equation

$$\frac{\partial^2 \psi}{\partial x_i \, \partial x_i} = \Box \, \psi = 0; \qquad (5.40)$$

cf. footnote 1 on p. 121.

We now take up the question of LORENTZ transformation of the various quantities under discussion. So far as the four-potential (A_i) is concerned, it will, of course, transform like a four-vector; this would give us the relevant transformation equations for A and φ. Of greater interest, however, is the transformation of the field tensor (F_{ik}), which, in turn, would give us the relevant equations for the field vectors E and B. By virtue of the transformation formulae (2.45), we obtain [making use of the α_{ik} appropriate to a general LORENTZ transformation without rotation, see (1.38), and the correspondence (5.38)]

$$\mathbf{E}' = \gamma \left[\mathbf{E} + \frac{1}{c}(\mathbf{v} \times \mathbf{B}) + \frac{(\mathbf{E} \cdot \mathbf{v}) \, \mathbf{v}}{v^2} \left(\frac{1}{\gamma} - 1 \right) \right], \qquad (5.41)$$

and

$$\mathbf{B}' = \gamma \left[\mathbf{B} - \frac{1}{c}(\mathbf{v} \times \mathbf{E}) + \frac{(\mathbf{B} \cdot \mathbf{v}) \, \mathbf{v}}{v^2} \left(\frac{1}{\gamma} - 1 \right) \right], \qquad (5.42)$$

with $\gamma = 1/\sqrt{1 - v^2/c^2}$. Equations inverse to the foregoing ones can be written down by interchanging the roles of the primed and unprimed quantities and also replacing v by $-$v.

Equations (5.41) and (5.42) make it evident that the division of the electromagnetic field into a pure electric field and a pure magnetic field, which we normally resort to for the sake of observational (and, hence, conceptual) convenience, is not an absolute process. The absence of one of these fields in one system of reference *does not* imply its absence in other systems as well.

In regard to the relative roles[1] of E and B in formulae (5.41) and (5.42), one

---

[1] It appears of interest to mention here that a set of six quantities, such as E and B, which transform as in (5.41) and (5.42) was, in the earlier stages of the development of the theory, usually referred to as a *six-vector*: *cf.* MINKOWSKI, 1908; also A. SOMMERFELD, *Electrodynamics*, Vol. III of his Lectures on Theoretical Physics (Academic Press, Inc., New York, 1952), Sec. 26. The use of such a term in this context must, however, be avoided, for, to us, it would mean a *vector* in a *six*-dimensional continuum [which the set (E, B) is not].

finds a significant difference which must be noted. That is, the sign of the second term on the right-hand side of (5.41) is positive while the corresponding sign in (5.42) is negative; consequently, these equations reproduce themselves under the substitutions $\mathbf{E} \rightarrow \mathbf{B}$, $\mathbf{B} \rightarrow -\mathbf{E}$, and not under the substitutions $\mathbf{E} \leftrightarrow \mathbf{B}$. This arises from the fact that there already exists a fundamentally significant difference in the manner in which $\mathbf{E}$ and $\mathbf{B}$ enter into the structure of the field tensor (5.38). From this structure it is evident that under a purely spatial rotation, $\mathbf{E}$ would behave like an ordinary *polar* vector, while $\mathbf{B}$ would behave like a three-tensor of rank 2 (which, however, being antisymmetric would have only three independent components), i.e. like an *axial* vector. It is, therefore, no surprise that the vectors $\mathbf{E}$ and $\mathbf{B}$ display a certain difference of behavior under a space-time rotation.

Finally, we write down transformation formulae for $\mathbf{E}$ and $\mathbf{B}$ under a special LORENTZ transformation without rotation. These follow from (5.41) and (5.42) on the substitution: $\mathbf{v} = (v, 0, 0,)$; we get

$$E'_x = E_x, \quad E'_y = \gamma \left[E_y - \frac{v}{c} B_z \right], \quad E'_z = \gamma \left[E_z + \frac{v}{c} B_y \right], \quad (5.43)$$

and

$$B'_x = B_x, \quad B'_y = \gamma \left[B_y + \frac{v}{c} E_z \right], \quad B'_z = \gamma \left[B_z - \frac{v}{c} E_y \right]. \quad (5.44)$$

The characteristic difference of signs in the two sets of equations is again conspicuous.

5.4. Lorentz invariance of the field equations

We shall now prove that the field equations (5.1) and (5.2) are invariant under LORENTZ transformation. For this, it would be sufficient to demonstrate that they can be put in the form of tensor equations. We must, therefore, utilize for this purpose the *field tensor* (F_{ik}), which is related to the field vectors $\mathbf{E}$ and $\mathbf{B}$ through the scheme (5.38), and the *four-vector* (s_i), which is related to the charge density ρ and the current density $\rho\mathbf{u}$ through the scheme (5.11). In terms of (F_{ik}) and (s_i), Eqs. (5.1) and (5.2) assume the form

$$\frac{\partial F_{ik}}{\partial x_l} + \frac{\partial F_{kl}}{\partial x_i} + \frac{\partial F_{li}}{\partial x_k} = 0, \quad i, k, l = 1, 2, 3, 4, \quad (5.45)$$

and

$$\frac{\partial F_{ik}}{\partial x_k} = 4\pi s_i, \quad i = 1, 2, 3, 4, \quad (5.46)$$

respectively. We shall now consider these equations at some length.

First of all, we note that the expression on the left-hand side of (5.45) is completely antisymmetric in the indices i, k and l; it would, therefore, reduce to triviality, $0 = 0$, unless i, k and l are all different. Further, any permutation among these indices leads us back to the original expression. We thus have only four independent equations, viz. those for which

$$(i, k, l) = (2, 3, 1), (3, 4, 2), (1, 4, 3) \text{ and } (2, 4, 1).$$

The first one among these coincides with (5.1a) and the remaining three with (5.1b). Thus, Eqs. (5.45), which may also be written as

$$\text{Curl } (F_{ik}) = 0, \tag{5.47}$$

represent *neither more nor less* than the four equations (5.1)—the so-called *Faraday tetrad*.

Equations (5.46) are clearly four in number; the first three among these coincide with (5.2b) and the fourth one with (5.2a). These equations, which may be written as

$$\text{Div } (F_{ik}) = 4\pi (s_i), \tag{5.48}$$

thus represent *exactly* the four equations (5.2)—the so-called *Ampere-Maxwell tetrad*.

Now, since the foregoing equations are all in the tensor form, their covariance under a LORENTZ transformation is self-evident.

Further, we note that a substitution from Eqs. (5.35) into (5.45) satisfies the latter identically, while the same substitution into (5.46) leads, with the use of (5.32), to Eqs. (5.34). Moreover, taking the Divergence of (5.46) and making use of the fact that (F_{ik}) is antisymmetric, we obtain (5.12). This proves beyond doubt the inner consistency of the formalism developed.

5.5.   The Lorentz force

In this section we shall calculate the electromagnetic forces experienced by electric charges moving in an electromagnetic field. We start with the simplest case of a single (charged) particle, carrying an electric charge e, moving with velocity $\mathbf{u}$ in a field specified by the tensor (F_{ik}). In the (instantaneous) rest system S^0 of the particle ($\mathbf{u}^0 = 0$), the force is known to be purely electrical:

$$\mathbf{F}^0 = e\mathbf{E}^0. \tag{5.49}$$

The four-force (F_i) in $S^0{}_2$ would therefore be, see (3.48),

$$(F_i)^0 = (e\mathbf{E}^0, 0). \tag{5.50}$$

In an arbitrary system S, however, the motion of the particle would be influenced by the field $\mathbf{B}$ as well, and we know that this influence is also dependent on the particle velocity $\mathbf{u}$. We must, therefore, construct a four-vector, out of the field tensor (F_{ik}) and the four-velocity (U_i), such that in S^0 it agrees with (5.50). The obvious choice seems to be the inner product

$$(F_{ik}\, U_k). \tag{5.51}$$

Now, in S^0, $(U_i)^0 = (0, 0, 0, ic)$; hence, from (5.51) and (5.38)

$$(F_{ik}\, U_k)^0 = ic(F_{i4})^0 = ic(-i\mathbf{E}^0, 0) = (c\mathbf{E}^0, 0). \tag{5.52}$$

A comparison of (5.50) and (5.52) shows that

$$(F_i)^0 = \frac{e}{c}\,(F_{ik}\, U_k)^0. \tag{5.53}$$

Now, (5.53) expresses the *equality* of two four-vectors in a particular system of reference S^0; this would obviously hold in all systems of reference. We therefore have, in the arbitrary system of reference S,

$$(F_i) = \frac{e}{c}\,(F_{ik}\, U_k), \tag{5.54}$$

which is the desired result.

Expressing (5.54), with the help of (3.32) and (5.38), in the standard form (3.48), we obtain

$$(F_i) = \left[\frac{\mathbf{F}}{\sqrt{1 - u^2/c^2}},\, \frac{i(\mathbf{F}\cdot\mathbf{u})/c}{\sqrt{1 - u^2/c^2}}\right], \tag{5.55}$$

with

$$\mathbf{F} = e\left[\mathbf{E} + \frac{1}{c}\,(\mathbf{u} \times \mathbf{B})\right], \tag{5.56}$$

the well known expression for the force experienced by a charged particle moving in an electromagnetic field, usually referred to as the *Lorentz force*. It must be emphasized at this point that the LORENTZ force, as obtained here, is a direct consequence of the relativity principle and does not require for its justification any extra physical, or axiomatic, premises.

We note that the four-force (F_i), as given by (5.54), is *orthogonal* to the four-velocity (U_i), for

$$F_i\, U_i = \frac{e}{c}\, (F_{ik}\, U_k\, U_i) = 0, \tag{5.57}$$

in view of the antisymmetric character of the tensor (F_{ik}). As discussed in Sec. 3.5, this orthogonality implies the constancy of the proper mass m_0 of the particle concerned. We can now write down the equations of motion of the particle in a covariant form:

$$\frac{dp_i}{d\tau} = m_0\, \frac{dU_i}{d\tau} = \frac{e}{c}\, F_{ik}\, U_k, \qquad i = 1, 2, 3, 4; \tag{5.58}$$

the first three equations here represent the ordinary equations of motion,

$$\frac{d\mathbf{p}}{dt} = m_0\, \frac{d}{dt}\left(\frac{\mathbf{u}}{\sqrt{1 - u^2/c^2}}\right) = e\left[\mathbf{E} + \frac{1}{c}\, (\mathbf{u} \times \mathbf{B})\right], \tag{5.59}$$

while the fourth one,

$$\frac{dE}{dt} = m_0 c^2\, \frac{d}{dt}\left[\frac{1}{\sqrt{1 - u^2/c^2}}\right] = e(\mathbf{E} \cdot \mathbf{u}), \tag{5.60}$$

equates the rate of increase of the particle energy with the power expended (by the field).

We now consider the more general case of a charge distribution. Assuming that the given distribution can be approximated by a *continuous* charge density function $\rho(x_i)$, we would have, instead of the four-force (5.54), a four-force density (f_i) given by

$$(f_i) = \frac{\rho^0}{c}\, (F_{ik}\, U_k), \tag{5.61}$$

where the (invariant) charge e has been replaced by the (invariant) proper charge density ρ^0. Recalling the definition of the four-current density (s_i), see (5.11), the relation between ρ and ρ^0, see (5.28), and the expression (3.32) for the four-velocity, we can write

$$(s_i) = \frac{\rho^0}{c}\, (U_i), \tag{5.62}$$

whence it follows that

$$(f_i) = (F_{ik}\, s_k). \tag{5.63}$$

The covariant form of the relevant equations of motion would then be

$$\mu^0 \frac{dU_i}{d\tau} = F_{ik} s_k, \qquad i = 1, 2, 3, 4, \qquad (5.64)$$

where μ^0 is the proper mass density of the distribution. The spatial and temporal parts of (5.64) are

$$\mu^0 \frac{d}{dt} \left(\frac{\mathbf{u}}{\sqrt{1 - u^2/c^2}} \right) = \rho^0 \left[\mathbf{E} + \frac{1}{c} (\mathbf{u} \times \mathbf{B}) \right] \qquad (5.65)$$

and

$$\mu^0 c^2 \frac{d}{dt} \left(\frac{1}{\sqrt{1 - u^2/c^2}} \right) = \rho^0 (\mathbf{E} \cdot \mathbf{u}), \qquad (5.66)$$

respectively;[1] the physical interpretation of these equations is straightforward.

5.6.   The electromagnetic energy-momentum tensor

We continue with the discussion, of the charge distribution, initiated in the preceding section. Assuming that the field (F_{ik}) in (5.63) arises from the distribution itself, we get, with the help of the field equations (5.46),

$$f_i = \frac{1}{4\pi} F_{ik} \frac{\partial F_{kl}}{\partial x_l}$$

$$= \frac{1}{4\pi} \left[\frac{\partial}{\partial x_l} (F_{ik} F_{kl}) - F_{kl} \frac{\partial F_{ik}}{\partial x_l} \right]. \qquad (5.67)$$

Now, $F_{kl} \dfrac{\partial F_{ik}}{\partial x_l}$ is obviously equal to $F_{lk} \dfrac{\partial F_{il}}{\partial x_k}$ (for, an interchange of the dummy variables k and l is quite insignificant); this, in turn, is equal to $F_{kl} \dfrac{\partial F_{li}}{\partial x_k}$ (because of the antisymmetric character of the field tensor). Hence, the second term in the brackets of (5.67) may be replaced by

$$\frac{1}{2} F_{kl} \left[\frac{\partial F_{ik}}{\partial x_l} + \frac{\partial F_{li}}{\partial x_k} \right],$$

[1] These facts were first recognized by POINCARE and were subsequently stated, in more clear terms by, MINKOWSKI.

which, in view of the field equations (5.45), is equal to

$$-\frac{1}{2} F_{kl} \frac{\partial F_{kl}}{\partial x_i} = -\frac{1}{4} \frac{\partial}{\partial x_i} (F_{kl} F_{kl}).$$

Equation (5.67) may, therefore, be written as

$$f_i = \frac{1}{4\pi} \left[\frac{\partial}{\partial x_m} (F_{ik} F_{km}) + \frac{1}{4} \delta_{im} \frac{\partial}{\partial x_m} (F_{kl} F_{kl}) \right]$$

$$= -\frac{\partial S_{im}}{\partial x_m}, \tag{5.68}$$

where

$$S_{im} = \frac{1}{4\pi} \left[F_{ik} F_{mk} - \frac{1}{4} \delta_{im} (F_{kl} F_{kl}) \right] \tag{5.69}$$

is the *electromagnetic energy-momentum tensor*. The foregoing result was first obtained by MINKOWSKI[1] in 1908.

We shall now examine the salient features of the tensor (5.69) and the tensor equation (5.68). First of all, we note that the tensor (S_{im}) is symmetric; it, therefore, has only ten independent components. However, this number reduces to nine, because the components S_{im} are bound by the identity

$$S_{ii} = \frac{1}{4\pi} \left[F_{ik} F_{ik} - \frac{1}{4} \delta_{ii}(F_{kl} F_{kl}) \right]$$

$$= \frac{1}{4\pi} [F_{ik} F_{ik} - F_{kl} F_{kl}] = 0; \tag{5.70}$$

we have used here the fact that $\delta_{ii} = 4$. Next, we consider the purely temporal part of the tensor (S_{im}), viz.

$$S_{44} = \frac{1}{4\pi} \left[F_{4k} F_{4k} - \frac{1}{4} (F_{kl} F_{kl}) \right]. \tag{5.71}$$

Now, $(F_{kl} F_{kl})$, which is an invariant of the field,[2] is equal to the sum of the

---

[1] For original references, see footnote on p. 100.

[2] The only other invariant of the field is the quantity $(e_{iklm} F_{ik} F_{lm})$; here, e_{iklm} is the completely antisymmetric unit tensor of rank 4; its components vanish unless i, k, l and m are all different, are equal to $+1$ if (i, k, l, m) is an even permutation of $(1, 2, 3, 4)$ and -1 if it is an odd permutation. It can be readily verified, with the help of the scheme (5.38), that this invariant, apart from a purely imaginary factor, is equal to $(\mathbf{E} \cdot \mathbf{B})$.

The invariance of the quantity $(\mathbf{E} \cdot \mathbf{B})$ incidentally shows that the transverse character of an electromagnetic wave is LORENTZ-invariant.

squares of all the (sixteen) components of the field tensor (F_{ik}); this comes out to be $2(B^2 - E^2)$. The quantity $F_{4k} F_{4k}$, on the other hand, stands for the sum of the squares of the four components comprising the last row in (5.38) and is equal to $-E^2$. Equation (5.71) then becomes

$$S_{44} = -\frac{1}{8\pi}(E^2 + B^2); \tag{5.72}$$

thus, the magnitude of the purely temporal part of the tensor (S_{im}) gives us the electromagnetic *energy density* of the field.

Further, we consider the mixed part of the tensor under discussion. We have here

$$S_{4p} = S_{p4} = \frac{1}{4\pi} F_{4k} F_{pk}, \qquad p = 1, 2, 3. \tag{5.73}$$

In view of the scheme (5.38), this turns out to be

$$(S_{4p}) = \frac{i}{4\pi}(\mathbf{E} \times \mathbf{B}) = \frac{i}{c}\mathbf{S}, \tag{5.74}$$

where $\mathbf{S}\left[= \dfrac{c}{4\pi}(\mathbf{E} \times \mathbf{B})\right]$ is the *Poynting vector* that gives us the rate of flow of energy in the field.

Finally, we have the purely spatial part:

$$S_{pq} = \frac{1}{4\pi}\left[F_{pk} F_{qk} - \frac{1}{2}\delta_{pq}(B^2 - E^2)\right], \qquad p, q = 1, 2, 3; \tag{5.75}$$

this turns out to be

$$
\begin{aligned}
S_{pq} &= \frac{1}{4\pi}\left[-\frac{1}{2}\delta_{pq}(B^2 - E^2) + \begin{cases}(B^2 - B_p^2 - E_p^2) & \text{if } p = q \\ (-B_p B_q - E_p E_q) & \text{if } p \neq q\end{cases}\right] \\
&= -\frac{1}{4\pi}\left[(B_p B_q + E_p E_q) + \delta_{pq}\left(\frac{B^2 - E^2}{2} - B^2\right)\right] \\
&= -t_{pq}, \tag{5.76}
\end{aligned}
$$

where

$$t_{pq} = \frac{1}{4\pi}\left[(B_p B_q + E_p E_q) - \delta_{pq}\frac{E^2 + B^2}{2}\right] \tag{5.77}$$

is the *Maxwell stress tensor*. Corresponding to the interpretation of θ_{pq}, defined by (3.80), this spatial tensor may be regarded as representing the *momentum current* in the field.[1] This lends a real physical meaning to the stress tensor which had hitherto been regarded as purely mathematical.

Recalling the equations of motion (3.72) for a continuous matter distribution, we note that if the forces experienced by the distribution are due to an electromagnetic field, we can write, see (5.68),

$$\text{Div}\ (\theta_{ik}) = (f_i) = -\text{Div}\ (S_{ik}),$$

that is,

$$\text{Div}\ (T_{ik}) = 0, \tag{5.78}$$

where

$$T_{ik} = \theta_{ik} + S_{ik}. \tag{5.79}$$

Being the sum of the kinetic and electromagnetic energy-momentum tensors, the tensor (T_{ik}) is called the *total energy-momentum tensor* of the system comprising the matter distribution as well as the electromagnetic field. Equations (5.78) constitute a concise and covariant statement of the laws of conservation of the total energy and the total momentum of this composite system.

5.7. Lagrangian and Hamiltonian functions of a particle in an electromagnetic field

Having established the relativistic equations of motion of a charged particle moving in an electromagnetic field, we are now in a position to examine the question of expressing them in the Lagrangian formulation. As is well known,[2] this formulation can be shown to follow directly from HAMILTON's *variational principle*, according to which the motion of a conservative system from time t_1 to time t_2 is such that the integral

$$I = \int_{t_1}^{t_2} L(q_j, \dot{q}_j, t)\, dt \tag{5.80}$$

---

[1] This interpretation of the spatial tensor (5.77) was first given by M. PLANCK, *Verh. d Deutschen phys. Ges.*, **6**, 728(1908); *Phys. ZS.* **9**, 828(1908).

[2] See, for instance, H. GOLDSTEIN, *Classical Mechanics* (Addison-Wesley Publishing Co. Inc., Reading, Mass. U.S.A., 1959), Chap. 2.

is *extremal* for the actual path of the motion; here, q_j and $\dot{q}_j$ ($j = 1, 2, \ldots, n$) are, respectively, the generalized coordinates and the corresponding 'velocities' that specify the mechanical state of the system at any time t, n is the number of degrees of freedom (3, in the case of a single particle), while L is a certain characteristic function of its arguments, the form of the function depending upon the details of the system under consideration. The function L is called the *Lagrangian* of the system.

From the extremal property of the integral (5.80) follow the so-called LAGRANGE's equations of motion :

$$\frac{d}{dt}\left(\frac{\partial L}{\partial \dot{q}_j}\right) - \frac{\partial L}{\partial q_j} = 0, \quad j = 1, 2, \ldots, n. \tag{5.81}$$

Further, one can show that if L does not depend explicitly on t we can define a *constant of motion H*, such that

$$H(p_j^*, q_j) = \sum_j (p_j^* \dot{q}_j) - L, \tag{5.82}$$

where p_j^* are defined by the relations

$$p_j^* = \frac{\partial L}{\partial \dot{q}_j}. \tag{5.83}$$

The function H is called the *Hamiltonian* of the system; variables p_j^* are the generalized 'momenta', conjugate to the generalized coordinates q_j, of the system. From the foregoing results we obtain the so-called HAMILTON's canonical equations of motion :

and

$$\left. \begin{array}{c} \dfrac{\partial H}{\partial p_j^*} = \dot{q}_j, \\[4mm] \dfrac{\partial H}{\partial q_j} = -\dot{p}_j^*, \end{array} \right\} \quad j = 1, 2, \ldots, n. \tag{5.84}$$

One is at liberty to use the differential equations (5.84), which are first-order but $2n$ in number, or the differential equations (5.81), which are second-order but only n in number. The physical content of the two sets of equations is the same; it is in regard to analytical details that they differ rather significantly. Finally, the physical significance of the Hamiltonian function is invariably the same, viz. the *total energy* of the system. However, the case of the Lagrangian function is not so plain; whereas in the simplest cases it represents the difference between the kinetic energy and the potential energy of the system (hence the name *kinetic-potential*), in general it is not so. Still, the theoretical importance of this function is great because it forms the very core of the basic variational principles which, in turn, are supposed to govern the most fundamental manifestations of Nature.

Let us now take up the question stated at the outset, viz. that of expressing the relativistic equations of motion of a charged particle moving in an electromagnetic field in terms of the Lagrangian formalism. Our first problem is to find an appropriate function such that it can serve as the Lagrangian for the motion concerned. In other words, we have to determine a function L such that the resulting equations of motion [the ones corresponding to Eqs. (5.81)] coincide with the actual equations of motion (5.59). Now, the latter ones, in terms of the field potentials, can be written as

$$m_0 \frac{d}{dt}\left(\frac{\mathbf{u}}{\sqrt{1 - u^2/c^2}}\right) = e\left[- \text{grad } \varphi - \frac{1}{c}\frac{\partial \mathbf{A}}{\partial t} + \frac{1}{c}(\mathbf{u} \times \text{curl } \mathbf{A})\right]. \quad (5.85)$$

The x-component of this equation is

$$m_0 \frac{d}{dt}\left(\frac{\dot{x}}{\sqrt{1 - u^2/c^2}}\right) = -e\frac{\partial \varphi}{\partial x} - \frac{e}{c}\frac{\partial A_x}{\partial t}$$

$$+ \frac{e}{c}\left[\dot{y}\left(\frac{\partial A_y}{\partial x} - \frac{\partial A_x}{\partial y}\right) - \dot{z}\left(\frac{\partial A_x}{\partial z} - \frac{\partial A_z}{\partial x}\right)\right]. \quad (5.86)$$

Since

$$\frac{dA_x}{dt} = \frac{\partial A_x}{\partial t} + \left(\dot{x}\frac{\partial A_x}{\partial x} + \dot{y}\frac{\partial A_x}{\partial y} + \dot{z}\frac{\partial A_x}{\partial z}\right), \quad (5.87)$$

(5.86) can be written in the form

$$\frac{d}{dt}\left[\frac{m_0\dot{x}}{\sqrt{1 - (\dot{x}^2 + \dot{y}^2 + \dot{z}^2)/c^2}} + \frac{e}{c}A_x\right]$$

$$= -e\frac{\partial \varphi}{\partial x} + \frac{e}{c}\left[\dot{x}\frac{\partial A_x}{\partial x} + \dot{y}\frac{\partial A_y}{\partial x} + \dot{z}\frac{\partial A_z}{\partial x}\right]. \quad (5.88)$$

Next, keeping in view the form of the equations (5.81), we re-write (5.88) as

$$\frac{d}{dt}\left\{\frac{\partial}{\partial \dot{x}}\left[-m_0 c^2 \sqrt{1-(\dot{x}^2+\dot{y}^2+\dot{z}^2)/c^2}+\frac{e}{c}\dot{x}A_x\right]\right\}$$

$$=\frac{\partial}{\partial x}\left[-e\varphi+\frac{e}{c}(\mathbf{u}\cdot\mathbf{A})\right]. \qquad (5.89)$$

Finally, adding to both sides of this equation certain quantities which are *identically* zero, we bring it to the desired form

$$\frac{d}{dt}\left\{\frac{\partial}{\partial\dot{x}}[L]\right\}=\frac{\partial}{\partial x}[L], \qquad (5.90)$$

with

$$L=m_0 c^2\,(1-\sqrt{1-u^2/c^2})+\frac{e}{c}\,(\mathbf{u}\cdot\mathbf{A})-e\varphi\,; \qquad (5.91)$$

the (constant) quantity $m_0 c^2$ has been included here so that the expression for L, in the *nonrelativistic* limit, reduces to the conventional form

$$L_{\text{N.R.}}=\frac{1}{2}m_0\,u^2+\frac{e}{c}\,(\mathbf{u}\cdot\mathbf{A})-e\varphi. \qquad (5.92)$$

Obviously, we shall also have equations similar to (5.90) arising from the y- and z-components of the vector equation (5.85); the set of LAGRANGE's equations then becomes complete, and the function (5.91) becomes the *Lagrangian function* of the problem under discussion. We then obtain for the conjugate momenta, see (5.83),

$$(p_j^*)=\left(\frac{m_0\mathbf{u}}{\sqrt{1-u^2/c^2}}+\frac{e}{c}\mathbf{A}\right); \qquad (5.93)$$

thus, the conjugate momenta p_j^* differ from the mechanical momenta p_j by the electromagnetic terms $\frac{e}{c}\,(A_x, A_y, A_z)$.

The Hamiltonian of the system would be, see (5.82), (5.91) and (5.93),

$$H=\left[\frac{m_0 u^2}{\sqrt{1-u^2/c^2}}+\frac{e}{c}\,(\mathbf{u}\cdot\mathbf{A})\right]$$

$$-\left[m_0 c^2\,(1-\sqrt{1-u^2/c^2})+\frac{e}{c}\,(\mathbf{u}\cdot\mathbf{A})-e\varphi\right]$$

$$=m_0 c^2\left(\frac{1}{\sqrt{1-u^2/c^2}}-1\right)+e\varphi, \qquad (5.94)$$

which is precisely equal to the total energy, kinetic plus potential, of the particle concerned. Now, in order to have the canonical form, the function H of (5.94) must be expressed as an explicit function of the coordinates q_j and the conjugate momenta p_j^* ; we then have

$$H = c\left[\left(\mathbf{p}^* - \frac{e}{c}\,\mathbf{A}\right)^2 + m_0^2\,c^2\right]^{1/2} - m_0 c^2 + e\varphi, \qquad (5.95)$$

which, with the help of the canonical equations (5.84), would also lead to the correct equations of motion of the particle. It may be noted here that the foregoing Hamiltonian differs from the one corresponding to particle motion in the absence of the electromagnetic field in that we have here $(\mathbf{p}^* - (e/c)\,\mathbf{A})$ instead of the simple $\mathbf{p}^*$ and $(H - e\varphi)$ instead of the simple H; it is obvious that these modifications disappear even in the presence of the field provided that the particle is uncharged.

Admittedly, throughout the foregoing formulation we have employed the language of the three-dimensional calculus; consequently, the resulting formulation is not a covariant one, though, undoubtedly, it is relativistically correct. In order to obtain a *covariant formulation,* we must first of all modify the form of the variational principle itself. The basic feature of this modification would be the replacement of the variable of integration t, in (5.80), by the invariant time parameter τ. Moreover, the integrand would be required to be LORENTZ-invariant so that the variational integral itself becomes invariant and, hence, the extremal principle holds in *all* inertial systems of reference. Clearly, the resulting equations of motion would themselves be covariant in form, coinciding with the equations

$$\frac{dp_i}{d\tau} = \frac{e}{c}\,F_{ik}\,U_k, \qquad i = 1, 2, 3, 4. \qquad (5.58)$$

Our problem then is to choose an invariant function $L'\,(x_i, U_i, \tau)$ such that the actual motion of the system is characterized by the fact that the integral

$$I' = \int_{\tau_1}^{\tau_2} L'\,(x_i, U_i, \tau)\,d\tau \qquad (5.96)$$

is extremal; in other words, the actual motion of the system is described by the (covariant) equations

$$\frac{d}{d\tau}\left(\frac{\partial L'}{\partial U_i}\right) - \frac{\partial L'}{\partial x_i} = 0, \qquad i = 1, 2, 3, 4. \qquad (5.97)$$

In analogy with the expressions (5.91) and (5.92), we try the following form for the covariant Lagrangian:

$$L' = \frac{1}{2} m_0 \, (U_i U_i) + \frac{e}{c} \, (U_k \, A_k).$$ (5.98)

From (5.97) and (5.98), we obtain

$$\frac{d}{d\tau} \left(m_0 U_i + \frac{e}{c} A_i \right) = \frac{e}{c} \frac{\partial}{\partial x_i} (U_k \, A_k),$$

that is,

$$\frac{d}{d\tau} (m_0 U_i) = \frac{e}{c} \left(\frac{\partial A_k}{\partial x_i} - \frac{\partial A_i}{\partial x_k} \right) U_k \quad \left(\because \frac{dx_k}{d\tau} \equiv U_k \right);$$

we thus get

$$\frac{d}{d\tau} (m_0 U_i) = \frac{e}{c} \, F_{ik} \, U_k,$$

in agreement with Eqs. (5.58).

We, therefore, conclude that the expression (5.98) is appropriate for the covariant form of the Lagrangian of the system. The four-vector

$$(p_i^*) \equiv \frac{\partial L'}{\partial U_i} = \left(m_0 U_i + \frac{e}{c} A_i \right)$$ (5.99)

is then the conjugate four-momentum of the particle, which differs from the mechanical four-momentum $(m_0 U_i)$ by the electromagnetic term (eA_i/c). Splitting (5.99) into spatial and temporal parts, we get

$$(p_i^*) = (p_i) + \frac{e}{c}(A_i)$$

$$= \left[\left(\mathbf{p} + \frac{e}{c} \mathbf{A} \right), \frac{i}{c} (E + e\varphi) \right]$$

$$= \left[\mathbf{p}^*, \frac{i}{c} H \right],$$ (5.100)

where $\mathbf{p}^*$ is the conjugate three-momentum, given by (5.93), while H is the conventional Hamiltonian (5.94), augmented by rest energy term $m_0 c^2$.

Finally, we determine the covariant form of the Hamiltonian of the system under study. We define, in analogy with (5.82),

$$H'(p_i^*, x_i, \tau) = p_i^* U_i - L'.\tag{5.101}$$

Substituting from (5.98) and (5.99), we get

$$H' = \frac{1}{2} m_0 U_i U_i.\tag{5.102}$$

This is, however, not in the canonical form; one must write instead

$$H' = \frac{1}{2m_0}\left(p_i^* - \frac{e}{c} A_i\right)^2.\tag{5.103}$$

It can now be verified that the foregoing expression for H' leads to the covariant form of HAMILTON's canonical equations, viz.

and

$$\left.\begin{array}{c} \dfrac{\partial H'}{\partial p_k^*} = \dfrac{dx_k}{d\tau}, \\[3mm] \dfrac{\partial H'}{\partial x_k} = -\dfrac{dp_k^*}{d\tau}. \end{array}\right\}\tag{5.104}$$

In this section we have considered the Lagrangian formulation, covariant as well as conventional, for the motion of a particle in an electromagnetic field. A corresponding formulation for the field itself is also of interest.[1]

5.8. Field due to a moving charge

Our next problem is concerned with calculating the electromagnetic field produced by an *arbitrarily* moving charge e. Let $P(x, y, z)$ be the point, and t the time, of observation of the field; we, then, speak of the event of observation $E(x, y, z, t)$. It is obvious that since the electromagnetic influences propagate at a finite speed (c, in free space), the effect observed at the space point P at time t must have left its place of origin, the point charge, at an *earlier* time t' such that

$$t' + \frac{R(t')}{c} = t;\tag{5.105}$$

---

[1] The interested reader may refer to W. PAULI, *loc. cit.*, Sec. 31; H. GOLDSTEIN, *loc. cit.*, Sec. 11-5. For an extensive treatment of this aspect, see L. D. LANDAU and E. M. LIFSHITZ, *Classical Theory of Fields, loc. cit*, Chaps. 3 and 4.

here, R is the distance between the point P and the position of the charge held at time t' — that is why we have indicated, rather explicitly, the dependence of R on t'

The value of t', for a given t, depends upon the state of motion of the charge with respect to the observer. In terms of space-time graphs, as discussed in Sec. 2.3, the value of t' will be given by the time coordinate of the point of intersection, $F(x'_i)$, of the time track of the given charge with the null half-cone, depicting the *past*,[1] drawn with the event of observation, $E(x_i)$, as its apex. The trajectory of the electromagnetic influence would, then, start from the world point F, proceed along a generator of the cone and pass through the world point E, corresponding to the time of observation t ; R is, then, the spatial separation between the world points E and F.

We first calculate the potentials A and φ. In the system S^0, with respect to which the charge was *at rest* at the instant of time t', the desired potentials at the space point P, at time t, would be

$$A^0 = 0, \qquad \varphi^0 = \frac{e}{R^0}, \tag{5.106}$$

where R^0 corresponds to a measurement in S^0. In the four-vector notation, Eqs. (5.106) would be written as

$$(A_i)^0 = \left(0, \, i \, \frac{e}{R^0}\right). \tag{5.107}$$

For passage to the system S, we have to construct, with the help of the four-vectors (U_i) and $(R_i = x_i - x'_i)$, a four-vector that reduces to (5.107) in S^0. Now, in the system S^0,

$$(U_i)^0 = (0, \, ic); \tag{5.108}$$

consequently, the four-vectors (A_i) and (U_i) are parallel in S^0. Obviously, they must be parallel in the system S too. Hence, our task is reduced to

---

[1] We do not intend to deal with the 'advanced' effects, corresponding to a time t' greater than t; this time would be determined by the equation

$$t' - \frac{R(t')}{c} = t,$$

and would be given by the time coordinate of the point of intersection of the time track of the given charge with the null half-cone depicting the *future*. For reasons of causality, such effects are normally rejected from consideration; see, however, J. A. WHEELER and R. P. FEYNMAN, *Revs. Mod. Phys.* **21**, 425 (1949).

determining an invariant such that in S^0 it is equal to $e/(cR^0)$; this invariant would provide a multiplying factor for the four-velocity (U_i) so as to yield the desired four-potential (A_i).

In this connection we note that the scalar product

$$
\begin{aligned}
(R_k \, U_k) &= \frac{\mathbf{R} \cdot \mathbf{u}}{\sqrt{1 - u^2/c^2}} - \frac{(t - t') \, c^2}{\sqrt{1 - u^2/c^2}} \\
&= \frac{\mathbf{R} \cdot \mathbf{u} - Rc}{\sqrt{1 - u^2/c^2}},
\end{aligned}
\tag{5.109}
$$

which is indeed an invariant, becomes, in S^0, simply $(-R^0 c)$. Hence, quite generally,

$$
(A_i) = - e(U_i)/(R_k \, U_k),
\tag{5.110}
$$

which is the desired result.[1]

Separating the spatial and temporal parts, we obtain

$$
\mathbf{A} = \frac{e \, \mathbf{u}/c}{R - \dfrac{\mathbf{R} \cdot \mathbf{u}}{c}}, \qquad
\varphi = \frac{e}{R - \dfrac{\mathbf{R} \cdot \mathbf{u}}{c}}.
\tag{5.111}
$$

Here, $\mathbf{R}$ is the radius vector directed from the position of the charge at time t' to the point of observation, while $\mathbf{u}$ is the velocity of the charge (at that very instant of time) with respect to the observer; hence,

$$
\mathbf{u} = - \frac{\partial \mathbf{R}}{\partial t'}.
\tag{5.112}
$$

Moreover, the parameters $\mathbf{R}$, R and $\mathbf{u}$ appearing on the right-hand sides of the foregoing equations correspond to the time t', and not to the time t.

The electromagnetic potentials (5.111) are the well known *Lienard-Wiechert potentials* due to a point charge moving arbitrarily. From these, one can calculate the field vectors $\mathbf{E}$ and $\mathbf{B}$ with the help of the relations

$$
\mathbf{B} = \operatorname{curl} \mathbf{A}
\tag{5.5a}
$$

and

$$
\mathbf{E} = -\operatorname{grad} \varphi - \frac{1}{c} \frac{\partial \mathbf{A}}{\partial t}.
\tag{5.5b}
$$

---

[1] In this form, the result was first obtained by H. MINKOWSKI : *Phys. ZS.* **10**, 104 (1909); see, in this connection, his article in *The Principle of Relativity, loc. cit.*

However, the field operations involved here imply differentiations with respect to the coordinates of the event E, not of the event F; we must, therefore, determine beforehand the derivatives $\partial t'/\partial x$, $\partial t'/\partial y$, $\partial t'/\partial z$ and $\partial t'/\partial t$, and only then proceed to substitute (5.111) into (5.5).

Carrying out a partial differentiation of (5.105) with respect to t, and making use of the result

$$R \frac{\partial R}{\partial t'} = \mathbf{R} \cdot \frac{\partial \mathbf{R}}{\partial t'} = -\mathbf{R} \cdot \mathbf{u}, \tag{5.113}$$

we obtain[1]

$$\frac{\partial t'}{\partial t} = \frac{R}{R - \dfrac{\mathbf{R} \cdot \mathbf{u}}{c}}. \tag{5.114}$$

Further, we have from (5.105)

$$\operatorname{grad} t' + \frac{1}{c} \left[\frac{\mathbf{R}}{R} + \frac{\partial \mathbf{R}}{\partial t'} \operatorname{grad} t' \right] = 0 ; \tag{5.115}$$

the first term in the brackets here comes from the explicit dependence of R on the coordinates x, y and z, while the second one arises because of its implicit dependence on t'. Making use of (5.113), we obtain from (5.115)

$$\operatorname{grad} t' = -\frac{\mathbf{R}/c}{R - \dfrac{\mathbf{R} \cdot \mathbf{u}}{c}}. \tag{5.116}$$

Substituting (5.111) into (5.5b) and making use of the formulae (5.114) and (5.116), we obtain, after a rather lengthy calculation,[2]

$$\mathbf{E} = \frac{e}{\left(R - \dfrac{\mathbf{R} \cdot \mathbf{u}}{c}\right)^3} \left[(1 - u^2/c^2)\, (\mathbf{R} - R\mathbf{u}/c) \right.$$

$$\left. + \frac{1}{c^2} \{ \mathbf{R} \times [(\mathbf{R} - R\mathbf{u}/c) \times \dot{\mathbf{u}}] \} \right], \tag{5.117}$$

---

[1] Note that the relations (5.105) and (5.114) enable us to write the potentials (5.111) in the interesting form:

$$\mathbf{A} = \varphi\,(\mathbf{u}/c), \qquad \varphi = \frac{e}{c\,(t - t')} \frac{\partial t'}{\partial t}.$$

[2] While carrying out the field operations here, one must remember the implicit dependence of the various quantities on t' as well as their explicit dependence, if any, on the coordinates x, y and z,

where

$$\dot{\mathbf{u}} = \frac{\partial \mathbf{u}}{\partial t'}. \tag{5.118}$$

Again the various quantities appearing on the right-hand side of the foregoing equation must correspond to the time t'.

Next, we substitute (5.111) into (5.5a) and make use of the formulae (5.114), (5.116) and the relations

$$\mathrm{curl}\ (\varphi \mathbf{u}) = \varphi\ (\mathrm{curl}\ \mathbf{u}) + (\mathrm{grad}\ \varphi) \times \mathbf{u} \tag{5.119}$$

and

$$\mathrm{curl}\ \mathbf{u} = (\mathrm{grad}\ t') \times \dot{\mathbf{u}}, \tag{5.120}$$

with the result

$$\mathbf{B} = \frac{e}{c^2 \left(R - \dfrac{\mathbf{R} \cdot \mathbf{u}}{c} \right)^3} \left[c\ (\mathbf{u} \times \mathbf{R}) \left(1 - \frac{u^2}{c^2} \right) \right.$$
$$\left. + (\dot{\mathbf{u}} \times \mathbf{R}) \left(R - \frac{\mathbf{R} \cdot \mathbf{u}}{c} \right) + \frac{1}{c}\ (\mathbf{u} \times \mathbf{R})\ (\mathbf{R} \cdot \dot{\mathbf{u}}) \right]. \tag{5.121}$$

Clearly, the vector $\mathbf{B}$ is perpendicular to $\mathbf{R}$. With the help of Eqs. (5.117) and (5.121) one can also show that

$$\mathbf{E} \cdot \mathbf{B} = 0\ ; \tag{5.122}$$

thus, the field vectors $\mathbf{E}$ and $\mathbf{B}$ are, quite generally (i.e. for any arbitrary motion of the charge and for any event of observation), mutually orthogonal. Consequently, $\mathbf{B}$ must be perpendicular to the plane of the vectors $\mathbf{R}$ and $\mathbf{E}$; indeed,

$$\mathbf{B} = \frac{1}{R}\ (\mathbf{R} \times \mathbf{E}). \tag{5.123}$$

One cannot fail to notice that the expression for $\mathbf{E}$, as well as for $\mathbf{B}$, consists of two distinct parts. The first part depends only on the velocity of the particle (not on its acceleration), and at large distances varies as $1/R^2$. The second part depends on acceleration as well, and at large distance varies as $1/R$. It is the second part alone which is responsible for the electromagnetic radiation emitted by the particle ;[1] see Problems 5.14 and 5.15.

---

[1] For a discussion of the radiation field arising from an accelerated charge, see W. HEITLER, *The Quantum Theory of Radiation* (Clarendon Press, Oxford, 1954), Chap. 1; W. K. H. PANOFSKY and M. PHILLIPS, *loc. cit.*, Chap. 19. See also L. D. LANDAU and E.M. LIFSHITZ, *The Classical Theory of Fields, loc. cit.*, Chap. 9,

We shall now discuss at some length the case of a *uniformly* moving charge; Eqs. (5.117) and (5.121) then become

$$E = \frac{e\,(1 - u^2/c^2)}{\left(R - \dfrac{R \cdot u}{c}\right)^3}\,(R - Ru/c) \tag{5.124}$$

and

$$B = \frac{e\,(1 - u^2/c^2)}{c\left(R - \dfrac{R \cdot u}{c}\right)^3}\,(u \times R), \tag{5.125}$$

respectively. We now have, in addition to the relations (5.122) and (5.123) which are quite generally valid, another relation, viz.

$$B = (u \times E)/c. \tag{5.126}$$

Thus, in this special case, the field vector E lies in the plane of the vectors R and u, and the field vector B is perpendicular to this plane.

We now make an important observation on the fields (5.124) and (5.125). Since R/c is the time taken by the electromagnetic influence in traversing the distance between the place of its origin and the place of observation, Ru/c would be the spatial displacement of the charge during this interval of time;[1] consequently, $R(t') - \dfrac{R(t')}{c}\,u$, which may be denoted by R^*, would be the position vector of the point of observation P with respect to the *instantaneous* position held by the charge. However, the field vector E, according to (5.124), is parallel to this vector (while B, according to (5.125), is perpendicular to it); thus, we have the most interesting result, viz. in spite of the fact that we have to consider only 'retarded' effects, the field vector E is *instantaneously* radial and the field vector B *instantaneously* transverse.

Further, one can verify that Eqs. (5.124) and (5.125) can be written as

$$E = \frac{eR^*}{r'^3\sqrt{1 - u^2/c^2}} \tag{5.127}$$

and

$$B = \frac{e\,(u \times R^*)}{cr'^3\sqrt{1 - u^2/c^2}}, \tag{5.128}$$

---

[1] Obviously, this is true only when the motion of the charge is *uniform*.

where

$$r' = \frac{R - \dfrac{\mathbf{R} \cdot \mathbf{u}}{c}}{\sqrt{1 - u^2/c^2}} = \left[\mathbf{R}^{*2} + \frac{(\mathbf{R}^* \cdot \mathbf{u})^2}{c^2 - u^2} \right]^{1/2} \tag{5.129}$$

The vector $\mathbf{R}^*$, as discussed above, is given by

$$\mathbf{R}^* = \mathbf{R} - \frac{R}{c}\mathbf{u}. \tag{5.130}$$

In this form, the expressions for $\mathbf{E}$ and $\mathbf{B}$ enable us to discuss, quite instructively, the field arising from a *rapidly* moving charge: $(1 - u^2/c^2) \ll 1$. We notice that because of the factor r'^3 in the denominator, the field is quite weak for finite values of $(\mathbf{R}^* \cdot \mathbf{u})$, see (5.129); however, as $(\mathbf{R}^* \cdot \mathbf{u}) \to 0$, the field becomes excessively strong (now, because of the factor $\sqrt{1 - u^2/c^2}$ in the denominator), and as $(\mathbf{R}^* \cdot \mathbf{u})$ again builds up its value the field again becomes very weak. Obviously, the strength of the field shoots up to its peak value when $(\mathbf{R}^* \cdot \mathbf{u}) = 0$, i.e. when the charge is *closest* to the point of observation. In the limit $u \to c$, this effect becomes exceptionally sharp ; the field is then 'vanishingly' small for all positions of the charge except when it is nearest to the point of observation, and at that instant the field is 'infinitely' strong. Clearly, this limiting case simulates rather well the salient features of the field associated with a sharp light pulse.

Problems

5.1. Two inertial frames, S and S', move with speed v relative to one another. Along the x-axis of S lies an infinitely long wire which is composed of stationary positive charges, and negative charges moving in the x-direction with speed v. Thus, a current flows through the wire although the net charge density in S is everywhere zero. What is the net charge density in S'? Does this result imply that total charge is not conserved in a LORENTZ transformation? Explain.

5.2. A classical point magnetic moment μ at rest has a vector potential

$$\mathbf{A} = \frac{\mu \times \mathbf{r}}{r^3},$$

and no scalar potential. Show that, if the magnetic moment moves with a velocity $\mathbf{v}$ ($v \ll c$), there is an electric dipole moment $\mathbf{p}$ associated with it:

$$\mathbf{p} = \frac{\mathbf{v}}{c} \times \mu.$$

What happens if v is not small compared to c?

5.3. Determine the system of reference S' in which a given electromagnetic field, characterized by the field vectors **E** and **B** in system S, appears as

(i) purely electric [if $(B^2 - E^2)$ is negative], or

(ii) purely magnetic [if $(B^2 - E^2)$ is positive].[1]

5.4. (a) Show that a system S', in which the field vectors **E**′ and **B**′ turn out to be parallel, is determined by the equation

$$\frac{v/c}{1 + v^2/c^2} = \frac{\mathbf{E} \times \mathbf{B}}{E^2 + B^2},$$

where **v** is the velocity of the system S' with respect to system S, **E** and **B** being the field vectors in the latter. Consider the limiting case of a plane electromagnetic wave.

(b) Is system S' determined in part (a) unique? Comment.

(c) If vectors **E** and **B** in S are mutually perpendicular, then $(\mathbf{E} \cdot \mathbf{B}) = 0$. Because of the invariance of $(\mathbf{E} \cdot \mathbf{B})$, one would expect that the vectors **E**′ and **B**′ in S' would also be mutually perpendicular. However, part (a) of this problem (which seeks systems in which **E**′ and **B**′ are parallel) leads to a definite answer in this case as well. How do we resolve this discrepancy?

5.5. In a certain reference system a static, uniform, electric field E_0 is parallel to the x-axis, and a static, uniform, magnetic induction $B_0 = 2E_0$ lies in the (x, y)-plane, making an angle θ with the x-axis. Determine the relative velocity of a reference system in which the electric and magnetic fields are parallel. What are the fields in that system for $\theta \ll 1$ and for $\theta \to (\pi/2)$?

5.6. NEWTON's equations of motion

$$m\mathbf{a}' = e\mathbf{E}'$$

hold for a small charged body of mass m and charge e in a coordinate system K' in which the body is momentarily at rest. Show that the LORENTZ force equations

$$\frac{d\mathbf{p}}{dt} = e\left(\mathbf{E} + \frac{\mathbf{u}}{c} \times \mathbf{B}\right)$$

follow directly from the LORENTZ transformation properties of accelerations and electromagnetic fields.

5.7. Study the motion of a charged particle in a *uniform* magnetic field. Derive expressions for the physical parameters of the resulting orbit in terms of the dynamical parameters of the particle and the intensity of the field.

5.8. Study the motion of a charged particle in a *uniform* electric field, the initial velocity of the particle being perpendicular to the field. Show that, in the limit $c \to \infty$, the resulting trajectory reduces to a parabola.

5.9. Discuss the motion of a charged particle in a *uniform* electromagnetic field in the following two cases:

(i) when **E** and **B** are parallel, and

---

[1] In view of the invariance of the scalar product $(\mathbf{E} \cdot \mathbf{B})$, see footnote 2 on p. 134, this requires that, at the point under consideration, the quantity $(\mathbf{E} \cdot \mathbf{B})$ be already vanishing.

(ii) when E and B are perpendicular, E being equal to B.[1]

5.10. (a) Show from HAMILTON'S principle that Lagrangians which differ only by a *total* time derivative of some function of the coordinates and time are equivalent, in the sense that they yield the same EULER-LAGRANGE equations of motion.

(b) Show explicitly that the gauge transformation (5.6) of the potentials in the charged-particle Lagrangian (5.91) merely generates another equivalent Lagrangian.

5.11. Since the potentials (5.111) are acceleration-independent, they should be derivable by applying a four-vector transformation to the potentials (5.106). Verify this.

5.12. Write down the expressions for E^0 and B^0 appropriate to the rest system S^0 of a uniformly moving charge and derive, by applying the relevant transformation equations, the expressions (5.127) and (5.128) for E and B appropriate to the system S.[2]

5.13. It is well known that the electric field E at a point P, whose perpendicular distance from an infinite linear charge distribution (of charge density ρ per unit length) is r, is $2\rho/r$ and is directed radially. Show, by applying the relevant LORENTZ transformation, that the magnetic induction B, at any given point, due to an infinitely long rectilinear current **i** is given by

$$\mathbf{B} = \frac{2(\mathbf{i} \times \mathbf{r})}{cr^2},$$

r being the perpendicular (vector) distance of the given point from the current.

5.14. (a) Making use of the *radiative* parts of the electromagnetic fields, (5.117) and (5.121), of an arbitrarily moving charged particle, calculate the instantaneous energy flux as given by the *Poynting vector* $c(\mathbf{E} \times \mathbf{B})/4\pi$. Integrating over a large sphere, show that the total instantaneous power radiated by the particle is given by the *Lienard formula*

$$P = -\frac{dW}{dt} = \frac{2e^2}{3c^3}\left[1 - \frac{u^2}{c^2}\right]^{-3}\left[\dot{\mathbf{u}}^2 - \frac{(\mathbf{u} \times \dot{\mathbf{u}})^2}{c^2}\right].$$

(b) For $u \ll c$, we have the nonrelativistic formula of LARMOR:

$$P = \frac{2e^2}{3c^3}\dot{\mathbf{u}}^2.$$

Show that LIENARD'S result can be obtained directly from LARMOR'S by applying appropriate LORENTZ transformations; see Problem 1.8.

(c) In view of the fact that the power radiated by the particle is LORENTZ-invariant, we should be able to construct a scalar (in the four-dimensional language) which directly determines P. Show that this scalar is

$$\frac{2e^2}{3m_0^2c^3}\ (F_i\,F_i),$$

---

[1] Evidently, since $(\mathbf{E} \cdot \mathbf{B})$ is an invariant (see footnote 2 on p. 134), the case of perpendicular fields with $E \neq B$ can be reduced, by a suitable choice of the reference system, to that of either a purely electric or a purely magnetic field; see Problem 5.3.

[2] Such a derivationfirst appeared in POINCARE'S paper of 1906, *loc. cit.*

where (F_i) is the force four-vector. Verify that this leads correctly to LIENARD's formula as well as to LARMOR'S.

5.15. (a) Show that the acceleration $\dot{\mathbf{u}}$ of a charged particle moving in an electromagnetic field, in terms of the quantities $\mathbf{u}$, $\mathbf{E}$ and $\mathbf{B}$, is given by

$$\dot{\mathbf{u}} = \frac{e}{m_0} \sqrt{1 - u^2/c^2} \left[\mathbf{E} + \frac{1}{c} (\mathbf{u} \times \mathbf{B}) - \frac{1}{c^2} (\mathbf{u} \cdot \mathbf{E}) \mathbf{u} \right]$$

in general,

$$\dot{\mathbf{u}} = \frac{\sqrt{1 - u^2/c^2}}{m_0} \left[\mathbf{F} - \frac{1}{c^2} (\mathbf{u} \cdot \mathbf{F}) \mathbf{u} \right].$$

(b) Using this result, establish the following formula for the 'total rate of emission of radiation' by the particle:

$$-\frac{dW}{dt} = \frac{2e^4}{3m_0^2 c^3} \left[1 - \frac{u^2}{c^2} \right]^{-1} \left[\left\{ \mathbf{E} + \frac{1}{c} (\mathbf{u} \times \mathbf{B}) \right\}^2 - \frac{1}{c^2} (\mathbf{u} \cdot \mathbf{E})^2 \right].$$

(c) In the special case when the field is purely magnetic ($\mathbf{E}=0$), examine the influence of emission on the orbit of the particle.

PART II
THE GENERAL THEORY

CHAPTER 6

GENERAL TRANSFORMATIONS IN THE SPACE-TIME CONTINUUM

6.1. The principle of covariance

In the preceding chapters we were concerned with a systematic development of physical theories in accordance with the principle of *special* relativity. As such, the validity of this principle was restricted to *inertial* systems of reference alone. Consequently, the resulting formulation was covariant only under LORENTZ transformations. This appears to imply that the laws of Nature assume their simplest form only in inertial systems of reference and may be unduly involved when referred to any system outside this class. We have thus given to the family of inertial systems of reference an especially privileged position among all conceivable systems of reference.

This attitude, however, appears contrary to the spirit of a truly relativistic viewpoint. No doubt the description of the large scale physical phenomena is generally simpler for an observer in an inertial system of reference than for one in a noninertial system, it appears physically unreasonable to expect that the fundamental laws governing physical phenomena would be different for the two observers. One is, therefore, faced with the question: 'Shouldn't the validity of the principle of relativity be extended to noninertial systems of reference as well'?

We first of all note that the foregoing extension would introduce far more radical changes into our concepts of space-time measurements than the ones introduced by the postulate of the 'constancy of the speed of light'. This postulate, as already seen in Chapter 1, resulted in demolishing the 'absolute' character of spatial and temporal measurements and brought out, quite inevitably, their essential 'relative' character; nevertheless, it did maintain the uniqueness of their meaning for each individual observer. The proposed generalization, however, demolishes even this aspect, as can be seen with the help of the following illustration.

Let K_0 be an inertial system of reference in which an event E is specified by the coordinates X, Y, Z and T; the first three coordinates give the position of the event in the physical space while the fourth one gives its time of occurrence. Let us also have a system of reference K which is in uniform rotation, with respect to the system K_0, with angular velocity ω about the common Z-axis. By symmetry, the Z-coordinate of the event will remain unchanged in the transformation $K_0 \rightarrow K$. However, for the other two spatial coordinates we have

$$X = R \cos \theta, \qquad Y = R \sin \theta \tag{6.1}$$

in K_0, and

$$x = r \cos (\theta - \omega T), \qquad y = r \sin (\theta - \omega T) \tag{6.2}$$

in K; we have switched over to cylindrical coordinates for obvious reasons. Further, at $T = 0$, the two systems of coordinates are assumed to coincide.

Let us now consider spatial measurements carried out by an observer in system K. Clearly, the measure of a radial line element, dr, will be the same as the corresponding measure, dR, in system K_0 (because the relevant measuring rod of K will be aligned orthogonal to its direction of motion with respect to K_0 and, hence, will not undergo any relativistic contraction); accordingly, $r = R$. On the other hand, a transverse line element will measure more in K than in K_0, because the corresponding measuring rod of K will undergo a relative contraction by the factor $\sqrt{1 - (r\omega)^2/c^2}$, $(r\omega)$ being the 'local' linear velocity in K_0; thus, the element of a circular arc, which measures $Rd\theta$ in K_0, will measure $rd\theta/\sqrt{1 - (r\omega)^2/c^2}$ in K. We, then, have for the measure of the line element connecting the points (r, θ) and $(r+dr, \theta + d\theta)$ in system K

$$dl^2 = dr^2 + \frac{r^2 d\theta^2}{1 - (r\omega)^2/c^2}. \tag{6.3}$$

Expression (6.3) for the line element dl^2 is obviously not of the Euclidean form. Consequently, the geometry of the space, as determined by measuring rods at rest in system K, is *non-Euclidean*. In particular, the ratio of the periphery of a circle ($dr=0$) to its radius ($d\theta=0$) is no longer 2π; it is instead

$$\frac{2\pi}{\sqrt{1 - (r\omega)^2/c^2}} > 2\pi. \tag{6.4}$$

To make things worse, this ratio is not even a constant; it is a space variable! Thus, if we increase the radius of the circle by a pre-assigned factor, its periphery increases by a different factor. Hence, for an observer in system K the space is neither homogeneous nor isotropic.[1]

Next, we consider temporal measurements. For this purpose, the observer in K would adopt clocks that are stationary in his own system. With respect to K_0, however, these clocks will be in motion and, in comparison with the clocks that are stationary in K_0, will run slower; the corresponding slowing factor will again be $\sqrt{1 - (r\omega)^2/c^2}$, r being the radial coordinate of the position held by a particular clock. A time interval T in K_0 would, therefore, appear as

$$t = T\sqrt{1 - (r\omega)^2/c^2} \tag{6.5}$$

in K; moreover, the same time interval will be recorded as different by different clocks of the system K. The temporal measure is, therefore, devoid of a unique meaning, even for an individual observer. It also follows that a clock in *uniform* motion with respect to K would run slower by a factor which itself changes with time, i.e. $dt/dT = f(T)$; consequently, time too loses its homogeneity.

These considerations show that transition from an inertial system of reference to a noninertial one entails a complete loss of objectivity in spatial and temporal measurements and, hence, in the allotment of space-time coordinates to an event. Now, by a noninertial system of reference, we have always meant a system in *arbitrary* motion with respect to an inertial system. A little reflection will show that however arbitrary the motion of our system may be, the resulting 'abnormalities' in space-time measurements will follow a somewhat systematic pattern (which, in principle, can be worked out if the necessary details of the motion of the system are known). Consequently, the totality of all conceivable noninertial systems of reference *does not* exhaust the set of all conceivable space-time transformations (in which, the aforementioned abnormalities could appear in all conceivable patterns, *not necessarily systematic*). Of course, in view of the fate that the space-time measurements have already met (on transition from inertial systems of reference to noninertial ones), it appears unnecessary to confine the principle of relativity even to arbitrarily moving systems; after all, there is no point in

---

[1] The anisotropy of space, for system K, is obvious because of the presence of a unique direction in space, specified by the axis of rotation.

fighting shy of a complete generalization of our principles, once the conceptual contents of our symbols have been so badly shaken.[1]

We are thus led to the general principle of relativity, viz. *all physical laws must be expressed in a form covariant with respect to arbitrary coordinate transformations*; in other words, *all systems of coordinates*[2] *must be regarded as equivalent in respect of the formulation of the physical laws.*

The foregoing principle is usually referred to as the *principle of* (*general*) *covariance* and the theory that follows from it the *general theory of relativity.*

At first sight it might appear that the extension of the principle of covariance to *all* conceivable systems of coordinates is a somewhat dubious step. However, one must note, as was emphasized by EINSTEIN in his concluding paper on the formulation of this theory,[3] that all physical measurements, in the last analysis, reduce to a determination of *coincidences* in the space-time continuum; nothing apart from these coincidences is observable. Consequently, any coordinate system which allocates to the world points (of the various physical events) coordinates in a *unique* and *continuous* manner is physically permissible. Space-time transformations among these systems— the so-called *Gaussian coordinate systems*—would leave intact the basic concepts of 'identity' and 'proximity' of events.

It will be assumed throughout this text that the coordinate systems employed are indeed of the Gaussian type.

6.2.   Arbitrary point transformations

We now consider arbitrary space-time transformations from a Galilean coordinate system, such as K_0, to a general Gaussian coordinate system K. Of special interest in this connection is the form the (invariant) quantity ds^2 assumes under such a transformation.

Let $X^i(i=1, 2, 3, 4)$ be the coordinates of a given event, as measured in the system K_0; we understand that $X^1 = X$, $X^2 = Y$, $X^3 = Z$ and $X^4 = cT$.[4]

---

[1] Henceforth, the four coordinates of an event would simply represent an arbitrary, but unambiguous, *numbering* of the event.

[2] By an *arbitrary* system of coordinates we do not simply mean a coordinate system adopted by an observer moving arbitrarily with respect to an inertial observer, but rather one obtained from a coordinate system adopted by an inertial observer through an *arbitrary* coordinate transformation.

[3] A. EINSTEIN, *Ann. der Phys.* **49**, 769(1916); English translation available in *The Principle of Relativity, loc. cit.*

[4] In view of the non-Euclidean form the line element is going to take, it is now futile to introduce the imaginary coordinate icT.

Then, the line element connecting two neighbouring events (X^i) and $(X^i + dX^i)$ would be

$$ds^2 = (dX^1)^2 + (dX^2)^2 + (dX^3)^2 - (dX^4)^2. \tag{6.6}$$

In an arbitrary system of coordinates K, these events will be specified by the coordinates (x^i) and $(x^i + dx^i)$ where x^i are certain arbitrary functions of X^i:

$$x^i = x^i(X^1, X^2, X^3, X^4), \qquad i = 1, 2, 3, 4. \tag{6.7}$$

The inverse transformations may be written as

$$X^i = X^i(x^1, x^2, x^3, x^4), \qquad i = 1, 2, 3, 4. \tag{6.8}$$

The coordinate differentials dX^i would be linear homogeneous functions of the coordinate differentials dx^i:

$$dX^i = \left(\frac{\partial X^i}{\partial x^k} \right) dx^k ; \tag{6.9}[1]$$

one would then have for (6.6)

$$ds^2 = g_{ik} \, dx^i \, dx^k, \tag{6.10}$$

where

$$g_{ik} = \frac{\partial X^1}{\partial x^i} \frac{\partial X^1}{\partial x^k} + \frac{\partial X^2}{\partial x^i} \frac{\partial X^2}{\partial x^k} + \frac{\partial X^3}{\partial x^i} \frac{\partial X^3}{\partial x^k} - \frac{\partial X^4}{\partial x^i} \frac{\partial X^4}{\partial x^k} = g_{ki}. \tag{6.11}$$

The geometrical significance of the set of quantities g_{ik} is evident from the place it occupies in the metric (6.10) of the space-time continuum. We find that the new metric (i.e. the expression for the quantity ds^2 in terms of the co-ordinate differentials dx^i) is no longer of the quasi-Euclidean type, (6.6), as in the Galilean coordinate systems; it is rather of a general quadratic form— the so-called *Riemannian* type—which characteristically corresponds to a 'curved' space-time; see the Appendix.

It is, however, clear that the form (6.10) of the metric encompasses (6.6) as well; this obtains when

$$\left. \begin{aligned} g_{ik} &= +1 \quad \text{for} \quad i = k = 1, 2, 3, \\ &= -1 \quad \text{for} \quad i = k = 4, \\ &= 0 \quad \text{for} \quad i \neq k. \end{aligned} \right\} \tag{6.12}$$

---

[1] The summation convention is assumed throughout.

This set of values is usually referred to as the 'normal' values of the g_{ik}. The degree of departure of the actual values of g_{ik} from the normal values (6.12) is a measure of the degree of departure of the actual (Riemannian) metric from the normal (quasi-Euclidean) one (which, in turn, determines the degree of 'curvature' of the space-time continuum, as observed in system K).

It thus appears that all laws of the special theory of relativity can be made *generally covariant* by a formal introduction of the g_{ik} [of course, in a manner that leaves (6.10) invariant]. It is clear that in this process no appeal may have to be made to the actual physical significance of these quantities. One is, thereby, led to suspect[1] that the postulate of general covariance has nothing to assert about the *physical content* of the generalized laws, and is concerned only with their *mathematical formulation*. The reflection is indeed genuine[2] and is redressed by the principle of equivalence, as a result of which the gravitational field in system K is determined *solely* and *uniquely* by the very quantities g_{ik}; see Sec. 7.2. These quantities are not given directly by matter itself but are obtained, rather indirectly, as solutions of certain field equations (see Sec. 7.5). For this reason the numbers g_{ik}, which are primarily *metrical*, may be looked upon as *physical quantities* as well.

In the foregoing discussion, we have considered transformations which are *quite* arbitrary, save for the requirements of uniqueness and continuity of the coordinates allotted to the various physical events. We, however, note that this arbitrariness must be somewhat curtailed otherwise the set of quantities g_{ik} may not be relevant to the *real* physical world. This curtailment arises from the fact that in K_0 no real motion can take place with a velocity v exceeding c. As can be seen from (6.6), this amounts to the restriction

$$ds^2 \leqslant 0, \tag{6.13}$$

the sign of equality holding in the limiting case of a signal propagating with the speed c; for particle motions,

$$ds^2 < 0. \tag{6.14}$$

Since ds^2 is an invariant, condition (6.14) would be quite generally true; we can, therefore, write from (6.10)

$$g_{ik}\, dx^i\, dx^k < 0. \tag{6.15}$$

[1] See, e.g. E. KRETSCHMANN, *Ann. der Phys.* **53**, 575 (1917).

[2] A. EINSTEIN, *Ann. der Phys.* **55**, 241 (1918).

Let us now imagine a clock, in system K_0, stationed at the space point (X^α), $\alpha = 1, 2, 3$; its time recording is given by, see (6.6),

$$dT = \frac{dX^4}{c} = \frac{\sqrt{-(ds)^2}}{c}. \qquad (6.16)$$

In an arbitrary system of coordinates K, this would correspond to a time interval dt given by, see (6.10),

$$dt = \frac{dx^4}{c} = \frac{\sqrt{(ds)^2/g_{44}}}{c}. \qquad (6.17)$$

However, ds^2 is negative [see condition (6.14)]; hence, for the expression (6.17) to give a physically realizable *time measure*, we must have

$$g_{44} < 0. \qquad (6.18)$$

Next, let us consider two events in space which, according to system K, took place simultaneously $(dx^4 = 0)$. Obviously, these events *cannot* be connected by any real physical signal, for that would require the signal to propagate with infinite speed. Consequently, the line element joining these events must not conform to the conditions (6.13)—(6.15); we must now have

$$g_{ik}\, dx^i\, dx^k > 0, \qquad (6.19)$$

whence, since $dx^4 = 0$, we obtain the inequality

$$g_{\alpha\beta}\, dx^\alpha\, dx^\beta > 0, \quad \alpha, \beta = 1, 2, 3. \qquad (6.20)$$

This must, of course, hold for arbitrary values of dx^α; hence, the quadratic form (6.20) must be *positive definite*.   The necessary and sufficient conditions for this to be true are that *all* the subdeterminants that can be constructed from the quantities $g_{\alpha\beta}$ be positive[1]:

$$g_{\alpha\alpha} > 0, \quad \begin{vmatrix} g_{\alpha\alpha} & g_{\alpha\beta} \\ g_{\beta\alpha} & g_{\beta\beta} \end{vmatrix} > 0 \quad \text{and} \quad \begin{vmatrix} g_{11} & g_{12} & g_{13} \\ g_{21} & g_{22} & g_{23} \\ g_{31} & g_{32} & g_{33} \end{vmatrix} > 0. \qquad (6.21)$$

From (6.18) and (6.21), it follows that

$$g = \begin{vmatrix} g_{11} & g_{12} & g_{13} & g_{14} \\ g_{21} & g_{22} & g_{23} & g_{24} \\ g_{31} & g_{32} & g_{33} & g_{34} \\ g_{41} & g_{42} & g_{43} & g_{44} \end{vmatrix} < 0. \qquad (6.22)$$

[1] For symbols appearing in the inequalities (6.21), the summation convention does not apply.

These are the conditions that g_{ik} must satisfy if they are to represent a coordinate system realizable by means of real physical objects. Throughout the text, this will be assumed to be the case.

It may be noted that in a system like K_0 each of the subdeterminants (6.21) equals $+1$, while the element (6.18) and the determinant (6.22) equal -1.

6.3.   Elements of general tensors

Having extended the principle of relativity to include arbitrary coordinate transformations in the space-time continuum, it is necessary to develop an equally general formalism of tensors, so that the fundamental laws of physics may be given a generally covariant expression.

As seen in the preceding section, an arbitrary point transformation leads to the Riemannian metric (6.10) instead of the normal metric (6.6) of the special theory; consequently, the general coordinate system K is essentially curvilinear. Accordingly, the formalism of Cartesian tensors (developed in Sec. 2.6), which was quite appropriate for *linear orthogonal* transformations characteristic of Galilean systems (like K_0), must now be replaced by one appropriate for the generalized situation under study.

Let us consider two systems of coordinates, K and K', in which physical events are specified by the coordinates x^i and x'^i, respectively. The functional relationship between the two sets of coordinates is assumed to be quite arbitrary (save for the restrictions of *reality*, discussed towards the end of the preceding section):

$$x'^i = x'^i(x^k), \qquad i = 1, 2, 3, 4. \tag{6.23}$$

We then have for the coordinate differentials

$$dx^i = \frac{\partial x'^i}{\partial x^k} dx^k, \tag{6.24}$$

where the coefficients $\partial x'^i/\partial x^k$ are certain space-time functions which depend upon the precise form of the relationships (6.23) [*cf.* the corresponding formulae (2.40), in the case of LORENTZ transformations, where the corresponding coefficients are *constants* of the transformation].

We now introduce the following definition :

An entity which, in every coordinate system, is characterized by four components A^i that transform *in the same manner* as the coordinate differentials

dx^i is called a *contravariant four-vector*. Thus

$$A'^i = \frac{\partial x'^i}{\partial x^k} A^k. \tag{6.25}$$

Next, we take up a *scalar* function φ, viz. one that remains invariant under coordinate transformations:

$$\varphi' = \varphi, \tag{6.26}$$

and examine how its derivatives transform. We have

$$\frac{\partial \varphi'}{\partial x'^i} = \frac{\partial \varphi'}{\partial \varphi} \frac{\partial \varphi}{\partial x^k} \frac{\partial x^k}{\partial x'^i} = \frac{\partial x^k}{\partial x'^i} \frac{\partial \varphi}{\partial x^k}. \tag{6.27}$$

We then introduce the following definition:

An entity which, in every coordinate system, is characterized by four components A_i that transform *in the same manner* as the derivatives $\partial \varphi / \partial x^i$ of a scalar φ is called a *covariant four-vector*. Thus

$$A'_i = \frac{\partial x^k}{\partial x'^i} A_k. \tag{6.28}$$

The difference in the transformation coefficients appearing in the formulae (6.25) and (6.28) must be noted carefully.

We now define a *tensor*[1] of rank 2. Here, we have three different types:

(i) a *contravariant* tensor (A^{ik}), whose components transform according to the formulae

$$A'^{ik} = \frac{\partial x'^i}{\partial x^l} \frac{\partial x'^k}{\partial x^m} A^{lm} ; \tag{6.29}$$

(ii) a *covariant*[2] tensor (A_{ik}), whose components transform according to the formulae

$$A'_{ik} = \frac{\partial x^l}{\partial x'^i} \frac{\partial x^m}{\partial x'^k} A_{lm} ; \tag{6.30}$$

---

[1] In the sequel, we shall use the term 'tensor' without specifying the relevant number of dimensions; it will be assumed that the term is meant to denote a *four-tensor*.

[2] This nomenclature may not be confused with the use of the same word in phrases such as 'covariant equations, covariant formalism, etc.', where it is intended to convey the sense of 'invariance' under coordinate transformations.

(iii) a *mixed* tensor (A_k^i), whose components transform according to the formulae

$$A'^i_k = \frac{\partial x'^i}{\partial x^l} \frac{\partial x^m}{\partial x'^k} A^l_m .$$
(6.31)

Obviously, the *outer product* of two contravariant vectors would be a contravariant tensor (of rank 2), that of two covariant vectors would be a covariant tensor and that of one contravariant vector and one covariant vector would be a mixed tensor.

Tensors of higher rank can be defined analogously. As a typical example, we define a mixed tensor, contravariant with rank s and covariant with rank p, usually designated as of rank $(s + p)$, to be an entity which, in every coordinate system, has $(4)^{s+p}$ components, $A^{i_1 i_2 \ldots i_s}_{k_1 k_2 \ldots k_p}$, that transform according to the formulae

$$A'^{i_1 i_2 \ldots i_s}_{k_1 k_2 \ldots k_p} = \frac{\partial x'^{i_1}}{\partial x^{l_1}} \cdots \frac{\partial x'^{i_s}}{\partial x^{l_s}} \cdot \frac{\partial x^{m_1}}{\partial x'^{k_1}} \cdots \frac{\partial x^{m_p}}{\partial x'^{k_p}}$$

$$\times A^{l_1 l_2 \ldots l_s}_{m_1 m_2 \ldots m_p} .$$
(6.32)

These formulae, though formidable in appearance, are a straightforward generalization of the ones like (6.25) for contravariant indices and the ones like (6.28) for covariant indices.

We shall now enumerate some of the salient properties of general tensors:

(i)   A linear combination of two tensors of the same rank (both with respect to contravariant and covariant indices) is a tensor of the same rank.

(ii)   The direct product of a tensor of rank $(s_1 + p_1)$ with a tensor of rank $(s_2 + p_2)$ is a tensor, of rank $(s_1 + s_2 + p_1 + p_2)$.

(iii)   Two tensors of the same rank are equal if, and only if, their corresponding components are one-to-one equal. The property of two tensors being equal is an invariant property; consequently, *a tensor equation is an invariant equation.*

(iv)   Closely connected with property (iii) is the following one : if a tensor is equal to zero, i.e. if all its components vanish, in any one system of coordinates, it will be so in all systems of coordinates.

(v) The symmetry properties of a general tensor are determined in the same manner as in the case of a Cartesian tensor (see Sec. 2.6). However, in the general case the considerations of symmetry apply only to indices of the *same* type, for otherwise the property concerned may not be generally covariant (because two indices of different type transform differently). Thus, we can speak of a tensor being symmetric (or antisymmetric) with respect to two particular contravariant indices or with respect to two particular covariant indices, but not with respect to one contravariant index and one covariant index.

We now introduce an especially important tensor, viz. a mixed tensor of rank 2 whose components in system K are defined as follows :

$$A_k^i = \begin{cases} 1 & \text{for} \quad i = k, \\ 0 & \text{for} \quad i \neq k. \end{cases} \tag{6.33}$$

Its components in system K' would be, see (6.31),

$$A_k'^i = \frac{\partial x'^i}{\partial x^l} \frac{\partial x^m}{\partial x'^k} A_m^l$$

$$= \frac{\partial x'^i}{\partial x^l} \frac{\partial x^l}{\partial x'^k} \qquad \text{(by virtue of 6.33)}$$

$$= \frac{\partial x'^i}{\partial x'^k}, \tag{6.34}$$

which have exactly the *same* values as were attributed to the components A_k^i in system K. Thus, the mixed tensor (A_k^i), as defined by (6.33), possesses the remarkable property that the values of its components are defined, once and for all, for *every* system of coordinates. We denote this tensor by the symbol (δ_k^i), the KRONECKER delta, and refer to it as the *fundamental mixed tensor*. One can verify that the corresponding quantities δ^{ik}, or δ_{ik}, do not possess the foregoing features : if they are defined to have the same components, as in (6.33), in *all* systems of coordinates, then they *cannot* be tensors, and if they are defined as tensors, having components (6.33) in any *one* system of coordinates, then they *do not*, in general, have the same components in other systems. It may also be mentioned that (δ_k^i) {is the *only* tensor of rank 2 which is 'symmetric' in its indices, though one of them is contravariant and the other covariant.

Next, we take up the process of *contraction*, as applicable to general tensors. Suppose we have a mixed tensor of rank $(s + p)$, as in (6.32). Then, equating any one of its contravariant indices with any one of the covariant indices and carrying out a summation over this common index, we are left with a (smaller) set of quantities, numbering $(4)^{s+p-2}$. A little reflection on the transformation formulae (6.32) shows[1] that the resulting set would *also* transform like a tensor, with rank $\overline{(s - 1 + p - 1)}$. Obviously, we can choose for this purpose any of the $(s \times p)$ pairs of *unlike* indices and effect contraction; in each case a new tensor of rank $\overline{(s - 1 + p - 1)}$ would result. It must, however, be noted that we never contract with respect to two indices of the same type, for the resulting set may not necessarily be a tensor. In view of this, our summation convention will henceforth apply *only* to a pair of unlike indices!

It is obvious that the process of contraction may be applied repeatedly, so long as the tensor to be contracted is of the mixed type. Moreover, we may combine the processes of multiplication and contraction to produce new tensors; thus, from the vectors (A^i) and (B_i) we obtain the *direct* or *outer* product $(A^i B_k)$ which, on contraction, yields the invariant $(A^i B_i)$; incidentally, this shows that an invariant may be regarded as a tensor of rank zero. Such a process is referred to as an *inner multiplication* of the given tensors and the resulting tensor the *inner product*. An especially interesting example of an inner product is provided by the following:

$$(A^i\, \delta_i^k) = (A^k); \tag{6.35}$$

the tensor (δ_k^i) thus plays the role of a *unit tensor*.

We now state an important law of tensor analysis:

A given set of quantities (of a requisite number) constitute a tensor (of the corresponding rank) if an inner product of this set with an *arbitrary* tensor is

---

[1] Suppose we contract the indices i_α and k_β in $A'^{i_1 i_2 \ldots i_s}_{k_1 k_2 \ldots k_p}$. Then, in the transformation formulae (6.32) the corresponding factors $\partial x'^{i_\alpha} / \partial x^{l_\alpha}$ and $\partial x^{m_\beta} / \partial x'^{k_\beta}$ would give, on contraction, the single factor $\partial x^{m_\beta}/\partial x^{l_\alpha}$, which is nothing but $\delta^{m_\beta}_{l_\alpha}$. This would imply an exactly similar contraction in the components $A^{l_1 l_2 \ldots l_s}_{m_1 m_2 \ldots m_p}$ as well. Thus, the transformation formulae for the contracted set would again be of the tensor type.

itself a tensor. This law is commonly referred to as the *quotient law*. To give an example, if we can show that

$$A^{ijk} \, B^p_{ij} = C^{kp},$$ (6.36)

where (B^p_{ij}) is an *arbitrary* tensor while (C^{kp}) is a tensor, it would follow that the quantities A^{ijk} themselves constitute a tensor. On the other hand, if (B^p_{ij}) were not completely arbitrary the foregoing conclusion could not be drawn. However, it can be shown[1] that if (B^p_{ij}) is symmetric in i and j but is otherwise arbitrary, then (6.36) implies that the quantities $(A^{ijk} + A^{jik})$ constitute a tensor. Of course, if it is further known that the set (A^{ijk}) is itself symmetric in i and j, its tensor character becomes established.

A fundamental application of the quotient law is in regard to the metric tensor (6.10). Here, we have

$$g_{ik}(dx^i \, dx^k) = ds^2,$$ (6.37)

where ds^2 is a tensor (of rank zero) while $(dx^i \, dx^k)$ is an arbitrary contravariant tensor of rank 2 *but symmetric in its indices*. Consequently, $(g_{ik} + g_{ki})$ would be a covariant tensor of rank 2. However, the set (g_{ik}), by 6.11, is itself symmetric; hence, it follows that the quantities g_{ik} constitute a tensor— the so-called *metric tensor*. This tensor is the most basic one in our study of the general theory of relativity; accordingly, it is also referred to as the *fundamental covariant tensor*.

Next, we introduce the concept of *reciprocal tensors*. Two tensors (A^{ik}) and (B_{ik}) are said to be reciprocal to each other if

$$A^{ik} \, B_{kl} = \delta^i_l \, ,$$ (6.38)

i.e. if their inner product is equal to the unit tensor. We then define the *fundamental contravariant tensor* (g^{ik}) as the one reciprocal to the tensor (g_{ik}):

$$g_{ik} \, g^{kl} = \delta^l_i \, .$$ (6.39)

Evidently, the element g^{kl} of the new tensor would be given by the cofactor of the element g_{kl} in the determinant $| \, g_{ik} \, |$, divided by g, the value of the

---

[1] See, e.g., B. SPAIN, *Tensor Calculus* (Oliver and Boyd Ltd., Edinburgh, 1956), Sec. 12.

determinant. From this it follows that the tensor (g^{ik}), like (g_{ik}), is also symmetric in its indices. Such tensors, which are symmetric as well as reciprocal, are referred to as *conjugate tensors*. It may be noted that in a system like K_0, where the g_{ik} have their normal values (6.12), the g^{ik} would also have the same values. Actually, it can be readily seen that if (g_{ik}) is diagonal, then (g^{ik}) would also be diagonal and its elements would be exactly the reciprocals of the corresponding elements of (g_{ik}).

Finally, we shall show that the distinction between the contravariant and covariant indices of a tensor disappears if we restrict ourselves to *linear orthogonal transformations*

$$x'^i = \alpha^i_k x^k + \beta^i, \tag{6.40}$$

where β^i are four constants (not necessarily forming a vector) while α^i_k are sixteen constants (not necessarily forming a tensor) satisfying the orthogonality conditions

$$\alpha^i_k \alpha^i_m = \delta_{km}, \tag{6.41}$$

where δ_{km} is the KRONECKER symbol (2.34). We obtain from (6.40)

$$\frac{\partial x'^i}{\partial x^k} = \alpha^i_k. \tag{6.42}$$

Next, we have the inverse relations

$$\alpha^i_k x'^i = \alpha^i_k (\alpha^i_m x^m + \beta^i)$$
$$= \delta_{km} x^m + \gamma_k = x_k + \gamma_k, \tag{6.43}$$

whence (since $x_k \equiv x^k$)

$$\frac{\partial x^k}{\partial x'^i} = \alpha^i_k. \tag{6.44}$$

Thus, we arrive at the result[1]

$$\frac{\partial x'^i}{\partial x^k} = \frac{\partial x^k}{\partial x'^i}. \tag{6.45}$$

---

[1] At first sight the equality of the coefficients $\partial x'^i/\partial x^k$ and $\partial x^k/\partial x'^i$ appears somewhat curious, because formally they appear to be rather reciprocal to each other. The point, however, is that in the case of the former coefficient partial differentiation has to be carried out under the constancy of all x^m except $m = k$, while in the case of the latter this has to be done under the constancy of all x'^m except $m = i$.

Now, the distinction between the two types of indices arises because of the difference in their mode of transformation, which, in turn, is due to the fact that the coefficients $\partial x'^i/\partial x^k$ and $\partial x^k/\partial x'^i$ are in general unequal. Thus, in the case of linear orthogonal transformations, for which we have the equalities (6.45), this distinction disappears; the formalism of general tensors then reduces *in toto* to that of Cartesian tensors (Sec. 2.6). It is now clear that for the purpose of the special theory of relativity, which deals only with linear orthogonal transformations, the scheme of Cartesian tensors suffices.

6.4. Covariant differentiation

Let us now consider a vector field, with components A^i at the point (x^i) and $A^i + dA^i$ at a neighbouring point $(x^i + dx^i)$. Ordinarily, the differentials dA^i would also constitute a vector and the derivatives $\partial A^i/\partial x^k$ a tensor. However, this is not true in the case of general curvilinear coordinates, the reason being that (dA^i) is the difference of two vectors, $(A^i + dA^i)$ and (A^i), of which one is located at the point $(x^i + dx^i)$ while the other is located at the point (x^i). Since the transformation coefficients $\partial x'^i/\partial x^k$ are, in general, functions of the coordinates, the difference of the two transformation formulae, one referring to the point $(x^i + dx^i)$ and the other to the point (x^i), which are individually vector-like, would not, in general, be vector-like. Actually, the variation of the transformation coefficients $\partial x'^i/\partial x^k$, from one point to the other, shows up in the transformation formulae for the dA^i and, thereby, destroys its (expected) vector character. Mathematically, we have from (6.25)

$$dA'^i = \frac{\partial x'^i}{\partial x^k} \, dA^k + \frac{\partial^2 x'^i}{\partial x^l \partial x^k} \, dx^l A^k. \tag{6.46}$$

It is obvious that due to the presence of the second term in (6.46) the transformation behavior of the differentials dA^i is not vector-like. One, of course, notes that in the case of linear transformations, viz. the ones among rectilinear coordinate systems, for which the derivatives $\partial^2 x'^i/(\partial x^l \, \partial x^k)$ are identically zero, the differentials dA^i *do* transform like the components of a vector.

From the foregoing discussion it follows that in order to obtain a vector as a difference of two vectors, in a general curvilinear coordinate system, it is necessary that the two vectors be located at the *same* point. In other words, we must somehow 'translate' one of the vectors, say (A^i), to the point of location, $(x^i + dx^i)$, of the other and then take their difference; this difference, being related to a *single* point, would indeed obey the transformation formulae appropriate to a vector. Now, the operation of translation of the vector (A^i), from its original location (x^i) to the new location $(x^i + dx^i)$, must be such that in the case of a rectilinear coordinate system the final vector difference coincides with the ordinary differential (dA^i); but this is exactly the difference we already have, even before the proposed process of translation. This means that the operation of translation must be such that in the case of a rectilinear coordinate system it does not *by itself* cause any change in the components of the vector; in other words, we must translate the vector *parallel to itself*.

However, in the case of a general curvilinear coordinate system the foregoing operation of translation *by itself* causes changes in the components of the vector (in the same way as, on the surface of the Earth, a stick, translated *parallel* to itself, automatically changes its orientation). Let us denote these changes by δA^i; we then have, at the point $(x^i + dx^i)$, the original vector $(A^i + dA^i)$ and the transported vector $(A^i + \delta A^i)$. Their difference,

$$(DA^i) = (dA^i - \delta A^i), \tag{6.47}$$

must then be a vector. We call this vector the *covariant differential* of the given vector (A^i).

The quantities δA^i would be proportionate with the values of the components A^i of the vector under translation and the components dx^i of the displacement. We write:

$$\delta A^i = -\Gamma^i_{kl} A^k \, dx^l, \tag{6.48}$$

where the three-index symbols Γ^i_{kl}—the so-called *Christoffel symbols of the second kind*—are certain functions of the coordinates, representing in a way the metrical properties of the space-time continuum as observed in the general coordinate system K. It is evident that in a rectilinear coordinate system these symbols identically vanish. The very fact that these symbols vanish in some of the coordinate systems but do not vanish in others shows that they do not form the components of a tensor; this is also clear from (6.48), according to which their inner product with the arbitrary tensor $(A^k dx^l)$, viz. $(-\delta A^i)$, is *not* a vector.

Next, let us consider a contravariant vector (A^i) and a covariant vector (B_i). Their inner product, $(A^i B_i)$, would be a scalar and, hence, on a parallel transport of the vectors (A^i) and (B_i), from the point (x^i) to the point

$(x^i + dx^i)$, would remain unchanged:

$$\delta(A^i B_i) = (\delta A^i) B_i + A^i (\delta B_i) = 0, \tag{6.49}$$

whence, in view of (6.48),

$$A^i(\delta B_i) = B_i \, \Gamma^i_{kl} \, A^k \, dx^l$$

$$\equiv B_k \, \Gamma^k_{il} \, A^i \, dx^l \tag{6.50}$$

(by an interchange of the dummy variables i and k). The relation (6.50) holds for arbitrary A^i; hence, we must have

$$\delta B_i = \Gamma^k_{il} \, B_k \, dx^l . \tag{6.51}$$

We thus have for the *covariant differentials* of the contravariant and covariant vectors

$$DA^i = dA^i - \delta A^i = \left(\frac{\partial A^i}{\partial x^l} + \Gamma^i_{kl} \, A^k \right) dx^l, \tag{6.52}$$

and

$$DB_i = dB_i - \delta B_i = \left(\frac{\partial B_i}{\partial x^l} - \Gamma^k_{il} \, B_k \right) dx^l. \tag{6.53}$$

The quantities in the parentheses of (6.52) and (6.53) are, by the quotient law, tensors; we call them the *covariant derivatives* of the respective vectors and denote them by the symbols $A^i_{;l}$ and $B_{i;l}$. Thus

$$A^i_{;l} \equiv \frac{\partial A^i}{\partial x^l} + \Gamma^i_{kl} \, A^k, \tag{6.54}$$

and

$$B_{i;l} \equiv \frac{\partial B_i}{\partial x^l} + \Gamma^k_{il} \, B_k. \tag{6.55}$$

The process of obtaining covariant derivatives from given vectors is called *covariant differentiation*.

Extension of this process to tensors of arbitrary rank is straightforward: we have to add, to the ordinary derivative, a term like the second one on the right-hand side of (6.54) for each of the contravariant indices and subtract a term like the second one on the right-hand side of (6.55) for each of the covariant indices. For the sake of illustration, we write down covariant derivatives of the three fundamental tensors:

$$g_{ij;l} = \frac{\partial g_{ij}}{\partial x^l} - \Gamma^k_{il}\, g_{kj} - \Gamma^k_{jl}\, g_{ik}, \tag{6.56}$$

$$g^{ij}{}_{;l} = \frac{\partial g^{ij}}{\partial x^l} + \Gamma^i_{kl}\, g^{kj} + \Gamma^j_{kl}\, g^{ik}, \tag{6.57}$$

and

$$\delta^i_{j;l} = \frac{\partial \delta^i_j}{\partial x^l} + \Gamma^i_{kl}\, \delta^k_j - \Gamma^k_{jl}\, \delta^i_k\,. \tag{6.58}$$

Relation (6.58) is a trivial one, because it merely equates zero with zero; we shall see in the sequel that expressions (6.56) and (6.57) are also identically vanishing, but they are nontrivial.

It need hardly be said that in a rectilinear coordinate system the process of covariant differentiation reduces to that of ordinary differentiation, and covariant derivatives reduce to ordinary derivatives.

We now establish an important symmetry property possessed by the symbols Γ^i_{kl}. For this, we note that since $(B_{i;l})$ is a tensor, so must be $(B_{i;l} - B_{l;i})$. We may choose (B_i) to be $(\partial\varphi/\partial x^i)$, where φ is an arbitrary scalar function. Then, we have from (6.55)

$$B_{i;l} - B_{l;i} = (\Gamma^k_{li} - \Gamma^k_{il})\, \frac{\partial\varphi}{\partial x^k}. \tag{6.59}$$

Now, by choosing a suitable coordinate system the symbols Γ^i_{kl} can be made to vanish (locally); hence, in such a system the left-hand side of (6.59) would also vanish. Being a tensor, it must then vanish in *all* coordinate systems. Consequently, the right-hand side of (6.59) must also vanish identically in *all* coordinate systems; hence, quite generally,

$$\Gamma^k_{li} = \Gamma^k_{il}. \tag{6.60}$$

Thus, the CHRISTOFFEL symbols of the second kind are symmetric in their lower indices; consequently, the number of independent symbols among these is only forty, and not sixty-four.

Next, we determine the relation between the CHRISTOFFEL symbols and the elements of the metric tensor.   For this purpose, we have to introduce one more concept.

A given vector (A) in rectilinear coordinates may be represented, in the general curvilinear coordinates, either by a contravariant vector (A^i) or by a covariant vector (A_i), the two representations being connected by the relations

$$A^i = g^{ik} A_k \quad \text{and} \quad A_i = g_{ik}A^k. \tag{6.61}$$

The first of these processes is referred to as the 'raising of the index', and the second as the 'lowering of the index'; one can see that by virtue of the reciprocality of the g^{ik} and the g_{ik} $(g^{ik}g_{im} = \delta^k_m)$ these two relations follow from one another.   Now, in a rectilinear coordinate system, the g_{ik} take their normal values (6.12); consequently,

$$A^\alpha = A_\alpha \ (\alpha = 1, 2, 3) \quad \text{and} \quad A^4 = -A_4. \tag{6.62}$$

Thus, in such a coordinate system, the two vectors have identical components (except for a change of sign in the fourth) and, hence, represent one and the same vector (A).   We may, therefore, look upon vectors (A^i) and (A_i), in a general coordinate system too, as representing one and the same (physical) vector ; such vectors are said to be *associate* to each other.   Thus, the processes of raising and lowering of indices, which are equally applicable to tensors of arbitrary rank, do not produce any change in the physical significance of the tensor concerned; they are, no doubt, of great analytical importance.

Now, (DA^i) is a contravariant vector and (DA_i) a covariant one ; they must, of course, represent one and the same thing, viz. that which, in the case of rectilinear coordinates, is represented by (dA).   They must, therefore, be connected through the relations

$$DA_i = g_{ik}(DA^k). \tag{6.63}$$

From (6.61), however,

$$DA_i = D(g_{ik} A^k) = (Dg_{ik}) A^k + g_{ik}(DA^k). \tag{6.64}$$

Equating (6.63) and (6.64), it follows, in view of the arbitrariness of the A^i,

$$Dg_{ik} = 0, \quad \text{i.e.} \quad g_{ik;l} = 0. \tag{6.65}$$

Thus, the g_{ik} are *constant under covariant differentiation.*   Equation (6.56) then gives (for the ordinary derivatives of the g_{ik})

$$\frac{\partial g_{ik}}{\partial x^l} = g_{jk} \, \Gamma^j_{il} + g_{ij} \, \Gamma^j_{kl}. \tag{6.66}$$

We now define another set of three-index symbols,

$$\Gamma_{i,kl} = g_{im} \, \Gamma^m_{kl}, \tag{6.67}$$

which are the so-called *Christoffel symbols of the first kind;*[1] obviously, they too are symmetric in the indices k and l. In terms of these symbols, (6.66) becomes

$$\frac{\partial g_{ik}}{\partial x^l} = \Gamma_{k,il} + \Gamma_{i,kl}. \tag{6.68}$$

Combining Eq. (6.68) with two others, obtained by permuting the indices i, k and l in cyclic order, and making use of the symmetry properties of the g_{ik} and the $\Gamma_{i,kl}$, we obtain

$$\Gamma_{i,kl} = \frac{1}{2} \left(\frac{\partial g_{ik}}{\partial x^l} + \frac{\partial g_{il}}{\partial x^k} - \frac{\partial g_{kl}}{\partial x^i} \right) \tag{6.69}$$

and hence

$$\Gamma^i_{kl} = g^{im} \, \Gamma_{m,kl}$$

$$= \frac{1}{2} g^{im} \left(\frac{\partial g_{mk}}{\partial x^l} + \frac{\partial g_{ml}}{\partial x^k} - \frac{\partial g_{kl}}{\partial x^m} \right). \tag{6.70}$$

In a rectilinear coordinate system, the g_{ik} are constants; accordingly, the CHRISTOFFEL symbols (of both kinds) are identically zero.

It may be noted that although the CHRISTOFFEL symbols are *not* tensors the notation for their indices is in keeping with the summation convention (which applies to *unlike* indices alone).

---

[1] In literature, one also comes across the following notation for the CHRISTOFFEL symbols: $[kl, i]$ for those of the first kind and $\{^i_{kl}\}$ for those of the second kind.

6.5. Motion of a 'free' particle in a curvilinear coordinate system. The variational principle

As is well known, the time track of a free particle, in a rectilinear co-ordinate system, is characterized by the equations

$$dU^i = 0, \qquad i = 1, 2, 3, 4, \tag{6.71}$$

where $U^i(= dx^i/ds)$ is the four-vector tangential to the track. These equations imply a straight line in the four-dimensional continuum, which in turn corresponds to a *uniform rectilinear* motion in the three-dimensional physical space. The corresponding motion in a curvilinear coordinate system is not given by (6.71); we must replace the ordinary differentials by covariant differentials. Accordingly, we have

$$DU^i = 0, \tag{6.72}$$

that is,

$$dU^i + \Gamma^i_{kl} U^k \, dx^l = 0; \tag{6.73}$$

see Eqs. (6.47) and (6.48). In view of the definition of the U^i, we finally obtain

$$\frac{d^2 x^i}{ds^2} + \Gamma^i_{kl} \frac{dx^k}{ds} \frac{dx^l}{ds} = 0, \tag{6.74}$$

which are the basic differential equations for the world line of a 'free' particle, when referred to a general coordinate system. This line is generally referred to as the *geodesic line*, or simply the *geodesic*, of the motion concerned.[1]

It is clear from these equations that the world line, in general, is not straight; consequently, the corresponding three-dimensional path of the particle may also be curved. This curvature, of course, arises from the second

---

[1] Equations (6.72) mean: $dU^i = \delta U^i$, that is, the changes in the components of the vector (U^i) in going from the point (x^i) to the point $(x^i + dx^i)$ are precisely the same as the changes caused by translating the vector, parallel to itself, from the point (x^i) to the point $(x^i + dx^i)$. Thus, the displacement of the tangent vector along the geodesic line is a parallel displacement! The argument can obviously be reversed to provide another way of *defining* the geodesic. It also provides an 'operational' meaning to the process of 'parallel transport' of an arbitrary vector.

term in (6.74), which in turn is due to the departure of the metric of the space-time continuum from the normal Euclidean type. The quantities Γ^i_{kl} thus represent the 'forces' that arise inherently in the system, the g_{ik} being the corresponding 'potentials' [see (6.70)]; they produce a curvature in the track of the *otherwise free* particle. It is to emphasize this aspect of the problem that we put the word free, in the title of this section, under quotation marks.

Equations (6.74) can also be obtained with the help of a *variational principle*. Let us require, in analogy with the case of a straight track corresponding to a force-free motion in a rectilinear system of coordinates (see Sec. 2.4), that the motion of a 'free' particle is, quite generally, charactierzied by the condition that the integral

$$I = \int_A^B ds, \tag{6.75}$$

between two given world points A and B, is *extremal* for the actual track. This means that the variation

$$\delta I = \delta \int_A^B ds = \int_A^B \delta \, (ds), \tag{6.76}$$

of the integral (6.75), if we change over from the actual track to an arbitrary neighboring track *between the same two world points*, must vanish, i.e.

$$\int_A^B \delta \, (ds) = \frac{1}{2} \int_A^B \frac{1}{ds} \, \delta \, (ds)^2 = 0. \tag{6.77}$$

Now

$$\delta \, (ds)^2 = \delta \, (g_{ik} \, dx^i \, dx^k)$$

$$= \left(\frac{\partial g_{ik}}{\partial x^l} \, \delta x^l \right) dx^i \, dx^k + g_{ik} \, \{2 \, dx^i \, \delta \, (dx^k)\}.$$

Hence, (6.77) becomes

$$\int_A^B \left[\frac{1}{2} \frac{dx^i}{ds} \frac{dx^k}{ds} \frac{\partial g_{ik}}{\partial x^l} \, \delta x^l + g_{ik} \, \frac{dx^i}{ds} \frac{d \, (\delta x^k)}{ds} \right] ds = 0 \, ; \tag{6.78}$$

we have used here the fact that $\delta \, (dx^k) = d(\delta x^k)$.

Integrating the second portion of this integral by parts, we get

$$g_{ik} \frac{dx^i}{ds} (\delta x^k) \Big|_A^B - \int_A^B \frac{d}{ds} \left(g_{ik} \frac{dx^i}{ds} \right) \delta x^k \, ds \, ;$$

the first part here vanishes identically because the two tracks join at the point A and B (so $\delta x^k = 0$ there), while in the second part we may replace the dummy variable k by l. Then, (6.78) becomes

$$\int_A^B \left[\frac{1}{2} \frac{dx^i}{ds} \frac{dx^k}{ds} \frac{\partial g_{ik}}{\partial x^l} - \frac{d}{ds} \left(g_{il} \frac{dx^i}{ds} \right) \right] \delta x^l \, ds = 0, \qquad (6.79)$$

whence, in view of the arbitrariness of the points A and B and of the variations δx^l, we get

$$\frac{1}{2} \frac{dx^i}{ds} \frac{dx^k}{ds} \frac{\partial g_{ik}}{\partial x^l} - g_{il} \frac{d^2 x^i}{ds^2} - \frac{dx^i}{ds} \frac{\partial g_{il}}{\partial x^k} \frac{dx^k}{ds} = 0. \qquad (6.80)$$

In the last term we may replace $\partial g_{il}/\partial x^k$ by $\frac{1}{2} [(\partial g_{il}/\partial x^k) + (\partial g_{kl}/\partial x^i)]$, because the remaining factors are symmetric in i and k and a summation over both of them has to be carried out. Equation (6.80) then gives, in view of (6.69),

$$g_{il} \frac{d^2 x^i}{ds^2} + \Gamma_{l,ik} \frac{dx^i}{ds} \frac{dx^k}{ds} = 0. \qquad (6.81)$$

On raising the index l, with the help of g^{rl}, we obtain

$$\frac{d^2 x^r}{ds^2} + \Gamma^r_{ik} \frac{dx^i}{ds} \frac{dx^k}{ds} = 0 \, ; \qquad (6.82)$$

these equations are the same as (6.74).

In a rectilinear coordinate system, $\Gamma^r_{ik} = 0$; we then have the usual equations

$$\frac{d^2 x^r}{ds^2} = 0, \qquad (6.83)$$

which represent a straight track. In such a system, *the geodesics reduce to straight lines.*

If we construct all the geodesics emanating from a world point P (x^i), they fall into three categories :

(i) those for which

$$g_{ik} \, dx^i \, dx^k > 0,\tag{6.84}$$

(ii) those for which

$$g_{ik} \, dx^i \, dx^k < 0,\tag{6.85}$$

(iii) those for which

$$g_{ik} \, dx^i \, dx^k = 0.\tag{6.86}$$

They span, respectively, the world regions (A and B) and the surface of the null cone (C), in a continuous manner. We refer to them, and to the associated directions, as space-like, time-like and null; cf. the corresponding discussion in Sec. 2.3. A real signal starting from P can proceed only in time-like directions or, at best, in the direction of a null geodesic ; in the latter case, it must be a light signal (or, in general, an electromagnetic pulse). It follows that the event corresponding to the world point P can have a causal connection only with those events whose world points lie in time-like directions (i.e. in region B) or, at best, with those whose world points lie in null directions.

Finally we remark that in the case of null geodesics the derivation from the variational principle breaks down because the integral (6.75), whose variation is to be studied, is then identically zero. However, Eqs. (6.72) and (6.73), which are a *direct* generalization of (6.71), hold in this case as well, provided that the vector (U^i) is defined as ($dx^i/d\lambda$) where λ is some *nonzero* invariant parameter of the track (not equal to s). Equations (6.74) would then be

$$\frac{d^2x^i}{d\lambda^2} + \Gamma^i_{kl} \frac{dx^k}{d\lambda} \frac{dx^l}{d\lambda} = 0.\tag{6.87}$$

Problems

6.1 Establish the inequalities (6.21) and (6.22).

6.2. The square of the interval between two neighbouring events (x^i) and ($x^i + dx^i$) is given by (6.10), which can also be written as

$$ds^2 = g_{\alpha\beta} \, dx^\alpha \, dx^\beta + 2g_{\alpha 4} dx^\alpha \, dx^4 + g_{44} \, (dx^4)^2,$$

$$\alpha, \beta = 1, 2, 3,$$

The spatial separation dl between the given events cannot be obtained from the foregoing expression merely by putting $dx^4 = 0$ (because, in a general metric, the 'proper' time at different points in space has a different dependence on the coordinate x^4).   Nevertheless, the desired spatial separation may be determined by sending a light signal from the site of one of the two events to the site of the other and *back*, and multiplying the total 'proper' time elapsed (in the two-way transit) by $c/2$.   Prove that this leads to the result

$$dl^2 = \gamma_{\alpha\beta} \, dx^\alpha \, dx^\beta$$

$$= \left(g_{\alpha\beta} - \frac{g_{\alpha 4} \, g_{\beta 4}}{g_{44}} \right) dx^\alpha \, dx^\beta .$$

The quantities $\gamma_{\alpha\beta}$ determine the geometrical properties of the relevant physical space.

6.3.   Derive the foregoing result by applying an arbitrary coordinate transformation to the expression ($dX^\alpha \ dX^\alpha$), where (X^i) are the coordinates in a system like K_0, and making use of Eqs. (6.11).   Show that the three-by-three matrices $\gamma_{\alpha\beta}$ and $g^{\alpha\beta}$ are reciprocal to each other (see Sec. 6.3 for the definition of the g^{ik}), and that the determinant γ is equal to (g/g_{44}), which is positive definite.

6.4.   Derive the foregoing expression for dl^2 by *maximizing* ds^2 (with dx^α fixed but dx^4 variable).   Interpret the procedure physically.

6.5.   (*a*) Perform the coordinate transformation

$$X = r \cos (\theta + \omega t), \qquad Y = r \sin (\theta + \omega t)$$
$$Z = z, \qquad\qquad T = t,$$

on the MINKOWSKI metric $ds^2 = dX^2 + dY^2 + dZ^2 - c^2 \, dT^2$.   Can the resulting metric be interpreted as one pertaining to a uniformly rotating frame of reference ?   Discuss the limitations of this interpretation.

(*b*) Discuss the nature of *spatial* length measurements in this metric and evaluate the Gaussian curvature K experienced in this situation. Show that, for $(\omega r) \ll c$, $K \simeq -3(\omega/c)^2$.

6.6.   The geometry of a hitherto plane surface S_0 is made non-Euclidean by effecting a transformation such that the new expression of the line element is

$$ds^2 = [1 + (r/2R)^2]^{-2} \, (dr^2 + r^2 \, d\varphi^2);$$

see Eq. (29) of the Appendix.   Show that the geodesics resulting from this metric are, in general, circles which are stereographic projections of the great circles of the corresponding spherical surface S in the three-dimensional Euclidean space.

Note that the radial lines ($\varphi = $ const.) in S_0 are stereographic projections of the great circles that cut S_0 perpendicularly and hence are special cases of the geodesics under study.

6.7.   Consider a three-dimensional space of *constant* curvature, with line element

$$dl^2 = \frac{dx^2 + dy^2 + dz^2}{\left[1 + \dfrac{x^2 + y^2 + z^2}{4a^2} \right]^2} .$$

A spherical surface (of radius ρ) in this space will have an area S and volume V which may be written as

$$S (\rho) = 4\pi\rho^2 [1 - A\rho^2 + \dots]$$

and

$$V(\rho) = \frac{4\pi}{3}\rho^3 [1 - B\rho^2 + \ldots],$$

respectively.   Determine A and B in terms of the radius of curvature, a, of the space.

6.8. Generalize the results of the preceding problem to an n-dimensional space of constant curvature and show that

$$A_n = \frac{n-1}{6a^2} \quad \text{and} \quad B_n = \frac{n(n-1)}{6(n+2)a^2},$$

where a is again the radius of curvature of the given space.

Note that for the case $n = 2$ these results lead to Eqs. (18) and (19) of the Appendix.

6.9. Consider a spherical surface, of constant curvature $(1/a^2)$, and a circular path of radius ρ on it (like a latitude on the surface of the Earth). Show that the 'parallel transport' of a vector around this circle causes a change Δ in the orientation of the vector, where

$$\Delta = 2\pi [1 - \cos(\rho/a)].$$

[Note that the radius ρ is measured along a geodesic connecting the centre of the circle with any point P on the perimeter.]

6.10. Show that the 'principal directions' in which the curvature of a given surface

$$z = F(x, y),$$

at a point $P(x, y)$, is maximum (in one case) and minimum (in the other), are mutually orthogonal.

6.11. For the polar metric,

$$dl^2 = d\rho^2 + \rho^2 d\varphi^2,$$

of a plane surface, work out the CHRISTOFFEL symbols Γ^i_{kl} (with $\rho = x^1$ and $\varphi = x^2$) and show that the lines $\varphi = $ const. satisfy the geodesic equations (6.74).

CHAPTER 7

GEOMETRIZATION OF GRAVITATION

7.1. The principle of equivalence

It has already been remarked during our considerations of arbitrary point transformations, see Sec. 6.2, that an unrestrained extension of the principle of relativity derives its physical content from the *principle of equivalence*, which brings forth a straightforward connection between the curved space-time continuum of an arbitrary system of coordinates on the one hand and the presence of a gravitational field in an otherwise flat (Euclidean) continuum on the other. We now propose to discuss, in reasonable detail, the main ideas conveyed by this principle, its empirical basis, its physical implications and, finally, its limitations.

We start with an *inertial* system of reference $\bar{K}$ in a *field-free* space. Obviously, the law of inertia would be valid in this system; in consequence, a material body which is not being acted upon by external forces would pursue a uniform rectilinear motion. Next, we imagine two systems of reference, K_1 and K_2, which differ from $\bar{K}$ in the following : (i) the system K_1 is still in the 'field-free' space but is accelerated relative to $\bar{K}$ with a constant acceleration f in the positive direction of the common z-axis, while (ii) the system K_2 is still 'inertial' but is embedded in a uniform gravitational field of constant intensity f acting in the negative direction of the z-axis. Clearly, the motion of material bodies would follow *exactly* the same pattern in these two systems of reference ; each and every motion would experience a constant acceleration f in the negative direction of the z-axis, irrespective of where and when the motion takes place and what sort of material the body is made of.

Thus, from the point of view of mechanics, systems K_1 and K_2 are equivalent; no dynamical experiment can distinguish between the so-called *fictitious* (inertial) force arising in system K_1 and the so-called *real* (gravitational) force prevailing in system K_2. EINSTEIN generalized this conclusion and asserted that 'no physical experiments *whatsoever* can ever distinguish between the aforementioned fields of forces'; to this assertion he gave the name *principle of equivalence*, according to which systems K_1 and K_2 are equivalent, *not only* from the point of view of mechanics but of all physics. Accordingly, there is no justification in speaking of a distinction between a force field 'generated' in a system like K_1 (by virtue of its noninertial character) and a corresponding uniform gravitational field. This constitutes the first step of the process which finally enables us to apply the formalism of generalized transformations to the study of physical phenomena taking place in a gravitational field (or, in other words, to develop a covariant theory of gravitation in terms of the geometrical features of a Riemannian space-time continuum).

Let us now consider the physical facts underlying the principle of equivalence. We note, first of all, that the basic feature of the equivalence of systems K_1 and K_2 is that the motion of *all* material bodies follows the same pattern whether it is observed in system K_1 or in K_2. Now, in K_1, different material bodies undergo the same acceleration, viz. $-f$ in the direction of the z-axis, because the cause thereof lies in the state of motion of the system itself and has nothing to do with the form, constitution or other physical properties of the bodies concerned. On the other hand, one may not expect *a priori* that an exactly similar situation would obtain in system K_2 as well. However, it is an experimental fact that in a gravitational field '*all bodies fall alike*'—a fact first clearly realized by GALILEO; consequently, in K_2 also, all motions take place with *the same* acceleration.

It follows that the fundamental fact on which the principle of equivalence rests is the empirical result that all material bodies respond *equally* to the influence of gravitation. This, in turn, accrues from a deep-seated fact that *the ratio of the inertial mass of a body to its gravitational mass is a universal constant*. By a suitable choice of units, this ratio can be made equal to unity; one can, then, state the foregoing fact in the form: *the inertial mass of a body is equal to its gravitational mass*. The bearing of this statement on the point under discussion will be evident from the following.

It is well known that the gravitational force experienced by a material body, in a given gravitational field, is proportional to the mass of the body. The question arises: which mass do we have in mind when we make this statement? Ordinarily, this mass is not expected to have any *direct* relationship with the inertial properties of the body. Hence, in view of the apparent vagueness of its meaning, let us call it the 'gravitational' mass of the body. Now, under the action of this force the body will experience an acceleration which, as we know, is inversely proportional to the 'inertial' mass of the body. Hence, the acceleration experienced by a given body under the action of a force of gravitation will be proportional to the ratio of the gravitational mass of the body to its inertial mass. It then follows that the observed fact of equal acceleration for *all* bodies implies the universal constancy of this ratio, and *vice versa*. The proportionality (or equality) of the inertial mass and the gravitational mass of a body, thus, forms the cornerstone of the principle of equivalence.

We shall now discuss the experimental evidence in favor of the foregoing conclusion. First of all, we have the classic experiments of Eötvös, performed towards the end of the last century, in which one (effectively) observed the resultant of: (i) the force of gravity experienced by a body suspended in the Earth's gravitational field, and (ii) the force of centrifuge arising from the rotation of the Earth. Now, the former is directly proportional to the gravitational mass of the body while the latter, arising because of the particular state of motion of the Earth system (with respect to an inertial system like $\bar{K}$), would be proportional to the inertial mass of the body.[1] Thus, the direction of the resultant force on the body will depend upon the ratio of the two masses.

Eötvös observed that, to 5 parts in 10^9, the direction of the resultant force was the *same* for all bodies experimented upon, thus establishing the universal constancy of the ratio of the two masses to a fairly high degree of accuracy.[2] Dicke and his collaborators at Princeton[3] have repeated the Eötvös experiment with greatly refined apparatus and have reduced the error to less than 1 part in 10^{11}.

Mention may also be made of an ingenious experiment conducted by Southerns,[4] at the suggestion of J. J. Thomson, who determined the time

---

[1] Actually, in the second case, it would be preferable to speak first of the centrifugal acceleration, which is the primary effect of the state of motion of the Earth system, and then invoke an appropriate 'inertial' force.

[2] For original references, see the Historical Introduction, p. 9.

[3] See R. H. Dicke in *Gravitation and Relativity*, ed. H.-Y. Chiu and W. F. Hoffmann (Benjamin, New York, 1964).

[4] L. Southerns, *Proc. Roy. Soc. London* A, **84**, 325 (1910). See also P. Zeeman, *Proc. Amsterdam Acad.* **20**, 542 (1917).

period of a (compound) pendulum,

$$T = 2\pi \sqrt{\frac{I}{Wh}},$$

to which identical bobs of different materials were attached. It was so arranged that the weight W of the oscillating system and the distance h between the centre of suspension and the centre of gravity were the same in all cases. Now, the moment of inertia I of the system would depend upon the inertial mass of the system; hence, if the ratio I/W (in other words, the ratio of the inertial mass of the system to its gravitational mass) were different in different cases, the time periods would also be different. Since, in the actual experiment, it was only the bob that was changed from one set of observations to the other variations in T, if any, would be a measure of the variations of the afore-mentioned ratio over different bobs.

The novel feature of the experiment conducted by SOUTHERNS was that it carried out a comparison between a radioactive salt such as uranium oxide and a non-radioactive one such as lead oxide, the latter being a product of the former. Now, the two samples differed from each other by a significant amount of internal potential energy, whose equivalence with inertial mass was already well recognized (see Sec. 3.3). Hence, if SOUTHERNS' experiment failed to give any significant variation in T, over the two cases mentioned here, it would establish not only the equality of the inertial mass and the gravitational mass but would also lend support to the belief that the internal potential energy of a radioactive substance possesses, besides an inertial mass, an equivalent gravitational mass as well.[1] The actual results were indeed like this; they proved that the mass-to-weight ratios for the two salts did not differ by more than 5 parts in 10^6.

There is, thus, sufficient empirical evidence in favor of the result that the inertial mass of a body is equal to its gravitational mass. To some extent, this result can be understood theoretically, though in a qualitative and somewhat speculative manner, on the basis of Machian considerations, according to which the inertia of any given body is an outcome of its interaction with other bodies in the universe which comes into play when the body in question is accelerated with respect to those bodies. Thus, one may look upon the inertia of a given body as a result of the influence caused by the huge masses scattered in the universe [because, from the point of view of an observer attached to the body, it is these huge masses that undergo

---

[1] This had been emphasized in 1907 by PLANCK; see the Historical Introduction, p. 9.

acceleration whenever the state of motion of the body (with respect to an inertial system of reference) changes]. We are, thus, led to believe that the inertial mass of a body, which is a measure of its inertia, is of *gravitational* origin.[1] It is, then, no wonder if *this* mass and the gravitational mass (which is a direct measure of the response of the body towards gravitational interactions) exhibit a universal proportionality.[2]

It is not possible to say at present whether this concept can be accepted *in toto* (especially when there is no quantitative support for the assertion that inertia is *entirely* due to gravitational interactions). An interesting conclusion of this concept is that if all the masses of the universe are removed from the scene and only the body under question remains, then it would no longer experience any gravitational interactions at all and, consequently, it should *cease to have inertia*. It is not easy to comprehend this situation though we do realize that in this case the question of the gravitational mass of the body would indeed become vague ; but whether one can put it equal to zero, as suggested by Machian considerations, is very difficult to say.

In passing, it appears worthwhile to remark that neither the principle of equivalence nor the equality of inertial and gravitational masses (nor even MACH's concept of inertia) precludes the possibility of coming across bodies which possess a *negative* mass, both inertial and gravitational.[3] Such bodies would behave, in a gravitational field, in the same way as ordinary positive-mass bodies do, because the ratio of their inertial and gravitational masses is exactly the same, *even in sign*, as that of ordinary bodies. Thus, an ordinary body would attract *all* bodies, whether their mass is positive or negative, whereas a negative-mass body would repel *all* bodies, whether their mass is positive or negative. It follows that whereas two like bodies of positive mass attract each other, those of negative mass would repel; on the other hand, if we have two unlike bodies, the positive one would attract the negative one whereas the negative one would repel the positive one, with the result that one would 'chase' the other.[4]

---

[1] See also R. H. DICKE, *The Theoretical Significance of Experimental Relativity* (Gordon and Breach, New York, 1964) ; *Physics Today* **20**, no. 1, 55 (1967).

[2] In this sense, it may be possible to account for the value of the constant of gravitation in terms of the structure of the universe !

[3] Bodies such as these are not to be confused with the ones supposed to be composed of *antimatter*, whose mass is positive.

[4] For further details on this point, see H. BONDI, *Rev. Mod. Phys.* **29**, 3, 423(1957).

7.2.   The principle of equivalence and a general gravitational field

In the preceding section we have seen how effectively the principle of equivalence brings out a complete physical correspondence between a homogeneous gravitational field and a field of force generated in a suitably accelerated system of reference. This equivalence enables us to *replace* a given homogeneous gravitational field by an equivalent 'fictitious' field (by resorting to a corresponding accelerated system), carry out the calculation of various physical phenomena in the latter, and apply the results thus obtained to events happening in the given field. This prescription is fine so long as the given field, in system K_2, is perfectly homogeneous because it is only then that an unambiguous choice of the corresponding noninertial system K_1 can be made.[1]

In a case such as this, the given field can be 'transformed away' from the whole of the physical space by adopting a 'freely falling' system K_0 in which, again by virtue of the principle of equivalence, the effect of the given field is *exactly* cancelled by an equal and opposite effect arising from the 'free fall' of the system; in other words, a state of weightlessness prevails in K_0. For all practical purposes, system K_0, though not really an inertial system, is *equivalent* to an inertial system in field-free space, such as the system $\bar{K}$ of the preceding section. In K_0, therefore, the law of inertia holds (and, with it, the whole formalism of the special theory of relativity). We may transform back from K_0 to K_2 by introducing a suitable acceleration, which is equivalent to switching on the original gravitational field once again; this is equivalent to going from $\bar{K}$ to K_1. The form taken by the mathematical expressions of the theory as a result of this transformation would, then, apply to system K_2 as well as to K_1; in other words, the resulting expressions would correspond to situations in the presence of a gravitational field in the (otherwise inertial) system K_2. The importance of the intermediate role played by system K_0 should be appreciated because it is only through K_0 that the formulae of the special theory of relativity make their way into the general theory.

Of course, the foregoing procedure is applicable only to a very special class of gravitational fields, viz. those which can be replaced by *equivalent* fields of

---

[1] Obviously, a variation of the field intensity with time does not create any new difficulties, provided that the homogeneity of the field does not get impaired at any instant of time; this would simply require the state of motion of the system K_1 to be made suitably time-dependent. Physically, however, this would entail an *instant action at a distance* !

force generated in a suitably accelerated system of reference or, in other words, those which can be *completely* 'transformed away' by adopting a 'freely falling' system. Such fields are usually referred to as *temporary* fields. However, the real gravitational fields encountered in Nature do not, *as a rule*, conform to such an over-simplifying requirement. They are invariably inhomogeneous, and their space dependence forbids the possibility of their being replaced by equivalent 'artificial' fields (because, in general, they cannot be *completely* transformed away by choosing any particular system like K_0). The principle of equivalence as such is inapplicable to these fields ; hence, the possibility of studying physical phenomena taking place in such a field, in terms of those taking place in a *single*, suitably chosen, noninertial system of reference, no longer exists. Such fields are usually referred to as *permanent* fields.

We can, however, still make some headway by applying the principle of equivalence to a given permanent field *locally*, i.e. by transforming it away from a small (strictly speaking, infinitesimal) region of space by introducing a system K_0 which is undergoing a local 'free fall'.[1] Now, the formulae of the special theory of relativity will be applicable *locally* in this system. A suitable transformation would, then, give results applicable in the presence of the field; their application, however, would be restricted to the (infinitesimally) small region of space from which the given field was removed by the adoption of the system K_0. The same will have to be done for every other region of space ; the totality of the resulting formulae would constitute the scheme appropriate to the given gravitational field.

Obviously, the foregoing procedure is formidably tedious ; in fact, limitations on the applicability of the principle of equivalence to a general gravitational field suggest that in the formulation of a *covariant* theory of gravitation no direct appeal may be made to this principle ; we may instead look upon physical situations arising in a *general gravitational field* as akin to the ones following from a *general coordinate transformation* from a system like K_0 to a system like K, say,[2] rather than of considering an infinity of transformations from systems like K_0 to 'suitably' accelerated systems like K_1, one for each infinitesimal region of space! A general gravitational field will then be characterized by the Riemannian metric

$$ds^2 = g_{ik}\, dx^i\, dx^k, \tag{6.10}$$

[1] This is based on the assumption that the given field can, in regions infinitesimally small, be regarded as effectively homogeneous.

[2] See Secs. 6.1 and 6.2.

where the functions g_{ik}, and their space-time dependence, would determine the complete nature of the given field; the special case (6.12) would correspond to the field-free case. Moreover, the process of adopting a system like K_0, in which the gravitational effects are *locally* absent, would correspond to effecting a coordinate transformation which reduces (6.10) to the quasi-Euclidean form

$$ds^2 = dX^\alpha\, dX^\alpha - dX^4\, dX^4, \qquad \alpha = 1, 2, 3, \tag{6.6}$$

locally ; such a system is usually referred to as the *geodesic coordinate system*. On the other hand, if the given field is homogeneous, the foregoing transformation would bring (6.10) to the form (6.6) *throughout* the space. Metric (6.10) thus covers the case of permanent fields as well as of temporary fields.

We must, however, point out that though the principle of equivalence seems to have been set aside the equality of inertial and gravitational masses (which is the very backbone of this principle) cannot be dispensed with, for otherwise it would not be possible to put the gravitational field in correspondence with the space-time metric; the latter being an outcome of mere coordinate transformations and, hence, being completely independent of the nature and constitution of the test particles employed in the field, the aforementioned correspondence is possible only if the gravitational field, too, possesses those properties. This, in turn, requires the equality of the two types of mass for *any* given body. This equality is the most vital physical fact for the establishment of the present theory ; in fact, one may even regard it as a fundamental physical postulate rather than an empirical result, for in the latter case the question 'to what accuracy?' would always arise.[1]

Finally, we have to find out the necessary means for determining the quantities g_{ik} which appear in the expression (6.10) of the metric. Evidently, this cannot be done by using relations (6.11) because the mathematical form of the coordinate transformation (6.7), or (6.8), is, in general, unknown. The solution, however, lies in determining the g_{ik} from the relevant field equations which link them with the sources of the field ; see Sec. 7.5.

7.3. Motion of a mass point in a gravitational field. The Newtonian approximation

We have already shown that the theory of gravitation may be developed in

---

[1] For a detailed discussion of this and related points, refer to V. Fock, *The Theory of Space, Time and Gravitation* (Pergamon Press, London, 1959), Secs. 61 and 96.

terms of the geometrical structure of the space-time continuum which, in general, is governed by the Riemannian metric

$$ds^2 = g_{ik}dx^i dx^k, \tag{7.1}$$

the functions g_{ik} determining the nature of the gravitational field. We can, then, apply the formalism of curvilinear coordinates and general tensors, developed in the preceding chapter, to study physical phenomena taking place in a gravitational field ; of course, the emphasis now would be on the physical content of the various symbols appearing in the analysis. The success of this scheme—the so-called *geometrization of gravitation*—can be gauged only when the conclusions drawn from it are compared with the observational data.

Of immediate interest in this connection is the problem of the motion of an otherwise free particle in a gravitational field. In view of the scheme under consideration, this motion would be regarded as that of a totally 'free' particle moving in a curved space-time continuum governed by the metric (7.1) ; the physical effects of gravitation would now appear as resulting from the departure of the actual metric of the continuum from the '*normal*' metric of the MINKOWSKI world. We then have, for the differential equations of the trajectory of the particle, the *geodesic equations*

$$d(dx^i) + \Gamma^i_{kl}\, dx^k\, dx^l = 0, \quad i = 1, 2, 3, 4\,; \tag{7.2}$$

see (6.73). Obviously, for Γ^i_{kl} vanishing, the foregoing equations would correspond to a uniform rectilinear motion, as one has in the absence of the gravitational field. However, in general, these equations would not lead to a rectilinear orbit ; we would rather obtain a curved one, characteristic of the motion of a particle in a field of force. The primary responsibility for bringing about this difference rests with the quantities Γ^i_{kl} which, accordingly, determine the magnitude of the resulting difference ; these quantities are, therefore, a measure of the *intensity* of the given field.

Now, the Γ^i_{kl} of (7.2) are related to the g_{ik} of (7.1) by relations

$$\Gamma^i_{kl} = g^{im}\, \Gamma_{m,kl} = \frac{1}{2}\, g^{im} \left(\frac{\partial g_{mk}}{\partial x^l} + \frac{\partial g_{ml}}{\partial x^k} - \frac{\partial g_{kl}}{\partial x^m} \right); \tag{7.3}$$

the g_{ik} may, therefore, be looked upon as the *potentials* of the field. The fact that the present theory requires ten 'potentials' against one in the New-tonian theory speaks not only of its mathematical complexity but also, perhaps, of its physical profundity. One does, of course, expect that in an appropriate limiting case the formalism of the present theory would reduce to that of the Newtonian theory, so that one could, in that limit, descri be the gravitational field in terms of a single potential. We may refer to this reduction as the *correspondence principle* in the relativistic theory of gravitation. To investigate this point, we consider the motion of a *slow-moving* mass point in a *weak* gravitational field.

We note, first of all, that the condition of slow motion implies that

$$dx^\alpha \ll dx^4 (= c \, dt), \quad \alpha = 1, 2, 3, \tag{7.4}$$

and the condition of weak field implies that the elements g_{ik} of the funda-mental tensor differ only *slightly* from the normal ones, viz. (6.12) :

$$g_{ik} = g_{ik}^{(0)} + \varepsilon_{ik}, \tag{7.5}$$

where

$$g_{ik}^{(0)} = \begin{cases} +1 & \text{for} \quad i = k = 1, 2, 3, \\ -1 & \text{for} \quad i = k = 4, \\ 0 & \text{for} \quad i \neq k, \end{cases} \tag{7.6}$$

and

$$\varepsilon_{ik} \ll 1. \tag{7.7}$$

Now, in view of (7.4), the first three of Eqs. (7.2) become

$$d^2x^\alpha = -\Gamma_{kl}^\alpha \, dx^k \, dx^l \simeq -\Gamma_{44}^\alpha \, (dx^4)^2, \quad \alpha = 1, 2, 3 \, ; \tag{7.8}$$

hence (since $x^4 = ct$)

$$\frac{d^2x^\alpha}{dt^2} \simeq -c^2 \, \Gamma_{44}^\alpha. \tag{7.9}$$

Next, in view of (7.5)–(7.7), we have from (7.3) and (7.9)

$$\frac{d^2x^\alpha}{dt^2} \simeq -\frac{c^2}{2} g^{\alpha m} \left(2 \frac{\partial g_{m4}}{\partial x^4} - \frac{\partial g_{44}}{\partial x^m} \right)$$

$$\simeq -\frac{c^2}{2}\,g^{\alpha m}\left(-\frac{\partial g_{44}}{\partial x^m}\right)$$

$$\simeq -\frac{c^2}{2}\,g^{\alpha\alpha}\left(-\frac{\partial g_{44}}{\partial x^\alpha}\right)$$

$$\simeq +\frac{c^2}{2}\frac{\partial g_{44}}{\partial x^\alpha};\tag{7.10}$$

here, the first step of reduction tacitly assumes that the field under consideration is practically static, while the second and third steps make use of the fact that $g^{\alpha m}\ll 1$ for $m\neq\alpha$ and $\simeq 1$ for $m=\alpha$.

Equations (7.10) are of the same form as in the Newtonian theory of gravitation, viz.

$$\frac{d^2 x^\alpha}{dt^2}=-\frac{\partial\varphi}{dx^\alpha},\tag{7.11}$$

φ being the Newtonian field potential. Correspondence between (7.10) and (7.11) becomes complete if we stipulate that

$$g_{44}=-2\frac{\varphi}{c^2}+\text{const.}$$

The value of the constant can be determined by considering the extreme case of a vanishing field, i.e. $g_{44}=-1$ when $\varphi=0$. The constant thus turns out to be -1. We then have[1]

$$g_{44}=-\left(1+2\frac{\varphi}{c^2}\right)\tag{7.12}$$

or

$$\varphi=-\frac{c^2}{2}(g_{44}+1).\tag{7.13}$$

We now see that, of all the g_{ik}, it is only g_{44} that appears in the Newtonian approximation;[2] one *can*, therefore, describe the field, in the limiting case

---

[1] Equation (7.12) shows that a weak field is characterized by the condition : $(\varphi/c^2)\ll 1$, i e. the gravitational potential energy of a material body is much smaller than its rest mass energy.

[2] Although deviations of the other g_{ik}, from their respective normal values, are of the same order of magnitude as of g_{44}.

under discussion, by a single parameter φ. Indirectly, this is due to the fact that, in the expression (7.1) of the metric, the element g_{44} occupies the most dominant position [see conditions (7.4)].

It must be remarked here that, in view of the correspondence established above, all observational evidence that supports the Newtonian theory of gravitation is, *ipso facto*, an evidence in support of the present theory.   In the case of strong fields and/or fast motions, the two theories are expected to lead to different results ; only then the question of accepting one theory in preference to the other can be put to a decisive test.

7.4.   Basic tensors of the gravitational field

We saw, in Sec. 6.4, that the process of (infinitesimal) parallel transport of a vector, in a continuum characterized by a non-Euclidean metric, causes inherent (differential) changes in its components ; the magnitudes of the respective changes depend on the CHRISTOFFEL symbols (of the second kind). One, then, expects that the *net* change caused by a *finite* displacement would further depend on the path pursued.   If so, the total change caused by an outward displacement from the world point A to another world point B along a particular path and an inward displacement from B to A along a *different* path would, in general, be nonvanishing.   In other words, the components of a vector, translated parallel to itself along a closed contour, would not generally regain their original values on returning to the starting point.   Our first problem in this section is to study this important aspect of the space-time geometry.

Let (A_i) be an arbitrary covariant vector which undergoes a parallel displacement, starting from the point (x^i), along a closed contour C.   The resulting changes in its components would be given by, see (6.51),

$$\Delta A_k = \oint_C \Gamma^i_{km} A_i \, dx^m. \tag{7.14}$$

Transforming the line integral into a surface integral, with the help of STOKES' theorem, we obtain

$$\Delta A_k = \frac{1}{2} \int_S \left[\frac{\partial(\Gamma^i_{km} A_i)}{\partial x^l} \, df^{lm} + \frac{\partial(\Gamma^i_{kl} A_i)}{\partial x^m} \, df^{ml} \right]$$

$$= \frac{1}{2} \int_S \left[\frac{\partial(\Gamma^i_{km} A_i)}{\partial x^l} - \frac{\partial(\Gamma^i_{kl} A_i)}{\partial x^m} \right] df^{lm}, \tag{7.15}$$

where

$$(df^{\,lm}) = (dx^l\,\delta x^m - dx^m\,\delta x^l) = -(df^{\,ml}) \qquad (7.16)$$

is the surface element spanned by the line elements (dx^i) and (δx^i). Carrying out differentiations in the integrand of (7.15) and effecting an interchange of dummy variables in some of the terms, we get

$$\Delta A_k = \frac{1}{2}\int_S\left[\frac{\partial \Gamma^i_{km}}{\partial x^l} - \frac{\partial \Gamma^i_{kl}}{\partial x^m} + \Gamma^n_{km}\,\Gamma^i_{nl} - \Gamma^n_{kl}\,\Gamma^i_{nm}\right]A_i\,df^{\,lm}; \quad (7.17)$$

in arriving at this form we have also made use of the fact that, since our vector is undergoing a parallel displacement,

$$\frac{\partial A_i}{\partial x^l} = \Gamma^n_{il}\,A_n, \qquad (7.18)$$

again by (6.51).

For an infinitesimal contour, we may write

$$\Delta A_k = \frac{1}{2}\,R^i_{klm}\,A_i\,df^{\,lm}, \qquad (7.19)$$

where

$$R^i_{klm} = \frac{\partial \Gamma^i_{km}}{\partial x^l} - \frac{\partial \Gamma^i_{kl}}{\partial x^m} + \Gamma^n_{km}\,\Gamma^i_{nl} - \Gamma^n_{kl}\,\Gamma^i_{nm}. \qquad (7.20)$$

Now, in (7.19), the quantities $df^{\,lm}$, which are constructed out of the co-ordinate differentials, and the quantities ΔA_k, which are the differences of the components of a vector *at one and the same point* (x^i), constitute tensors; the same is true of the quantities A_i. Hence, the set (R^i_{klm}), by the quotient law, must also be a tensor. This (mixed) tensor of rank 4 is called the *Riemann-Christoffel tensor*[1] or the *curvature tensor*; it is formed exclusively from the fundamental tensor g_{ik} and its first and second derivatives.

We obtain a similar result for an arbitrary contravariant vector, say (A^i). For this purpose, we note that the *scalar product* $(A^k B_k)$ of the given vector with an arbitrarily chosen covariant vector (B_i) would remain unchanged as

---

[1] This tensor was discovered by RIEMANN in a memoir sent to the Paris Academy in 1861 [and was published posthumously in his *Werke* (1892), p. 401]; it was subsequently employed by CHRISTOFFEL in his work : *J. für. Math.* 70, 46, 241(1869).

the two vectors are made to undergo a parallel displacement. Hence, we must have

$$0 = \Delta(A^k B_k) = A^k \, \Delta B_k + B_k \, \Delta A^k$$

$$= A^k \left[\frac{1}{2} \, R^i_{klm} \, B_i \, df^{lm} \right] + B_k \, \Delta A^k$$

$$= B_k \left[\frac{1}{2} \, R^k_{ilm} \, A^i \, df^{lm} + \Delta A^k \right].$$

In view of the arbitrariness of the components B_k, it follows that

$$\Delta A^k = - \frac{1}{2} \, R^k_{ilm} \, A^i \, df^{lm}. \tag{7.21}$$

We now consider another important fact of tensor calculus, in which the curvature tensor figures prominently. We examine the process of *successive* covariant differentiation of a given vector which, as we shall see here, is not generally commutative. One obtains, with the help of (6.55),

$$A_{i;k;l} - A_{i;l;k} = \left[\frac{\partial A_i}{\partial x^k} - \Gamma^m_{ik} A_m \right]_{;l} - \left[\frac{\partial A_i}{\partial x^l} - \Gamma^n_{il} A_n \right]_{;k}$$

$$= \left[\frac{\partial^2 A_i}{\partial x^l \partial x^k} - \frac{\partial}{\partial x^l} (\Gamma^m_{ik} A_m) - \Gamma^s_{il} \left(\frac{\partial A_s}{\partial x^k} - \Gamma^m_{sk} A_m \right) \right]$$

$$- \left[\frac{\partial^2 A_i}{\partial x^k \partial x^l} - \frac{\partial}{\partial x^k} (\Gamma^n_{il} A_n) - \Gamma^s_{ik} \left(\frac{\partial A_s}{\partial x^l} - \Gamma^n_{sl} A_n \right) \right]. \tag{7.22}$$

Now, the first terms in the brackets cancel because the process of ordinary successive differentiation is indeed commutative; two other pairs of terms cancel when the second terms here are opened up. We are left with

$$A_{i;k;l} - A_{i;l;k} = \left[-A_m \frac{\partial \Gamma^m_{ik}}{\partial x^l} + \Gamma^s_{il} \Gamma^m_{sk} A_m \right]$$

$$- \left[- A_n \frac{\partial \Gamma^n_{il}}{\partial x^k} + \Gamma^s_{ik} \Gamma^n_{sl} A_n \right]$$

$$= A_n \left[\frac{\partial \Gamma^n_{il}}{\partial x^k} - \frac{\partial \Gamma^n_{ik}}{\partial x^l} + \Gamma^s_{il} \, \Gamma^n_{sk} - \Gamma^s_{ik} \, \Gamma^n_{sl} \right]$$

$$= A_n \, R^n_{ikl} \, , \tag{7.23}$$

by (7.20). Similarly, one can show that

$$A^i_{\;;k;l} - A^i_{\;;l;k} = - A^n \, R^i_{nkl} \, . \tag{7.24}$$

As pointed out in Sec. 6.4, the CHRISTOFFEL symbols Γ^i_{kl} can be made to vanish throughout the space-time, provided the latter is of the Euclidean type (of course, we do so by adopting a Cartesian coordinate system). This would imply that the quantities R^i_{klm} also vanish identically. Now, the set (R^i_{klm}) is a tensor; hence, its vanishing in any one coordinate system would insure its vanishing in every other coordinate system. Thus, a Euclidean continuum is characterized by the fact that the curvature tensor vanishes *everywhere*. The converse of this statement also holds, i.e. if the curvature tensor of a given continuum can be made to vanish *everywhere*, the continuum must be of the Euclidean type. The truth of this statement may be seen as follows.

Over an infinitesimal region of *any* given continuum one can employ an effectively Cartesian system, like K_0 (of Sec. 7.2), in which the quantities Γ^i_{kl} are locally vanishing.[1] Now, if it is given that the curvature tensor (R^i_{klm}) is zero throughout the continuum (which indeed implies the vanishing of the Γ^i_{kl} throughout), then any arbitrary vector can be translated parallel to itself, from *any* initial point to *any* final point and that too along *any* path, without causing any changes in its components due to the geometry of the continuum. This means that the coordinate system K_0, which was (effectively) Cartesian for the infinitesimal region we started with, can be given a parallel displacement from this region to any other region of the continuum, without impairing the orientations of its axes and, hence, its Cartesian

---

[1] It should be noted at this point that the *local* vanishing of the Γ^i_{kl} does not necessarily imply the vanishing of the R^i_{klm}, because the latter also depend on the derivatives of the Γ^i_{kl}— and these, even in K_0, would generally be nonzero.

character.   Thus, the whole continuum can be covered by a single Cartesian system of coordinates; the continuum must, therefore, be 'flat', i.e. without any curvature. Hence the *overall vanishing* (*or nonvanishing*) *of the components of the curvature tensor is a necessary and sufficient condition for the flatness* (*or otherwise*) *of the continuum.*

We shall now consider some of the important properties of the curvature tensor.   First of all, from its very definition (7.20), it follows that

$$R^i_{kml} = -R^i_{klm},$$
(7.25)

that is, the curvature tensor is antisymmetric in the second and third co-variant indices.   Next, it can be verified, again from (7.20), that

$$R^i_{klm} + R^i_{lmk} + R^i_{mkl} = 0;$$
(7.26)

to establish this result, one has to make use of the symmetry properties of the CHRISTOFFEL symbols as well.   Further, we have the following important identities—the so-called *Bianchi identities*:

$$R^n_{ikl;m} + R^n_{ilm;k} + R^n_{imk;l} = 0.$$
(7.27)

To prove these identities, it is useful to employ a (local) geodesic coordinate system; proved in this system, the identities, being tensor equations, would hold in any coordinate system. Now, in a geodesic coordinate system, the CHRISTOFFEL symbols vanish and, hence, the process of covariant differentiation reduces to that of ordinary differentiation; consequently, in this system

$$R^n_{ikl;m} = \frac{\partial^2 \Gamma^n_{il}}{\partial x^m \partial x^k} - \frac{\partial^2 \Gamma^n_{ik}}{\partial x^m \partial x^l}.$$
(7.28)

Substituting this expression, and similar ones for the other two derivatives, into the left-hand side of (7.27), we find that all the terms get cancelled, thus establishing the identities under question.

In terms of the mixed curvature tensor, we can also define a *purely covariant curvature tensor*, viz.

$$R_{iklm} = g_{in} R^n_{klm}.$$
(7.29)

Substituting from (7.20), we get

$$R_{iklm} = \left[\frac{\partial}{\partial x^l} (g_{in} \ \Gamma^n_{km}) - \frac{\partial g_{in}}{\partial x^l} \ \Gamma^n_{km} \right]$$

$$- \left[\frac{\partial}{\partial x^m} (g_{in} \Gamma^n_{kl}) - \frac{\partial g_{in}}{\partial x^m} \ \Gamma^n_{kl} \right]$$

$$+ \ g_{in} \ [\Gamma^s_{km} \ \Gamma^n_{sl} - \Gamma^s_{kl} \ \Gamma^n_{sm}],$$

which, with the help of (6.67) and (6.68), reduces to

$$R_{iklm} = \frac{\partial}{\partial x^l} \ \Gamma_{i,km} - \frac{\partial}{\partial x^m} \ \Gamma_{i,kl} + \Gamma^s_{kl} \ \Gamma_{s,im} - \Gamma^s_{km} \ \Gamma_{s,il}.$$

A further reduction, with the help of (6.69) and (6.70), leads to the important formula[1]

$$R_{iklm} = \frac{1}{2} \left[\frac{\partial^2 g_{im}}{\partial x^k \partial x^l} + \frac{\partial^2 g_{kl}}{\partial x^i \partial x^m} - \frac{\partial^2 g_{il}}{\partial x^k \partial x^m} - \frac{\partial^2 g_{km}}{\partial x^i \partial x^l} \right]$$

$$+ \ g^{st} \ (\Gamma_{t,kl} \ \Gamma_{s,im} - \Gamma_{t,km} \ \Gamma_{s,il}). \tag{7.30}$$

From this result it follows that

$$\left. \begin{array}{l} R_{ikml} = -R_{iklm} \\[4pt] R_{kilm} = -R_{iklm} \\[4pt] R_{lmik} = +R_{iklm}, \end{array} \right\} \tag{7.31}$$

that is, the tensor under discussion is antisymmetric with respect to an interchange among the first two or among the last two indices whereas it is symmetric with respect to an interchange among the two pairs of indices, taken together.

Further, in view of (7.26) and (7.29),

$$R_{iklm} + R_{ilmk} + R_{imkl} = 0; \tag{7.32}$$

it then follows from (7.31) that similar results hold if we permute cyclically

---

[1] The second term on the right-hand side of (7.30) can also be written as

$$g_{st} \ (\Gamma^s_{kl} \ \Gamma^t_{im} - \Gamma^s_{km} \ \Gamma^t_{il}).$$

any three of the four indices i, k, l and m, keeping the fourth one in its place. Finally, we obtain from the BIANCHI identities (7.27), by an inner multiplication with g_{sn}, the identities

$$R_{sikl;m} + R_{silm;k} + R_{simk;l} = 0. \tag{7.33}$$

From the curvature tensor(s) we can construct, by contraction, a tensor of rank 2. At first sight, there appear to be more than one different possibilities of doing this; however, on closer scrutiny we find that there is only one independent result obtainable in this manner. This is so because, firstly, the tensor

$$R^i_{ilm} = \delta^k_i R^i_{klm} = g^{kt} g_{ti} R^i_{klm} = g^{kt} R_{tklm}$$

vanishes identically (because of the symmetry of g^{kt} and the antisymmetry of R_{tklm} in the indices k and t) and, secondly, the tensors

$$R^i_{kim} \quad \text{and} \quad R^i_{kli}$$

are equal and opposite (because of the antisymmetry of R^i_{klm} in the indices l and m). We have, therefore, only one of these two tensors to consider, say

$$R_{km} = R^i_{kim} = g^{il} R_{iklm} . \tag{7.34}$$

Applying the relevant contraction, we obtain from (7.20)

$$R_{km} = \frac{\partial \Gamma^i_{km}}{\partial x^i} - \frac{\partial \Gamma^i_{ki}}{\partial x^m} + \Gamma^n_{km} \Gamma^i_{ni} - \Gamma^n_{ki} \Gamma^i_{nm} ; \tag{7.35}$$

this is called the *Ricci tensor* or the *contracted curvature tensor*. We note that the second term on the right-hand side of (7.35) is, by virtue of (6.70),

$$-\frac{\partial}{\partial x^m} \left[\frac{1}{2} g^{ir} \left(\frac{\partial g_{rk}}{\partial x^i} + \frac{\partial g_{ri}}{\partial x^k} - \frac{\partial g_{ki}}{\partial x^r} \right) \right] = -\frac{1}{2} \frac{\partial}{\partial x^m} \left[g^{ir} \frac{\partial g_{ri}}{\partial x^k} \right]$$

$$= -\frac{1}{2} \frac{\partial}{\partial x^m} \left(\frac{1}{g} \frac{\partial g}{\partial x^k} \right) = -\frac{\partial^2 (\ln \sqrt{-g})}{\partial x^m x \partial^k} ; \tag{7.36}$$

here, $g = |g_{ik}|$, so the quantity $(g^{ir} g)$ is the cofactor of g_{ir}. This term is, therefore, symmetric in m and k; the same being true of other terms in (7.35), the tensor (R_{km}) is symmetric.

Finally, we contract the RICCI tensor and obtain the so-called *curvature invariant* or the *scalar curvature*, viz.

$$R = g^{km} R_{km} = g^{km} g^{il} R_{iklm}. \tag{7.37}$$

In a Euclidean continuum, R is zero everywhere.

7.5. Einstein's field equations and the Poisson approximation

As remarked in Sec. 7.2, the elements g_{ik} of the fundamental tensor appropriate to a general gravitational field cannot, in practice, be determined by means of coordinate transformations. One must set up a covariant scheme of field equations which, on integration under specified boundary conditions, may lead to the desired analytical expressions for the g_{ik}. This is demanded not only by practical considerations but also by the obvious requirement that a physical theory which deals with manifestations in a *field* has to be based on a scheme of differential equations in terms of the basic parameters of the field. Such, for instance, was the case with NEWTON's theory of gravitation, for which one had the famous POISSON equation

$$\nabla^2 \varphi = 4\pi k \mu , \tag{7.38}$$

φ being the gravitational potential, μ the mass density in space and k the universal constant of gravitation. Our aim, then, is to evolve the covariant counterpart of the foregoing equation.

First of all, we note that the gravitational field is an outcome of the mass distribution or, more correctly, the mass-energy distribution in the space-time. Accordingly, the geometrical properties of the space-time continuum (which are, in essence, a measure of the gravitational field) would also be determined by the same distribution; one can even say, in view of MACH's concept of inertia (see Sec. 7.1), that the geometrical properties of the continuum actually *accrue* from the given mass-energy distribution.[1] The field equations, therefore, must somehow relate the curvature tensor of the

---

[1] This hypothesis was designated by EINSTEIN as *Mach's principle*, since he took it to be a generalization of the Machian concept of inertia; see *Ann. der Phys.* **55**, 241 (1918).

field (characterizing the effect) with the given energy-momentum tensor (characterizing the cause) in a form that reduces, in the correspondence limit, to the POISSON equation (7.38).

Now, the energy-momentum tensor (T_k^i) of a closed system [for instance, the one comprising the matter distribution and the (force) field together] is bound by the conservation laws

$$T_{k;i}^i = 0, \qquad k = 1, 2, 3, 4, \tag{7.39}$$

which represent the covariant generalization of Eqs. (5.78). Hence, the number of *really* independent components of this tensor is four less than the number characteristic of a symmetric tensor [such as (T^{ik}) is], i.e. six. At the same time, we need exactly six field equations, for out of the ten elements g_{ik} of the metric tensor four can be made *constant* by means of a suitable coordinate transformation [such as (6.23)]. Thus, the energy-momentum tensor seems to suffice for the task at hand.

The corresponding field tensor, which is of the same rank as the energy-momentum tensor and whose symmetry properties are also the same, is the RICCI tensor (R_{ik}), given by (7.35). One may, therefore, feel tempted to postulate a proportionality between the tensors (R_{ik}) and (T_{ik}). However, unlike (T_{ik}), the tensor (R_{ik}) does not satisfy conservation laws of the type (7.39), unless, of course, the scalar curvature R vanishes. This, on the other hand, cannot be admitted because the scalar T which, according to the foregoing postulation, would be proportional to R, may not necessarily vanish.[1] It is, therefore, essential to discover a field tensor such that not only are its symmetry properties the same as the RICCI tensor but it also satisfies the conservation laws.

We recall in this connection the BIANCHI identities (7.27) :

$$R_{ikl;m}^n + R_{ilm;k}^n + R_{imk;l}^n = 0. \tag{7.40}$$

Contracting (7.40) with respect to the indices n and l, we get, in view of (7.34),

$$-R_{ik;m} + R_{im;k} + R_{imkn}^n = 0; \tag{7.41}$$

---

[1] It should be noted that T in the case of an electromagnetic field does indeed vanish; in the case of matter distribution, it would be directly proportional to the rest energy density.

now, raising the index i, i.e. making an inner product with (g^{ri}) and then writing i for r, we obtain

$$-R^i_{k;m} + R^i_{m;k} + R^{ni}_{mk;n} = 0. \tag{7.42}$$

Next, contracting with respect to i and k, we get

$$-R_{;m} + R^i_{m;i} + R^n_{m;n} = 0, \tag{7.43}$$

that is,

$$-\delta^i_m R_{;i} + 2R^i_{m;i} = 0. \tag{7.44}$$

Thus, we finally have

$$[R^i_m - \frac{1}{2} \delta_m R]_{;i} = 0. \tag{7.45}$$

Introducing the *Einstein tensor* (G^i_m), defined by

$$G^i_m = R^i_m - \frac{1}{2} \delta^i_m R, \tag{7.46}$$

(7.45) can be written as

$$G^i_{m;i} = 0. \tag{7.47}$$

The last result shows that the tensor (G^i_m), or its purely covariant form

$$G_{ik} = g_{in} G^n_k = g_{in} \left(R^n_k - \frac{1}{2} \delta^n_k R \right)$$

$$= R_{ik} - \frac{1}{2} g_{ik} R, \tag{7.48}$$

which is symmetric in i and k, satisfies all the requirements stipulated above.

It is now perfectly plausible to postulate for the desired field equations

$$G_{ik} = R_{ik} - \frac{1}{2} g_{ik} R = \gamma T_{ik}, \tag{7.49}$$

where γ is a *scalar* multiplier, which is expected to be a universal constant. Equations (7.49) are bound by four conditions, viz.

$$G^i_{k;i} = \gamma T^i_{k;i} = 0. \tag{7.50}$$

Thus, we have just enough equations for the determination of the six non-constant elements of the tensor (g_{ik}).

In the mixed form, our field equations would be

$$G^i_k = R^i_k - \frac{1}{2} \delta^i_k R = \gamma T^i_k ; \tag{7.51}$$

it follows, on contraction, that

$$G = G^i_i = R - \frac{1}{2} \cdot 4R = - R = \gamma T. \tag{7.52}$$

Substituting from (7.52) into (7.51), we obtain

$$R^i_k = \gamma \left(T^i_k - \frac{1}{2} \delta^i_k T \right); \tag{7.53}$$

on lowering the index i, we get

$$R_{ik} = \gamma \left(T_{ik} - \frac{1}{2} g_{ik} T \right). \tag{7.54}$$

It is in the form (7.53) or (7.54) that the field equations are generally used for analytical purposes.

A special case of the foregoing may be noted. In empty space, $T_{ik} = 0$; hence, our equations become

$$R_{ik} = 0. \tag{7.55}$$

At first sight, this gives the impression that the corresponding space-time continuum is flat. This, however, may not be true because the real conditions for flatness are far more stringent, viz. $R^i_{klm} = 0$ (see the preceding section). The underlying point will be amply illustrated by the case studied in Sec. 8.2.

Our next problem is concerned with the determination of the coupling constant γ which appears in the field equations as a constant of proportionality. This is done by requiring that in the correspondence limit, i.e. for weak fields and slow motions, our equations reproduce the POISSON equation (7.38). Assuming, for this purpose, a pure matter distribution, we have for the energy-momentum tensor

$$T^{ik} = \mu^0 U^i U^k, \tag{7.56}$$

where μ^0 is the rest (or proper) mass density and (U^i) $[= (dx^i/d\tau)]$ is the vector representing the state of motion of the matter. The scalar T turns out to be

$$T = T^i_i = g_{ik} T^{ik} = \mu^0(g_{ik} U^i U^k)$$

$$= \mu^0 \frac{g_{ik} dx^i dx^k}{(d\tau)^2} = -\mu^0 c^2. \tag{7.57}$$

At the same time the purely covariant energy-momentum tensor would be given by

$$T_{ik} = g_{il} g_{km} T^{lm}$$

$$= -\mu^0 c^2 g_{il} g_{km} \frac{dx^l dx^m}{(ds)^2}. \tag{7.58}$$

Since the motion of the matter is assumed to be slow, $dx^\alpha \ll dx^4$ ($\alpha = 1, 2, 3$) and $(ds)^2 \simeq g_{44}(dx^4)^2$; consequently,

$$T_{ik} \simeq -\mu^0 c^2 g_{i4} g_{k4}/g_{44}. \tag{7.59}$$

Since the field is also assumed to be weak, the actual g_{ik} would not be very different from their normal values. Thus, the only significant component among the various T_{ik} would be the one corresponding to $i = k = 4$, for which we may write

$$T_{44} \simeq \mu^0 c^2. \tag{7.60}$$

Correspondingly, the only significant component of the tensor (R_{ik}), see (7.54), would be R_{44}. Substituting from (7.57) and (7.60) into (7.54), we get

$$R_{44} \simeq \frac{1}{2}\,\gamma\,\mu^0 c^2 \simeq \frac{1}{2}\,\gamma\,\mu c^2; \tag{7.61}$$

for obvious reasons, we have neglected the difference between μ and μ^0.

Further, we have for R_{44}, see (7.35),

$$R_{44} = \frac{\partial \Gamma^i_{44}}{\partial x^i} - \frac{\partial \Gamma^i_{4i}}{\partial x^4} + \Gamma^n_{44}\,\Gamma^i_{ni} - \Gamma^n_{4i}\,\Gamma^i_{n4}. \tag{7.62}$$

The last two terms can be neglected because they are both of the second order of magnitude, while the second one and the fourth part of the first one can be neglected because they are derivatives with respect to $x^4 (=ct)$ and, hence, are negligible in comparison with the space derivatives $\partial \Gamma^\alpha_{44}/\partial x^\alpha$ ($\alpha = 1, 2, 3$).[1] We can, therefore, write

$$R_{44} \simeq \frac{\partial \Gamma^\alpha_{44}}{\partial x^\alpha}, \tag{7.63}$$

which, by means of steps similar to the ones leading from (7.9) to (7.10), becomes

$$R_{44} \simeq -\frac{1}{2}\,\frac{\partial^2 g_{44}}{\partial x^\alpha\,\partial x^\alpha} \simeq \frac{1}{c^2}\,\frac{\partial^2 \varphi}{\partial x^\alpha\,\partial x^\alpha}$$

$$= \frac{1}{c^2}\,\nabla^2\varphi; \tag{7.64}$$

here, use has also been made of Eq. (7.12). From (7.61) and (7.64), we finally obtain

$$\nabla^2\varphi \simeq \frac{1}{2}\,\gamma c^4 \mu. \tag{7.65}$$

A comparison of (7.65) with (7.38) shows that we must take:

$$\gamma = 8\pi k/c^4. \tag{7.66}$$

---

[1] For a *static* field, these terms are identically zero.

Thus, the field equations (7.53) and (7.54) become

$$R_k^i = 8\pi \frac{k}{c^4} \left(T_k^i - \frac{1}{2} \delta_k^i T \right) \tag{7.67}$$

and

$$R_{ik} = 8\pi \frac{k}{c^4} \left(T_{ik} - \frac{1}{2} g_{ik} T \right), \tag{7.68}$$

respectively.

It may be pointed out that the very fact that EINSTEIN's field equations reduce, in the correspondence limit, to the POISSON equation (or, in the case of matter-free space, to the LAPLACE equation $\nabla^2 \varphi = 0$) shows that the *inverse square law* of gravitational force is automatically implied here (of course, only in the correspondence limit).

7.6. Electromagnetic field in a Riemannian space-time

We shall now discuss the generalization of the electromagnetic field equations of the special theory of relativity, as given in Chapter 5, to arbitrary coordinate systems ; after the generalization, these equations would become applicable even in the presence of a gravitational field. The main step to be taken here is the introduction of the metric tensor (g_{ik}) and the covariant differential operations (characteristic of a non-Euclidean continuum) into the tensor formalism of the theory already developed in Chapter 5. Thus, we must now appreciate the distinction between the contravariant and covariant indices of the various tensors and, moreover, replace the processes of ordinary differentiation by their covariant counterparts.

First of all, we note that the relations (5.35), defining the electromagnetic field tensor (F_{ik}) in terms of the field potential (A_i), now become

$$F_{ik} = A_{k;i} - A_{i;k} ; \tag{7.69}$$

however, from (6.55) it follows that these relations may equally well be written as

$$F_{ik} = \frac{\partial A_k}{\partial x^i} - \frac{\partial A_i}{\partial x^k} \tag{7.70}$$

which are of the same form as (5.35), with the only difference that the co-ordinate indices must now be contravariant. The correspondence between the tensor (F_{ik}) and the physical fields **E** and **B** remains unchanged. Further, the first tetrad of MAXWELL's equations, (5.45), now becomes

$$F_{ik;l} + F_{kl;i} + F_{li;k} = 0. \tag{7.71}$$

However, again, in view of the relations

$$B_{ik;l} = \frac{\partial B_{ik}}{\partial x^l} - \Gamma^j_{il} B_{jk} - \Gamma^j_{kl} B_{ij} \tag{7.72}$$

and the antisymmetric nature of the tensor (F_{ik}), (7.71) may be written in a form similar to (5.45), viz.

$$\frac{\partial F_{ik}}{\partial x^l} + \frac{\partial F_{kl}}{\partial x^i} + \frac{\partial F_{li}}{\partial x^k} = 0. \tag{7.73}$$

Next, in order to give a covariant form to the second tetrad of MAXWELL's equations, (5.46), we must first of all determine the current density four-vector appropriate to the generalized situation under discussion. We follow, for this purpose, the procedure indicated at the end of Sec. 5.2. Thus, let it be postulated that the charge element $\rho \delta V$ (where ρ is the charge density: $de/\delta V$) is invariant under arbitrary coordinate transformations. Now, the four-dimensional volume element $(\delta V dx^4)$ is *not* invariant under general transformations; on the other hand, one can show that the quantity

$$(\sqrt{-g} \; \delta V \, dx^4) \tag{7.74}$$

is, quite generally, invariant.[1]   Hence, the quantity

$$\frac{\rho}{\sqrt{-g} \; dx^4} \tag{7.75}$$

---

[1] It is evident that, in view of the tensor character of the set (g_{ik}),

$$g'_{ik} = \frac{\partial x^l}{\partial x'^i} \frac{\partial x^m}{\partial x'^k} g_{lm},$$

whence, taking the determinants, we get

$$g' = | g'_{ik} | = \left| \frac{\partial x^l}{\partial x'^i} \frac{\partial x^m}{\partial x'^k} g_{lm} \right|$$

$$= \left| \frac{\partial x^l}{\partial x'^i} \right| \left| \frac{\partial x^m}{\partial x'^k} \right| \left| g_{lm} \right|,$$

by the well known law of multiplication of determinants.   It follows that

$$g' = \left| \frac{\partial x^l}{\partial x'^i} \right|^2 g.$$

Now, the transformation of the volume element is given by

$$(\delta V dx^1) = \left| \frac{\partial x^l}{\partial x'^i} \right| (\delta V' dx'^4).$$

Consequently,

$$(\delta V dx^4) = (g'/g)^{1/2} (\delta V' dx'^4),$$

whence follows the statement made in the text.

would also be invariant. Consequently, the set

$$(s^i) = \frac{\rho \, (dx^i)}{\sqrt{-g} \, dx^4} = \frac{\rho}{\sqrt{-g}} \left(\frac{\mathbf{u}}{c}, 1 \right)$$ (7.76)

would be a four-vector; this represents the desired generalization of the corresponding set of quantities (5.11). The equation of continuity (5.12) would, then, be replaced by

$$s^i{}_{;i} = 0,$$ (7.77)

which, in view of (6.54), becomes

$$\frac{\partial s^i}{\partial x^i} + \Gamma^i_{ki} \, s^k = 0.$$ (7.78)

Next, following the sequence of steps involved in arriving at (7.36), we get

$$\frac{\partial s^i}{\partial x^i} + \left[\frac{\partial}{\partial x^k} \, (\ln \sqrt{-g}) \right] s^k = 0,$$

that is,

$$\frac{1}{\sqrt{-g}} \, \frac{\partial}{\partial x^i} (\sqrt{-g} \, s^i) = 0.$$ (7.79)

Equations (5.46) can now be written in the covariant form :

$$F^{ik}{}_{;k} = 4\pi s^i,$$ (7.80)

that is,

$$\frac{1}{\sqrt{-g}} \, \frac{\partial}{\partial x^k} (\sqrt{-g} \, F^{ik}) = 4\pi s^i.$$ (7.81)

Further, we note that the LORENTZ condition (5.32) becomes

$$A^i{}_{;i} = 0.$$ (7.82)

Substituting from (7.69) into (7.80), we get[1]

$$A^{k;i}{}_{;k} - A^{i;k}{}_{;k} = 4\pi s^i,$$ (7.83)

---

[1] An index of covariant differentiation appearing as a superscript implies that it has been raised *after* the process of (covariant) differentiation.

whence we obtain, by first lowering the index i and then making use of (7.24), (7.25), (7.34) and (7.82),

$$A_{i;k}{}^{;k} - R_{in} A^n = -4\pi s_i, \tag{7.84}$$

which represents the generalization of (5.34).

Finally, for the geodesic equations of a charged particle moving in an electromagnetic field, in the presence of a gravitational field, we have

$$\frac{d^2 x^i}{ds^2} + \Gamma^i_{kl} \frac{dx^k}{ds} \frac{dx^l}{ds} = \frac{e}{m_0 c^2} F^i_k \frac{dx^k}{ds}; \tag{7.85}$$

see Eqs. (5.58) and (6.72) – (6.74). One can write down analogously the equations of motion for a charge distribution as well.

In this section we have considered the presence of a gravitational field along with an electromagnetic field. However, there was one essential difference between the two fields: whereas the former appeared as an *inherent* feature of the space-time continuum, the latter appeared as an *external* manifestation superposed on the continuum. In other words, while the gravitational field was taken care of by the metrical properties of the continuum itself, the electromagnetic field had to be described in terms of parameters external to the continuum. This is very different from the so-called 'unified' point of view, according to which one would like to treat the electromagnetic field, and for that matter any fundamental field of physics, in the same spirit as the gravitational field. Efforts towards evolving such an all-embracing theory have been going on ever since the development of the general theory of relativity; however, a truly 'unified' picture has not yet emerged.

Problems

7.1. (a) Perform the coordinate transformation

$$X = x \cosh \frac{gt}{c} + \frac{c^2}{g} \left(\cosh \frac{gt}{c} - 1 \right),$$

$$Y = y, \quad Z = z,$$

$$T = \frac{c}{g} \sinh \frac{gt}{c} + \frac{x}{c} \sinh \frac{gt}{c}$$

on the MINKOWSKI metric $ds^2 = dX^2 + dY^2 + dZ^2 - c^2 dT^2$. Can the resulting metric be interpreted as one pertaining to a *uniformly* accelerated observer? Discuss the limitations of this interpretation.

(b) Examine the foregoing problem in the correspondence limit.

7.2. Proceeding as in the case of (7.23), or otherwise, prove the relations (7.24).

7.3. Complete the analytical steps leading from (7.29) to (7.30).

7.4. Prove the relations (7.32) by setting up a (local) geodesic system of coordinates.

7.5. Show that the curvature tensor R_{iklm} in a four-dimensional continuum possesses, in view of its symmetry properties (7.31) and the relations (7.32), only 20 independent components. Further show that in an n-dimensional continuum, where each index runs from 1 to n, this number is $n^2 (n^2 - 1)/12$.

Note that on a two-dimensional surface the curvature tensor is completely defined by a single component, R_{1212} say.

7.6. In a two-dimensional space the line element dl^2 is given by

$$dl^2 = du^2 + G^2(u, v) \, dv^2.$$

Show that the Gaussian curvature K and the curvature invariant R of this space are given by

$$K = -\frac{1}{2} R = -\frac{R_{1212}}{G^2} = \frac{1}{G} \frac{\partial^2 G}{\partial u^2}.$$

Apply these results to the relevant subspace of Problem 6.5.

7.7. Show that the hypersurface

$$x^1 = a \cos \chi, \qquad\qquad x^2 = a \sin \chi \cos \theta,$$

$$x^3 = a \sin \chi \sin \theta \cos \varphi, \qquad x^4 = a \sin \chi \sin \theta \sin \varphi,$$

in a Euclidean space of dimensionality 4 is itself a three-dimensional space of constant curvature.

CHAPTER 8

EXPERIMENTAL TESTS OF EINSTEIN'S THEORY OF GRAVITATION

8.1. Centrally symmetric fields

In the preceding chapter we developed a theory of gravitation in terms of the geometrical properties of space-time continuum and checked that in the limit of weak fields and slow motions this theory reduced to the Newtonian theory of gravitation. This means that the vast observational data, both terrestrial and astronomical, normally cited as evidence in support of the Newtonian theory of gravitation, is automatically an evidence for the new theory as well. However, differences arise when the basic assumptions, viz. that the field is weak and the motion is slow, which correspond to the Newtonian limit, fail to hold. In such cases, it is possible to subject the various theoretical results to appropriate experimental tests and thereby decide the merits of one theory over the other.

In this connection we consider a typical class of gravitational fields, viz. the ones that possess central symmetry. Such fields arise from a centrally symmetric distribution of matter, whose motion is also centrally symmetric (that is, the velocity, at each point of space and at each instant of time, is radial). In view of this symmetry, the expression for the space-time metric, at any instant of time, should be the same for all points of space located symmetrically with respect to (that is, equidistant from) the centre of the distribution : the latter point is, obviously, the most natural choice for the origin of the spatial coordinates appropriate to this study. Now, the measure of the spatial distance in a Euclidean space is the well known *radius vector r*; however, in a non-Euclidean space there does not exist any quantity which possesses all the conventional properties of the Euclidean radius vector. Actually, in such a space the choice of a 'radius vector' is quite arbitrary (see Sec. 6.1)—so much so that the various parameters of the non-Euclidean geometry acquire physical meaning only through the (relativistic) *correspondence principle*.

Now, so far as the choice of a suitable coordinate system is concerned, it is most natural to adopt the 'spherical polar' coordinates (r, θ, φ). The most general form of the metric, in view of the spatial symmetry, would then be

$$ds^2 = A \, dr^2 + Br^2 \, (d\theta^2 + \sin^2 \theta \, d\varphi^2) + C \, dr \, dt + D \, dt^2, \qquad (8.1)$$

where A, B, C and D are certain functions of the 'radius' r and the 'time' t; the coordinates θ and φ, due to the special nature of the problem, (can be made to) stay aloof from the non-Euclidean aspects of the metric. Next, because of the inherent arbitrariness in the choice of a space-time coordinate system in general relativity, we may carry out transformations of the type

$$r' = r' \, (r, t) \qquad \text{and} \qquad t' = t' \, (r, t), \qquad (8.2)$$

which, evidently, do not impair the central symmetry of the metric. However, by means of such a transformation, we can make any two of the four quantities A, B, C and D, appearing in (8.1), take values that suit our convenience. We arrange the transformation (8.2) such that it makes

$$B = 1 \qquad \text{and} \qquad C = 0. \qquad (8.3)$$

The metric (8.1) then becomes

$$ds^2 = A \, dr^2 + r^2 \, (d\theta^2 + \sin^2 \theta \, d\varphi^2) + D \, dt^2, \qquad (8.4)$$

where we have removed the primes appearing over the various symbols. We now note that the 'instantaneous' *spatial* geometry in a centrally symmetric field is determined by the (spatial) metric

$$dl^2 = A \, dr^2 + r^2 \, (d\theta^2 + \sin^2 \theta \, d\varphi^2) ; \qquad (8.5)$$

obviously, a purely radial element in this case measures $A^{1/2}$ times the corresponding measure in a Euclidean space, while a purely transverse element measures the same as the corresponding Euclidean measure. Thus, while the circumference of a circle passing through the point (R, θ, φ), with centre at the origin, will be equal to the normal value $2\pi R$, the centre-to-rim distance will be equal to $\int_0^R A^{1/2}\,dr.$ The ratio of the two would, therefore, be

$$2\pi R \div \int_0^R A^{1/2}\,dr \neq 2\pi\,; \tag{8.6}$$

cf. the corresponding situation in (6.4).

It is also evident from the expression on the right-hand side of (8.4) that the 'proper' times of the various clocks stationed at different points in space (at which D, in general, will have different values) will no longer agree with one another ; *cf.* the corresponding situation in (6.5). A detailed study of this question will be taken up in Secs. 8.5 and 8.6.

We now make the customary substitutions

$$A \equiv e^{\lambda} \quad \text{and} \quad D \equiv -c^2\,e^{\nu}\,, \tag{8.7}$$

where λ and ν are as yet unknown functions of r and t, while c is the speed of light in a system like K_0 (see Sec. 7.2) ; thus, we can write

$$ds^2 = e^{\lambda}\,dr^2 + r^2(d\theta^2 + \sin^2\theta\,d\varphi^2) - e^{\nu}\,c^2\,dt^2. \tag{8.8}$$

A comparison of the metric (8.8) with the corresponding one appropriate to the special theory of relativity, viz.

$$ds^2 = dr^2 + r^2(d\theta^2 + \sin^2\theta\,d\varphi^2) - c^2\,dt^2, \tag{8.9}$$

shows that the latter corresponds to the special case

$$\lambda = \nu = 0. \tag{8.10}$$

Expression (8.8) may preferably be written in the standard form

$$ds^2 = g_{ik}\,dx^i\,dx^k, \tag{8.11}$$

with $(x^i) \equiv (r,\,\theta,\,\varphi,\,ct)$,

and

$$g_{11} = e^{\lambda}\,,\quad g_{22} = r^2,\quad g_{33} = r^2\sin^2\theta,\quad g_{44} = -e^{\nu}$$

$$g_{ik} = 0 \quad \text{for} \quad i \neq k\,; \tag{8.12}$$

it follows that

$$g = | g_{ik} | = - r^4 \sin^2 \theta \cdot e^{\lambda+\nu}. \tag{8.13}$$

The corresponding elements of the fundamental *contravariant* tensor (g^{ik}) would then be, see Sec. 6.3,

$$\left. \begin{array}{l} g^{11} = e^{-\lambda}, \quad g^{22} = r^{-2}, \quad g^{33} = (r \sin \theta)^{-2}, \quad g^{44} = - e^{-\nu} \\ \\ \text{and} \\ \\ g^{ik} = 0 \quad \text{for} \quad i \neq k. \end{array} \right\} \tag{8.14}$$

8.2. The Schwarzschild metric

We shall now discuss the field arising from a *single* mass, located at the origin of the spatial coordinates, having a spherically symmetric distribution; this field was first investigated in 1916 by K. SCHWARZSCHILD[1] and after him, independently, by J. DROSTE.[2] Confining our considerations to the *exterior* of the central body, we have for the energy-momentum tensor[3]

$$T_{ik} = 0 ; \tag{8.15}$$

the field equations for the region surrounding the body would then be, see Eqs. (7.48), (7.52) and (7.54),

$$G_{ik} = R_{ik} = 0. \tag{8.16}$$

Note that throughout this region, $T = 0$ and, hence, $R = 0$; the EINSTEIN tensor (G_{ik}) and the RICCI tensor (R_{ik}) are, therefore, identical. Further, the two tensors are also vanishing. This, however, does not mean that the continuum under study is flat, for the actual conditions for flatness are far more stringent than (8.16); that actually requires (see Sec. 7.4) the vanishing of the curvature tensor R^i_{klm}, which *does not* follow from (8.16). In fact, it will be seen in the sequel that the SCHWARZSCHILD solution does indeed correspond to a curvature in the space-time continuum.

---

[1] K. SCHWARZSCHILD, *Berlin Sitzungsberichte* (1916), p. 189.

[2] J. DROSTE, *Versl. gewone. Vergad. Akad. Amsterdam* 25, 163 (1916).

[3] For the *interior* solution, see H. P. ROBERTSON and T. W. NOONAN, *loc. cit.*, Sec. 9.7.

Now, the metric of the field under study should clearly be centrally symmetric ; hence, it may be characterized by the expression (8.8) of the preceding section, i.e. by the standard expression (8.11), with the metric tensor defined by (8.12). Our problem, then, consists of determining the space-time functions $\lambda(r, t)$ and $\nu(r, t)$ in terms of the parameters of the field. For this purpose, we carry out an integration of the field equations (8.16) under appropriate boundary conditions.

Our first task is to write out Eqs. (8.16) as differential equations for the functions λ and ν. This is accomplished by the following sequence of steps. First, we obtain from the metric tensors (8.12) and (8.14), with the help of relations (7.3), the following expressions for the various CHRISTOFFEL symbols Γ^i_{lk} :

$$\Gamma^1_{11} = \frac{1}{2} \frac{\partial \lambda}{\partial r}, \qquad\qquad \Gamma^1_{22} = - r\, e^{-\lambda},$$

$$\Gamma^1_{33} = - r \sin^2 \theta\ e^{-\lambda}, \quad \Gamma^1_{44} = \frac{1}{2}\, e^{\nu-\lambda} \frac{\partial \nu}{\partial r},$$

$$\Gamma^1_{14} = \Gamma^1_{41} = \frac{1}{2c} \frac{\partial \lambda}{\partial t},$$

$$\Gamma^2_{12} = \Gamma^2_{21} = \Gamma^3_{13} = \Gamma^3_{31} = \frac{1}{r}, \tag{8.17}$$

$$\Gamma^2_{33} = - \sin \theta \cos \theta, \qquad \Gamma^3_{23} = \Gamma^3_{32} = \cot \theta,$$

$$\Gamma^4_{11} = \frac{1}{2c}\, e^{\lambda-\nu} \frac{\partial \lambda}{\partial t}, \qquad \Gamma^4_{44} = \frac{1}{2c} \frac{\partial \nu}{\partial t},$$

and

$$\Gamma^4_{14} = \Gamma^4_{41} = \frac{1}{2} \frac{\partial \nu}{\partial r};$$

note that the symbols not appearing in Eqs. (8.17) are, in the case under study, identically zero. Next, with the help of the foregoing expressions and relations (7.35), we obtain expressions for the various components of the RICCI tensor (R_{ik}). It is, however, preferable to compute the mixed tensor (R^i_k) :

$$R^i_k = g^{im}\, R_{mk}. \tag{8.18}$$

By (8.16), this computation would hold for the mixed tensor (G^i_k) as well. We obtain the following nonvanishing components :

$$G_1^1 = e^{-\lambda} \left(\frac{1}{r} \frac{\partial \nu}{\partial r} + \frac{1}{r^2} \right) - \frac{1}{r^2},$$

$$\begin{aligned} G_2^2 = G_3^3 &= \frac{1}{2} e^{-\lambda} \left[\frac{\partial^2 \nu}{\partial r^2} + \frac{1}{2} \left(\frac{\partial \nu}{\partial r} \right)^2 \right. \\ &\quad + \frac{1}{r} \left(\frac{\partial \nu}{\partial r} - \frac{\partial \lambda}{\partial r} \right) - \frac{1}{2} \frac{\partial \nu}{\partial r} \frac{\partial \lambda}{\partial r} \Big] \\ &\quad - \frac{1}{2c^2} e^{-\nu} \left[\frac{\partial^2 \lambda}{\partial t^2} \quad \frac{1}{2} \left(\frac{\partial \lambda}{\partial t} \right)^2 - \frac{1}{2} \frac{\partial \lambda}{\partial t} \frac{\partial \nu}{\partial t} \right], \end{aligned}$$

$$G_4^4 = e^{-\lambda} \left(\frac{1}{r^2} - \frac{1}{r} \frac{\partial \lambda}{\partial r} \right) - \frac{1}{r^2},$$

and

$$G_4^1 = \frac{1}{cr} e^{-\lambda} \frac{\partial \lambda}{\partial t}.$$

$$(8.19)$$

Equating each of the foregoing expressions to zero, as required by the field equations

$$G_k^i = R_k^i = 0, \tag{8.16'}$$

we obtain four differential equations; however, a simplification among these equations leaves only three independent ones, namely

$$\frac{\partial \nu}{\partial r} = \frac{e^\lambda - 1}{r}, \tag{8.20}$$

$$\frac{\partial \lambda}{\partial r} = \frac{1 - e^\lambda}{r}, \tag{8.21}$$

and

$$\frac{\partial \lambda}{\partial t} = 0. \tag{8.22}$$

Adding (8.20) and (8.21), we get

$$\frac{\partial (\nu + \lambda)}{\partial r} = 0 ; \tag{8.23}$$

hence, $(\nu + \lambda)$ is a function of t alone :

$$(\nu + \lambda) = f(t), \quad \text{say.} \tag{8.24}$$

Now, by virtue of (8.22), λ is independent of t; hence, the t-dependence in (8.24) is due solely to the parameter ν, and is given by $\partial \nu/\partial t = \partial f/\partial t$, *independently* of r. It is, however, obvious from the analytical form of the metric (8.8) that such a dependence (of ν) on t can be eliminated by means of a *purely temporal* transformation (which we are always at liberty to do).[1] Denoting the transformed time coordinate again by t, the metric remains the same as before but now ν, as well as λ, is time-independent. This means that a centrally symmetric gravitational field in vacuum can always be given a time-independent description; in other words, such a field is *automatically static*.

Now the dependence of λ and ν on r is such that, see (8.23),

$$\frac{\partial \lambda}{\partial r} = -\frac{\partial \nu}{\partial r}. \tag{8.25}$$

Integrating (8.21) and keeping in mind (8.25), we get[2]

$$e^{-\lambda} = e^{\nu} = 1 + \frac{\text{const}}{r}. \tag{8.26}$$

It may be noted that as $r \to \infty$, both λ and ν tend to zero; hence, the metric (8.8) tends to the quasi-Euclidean limit (8.9) as the distance from the central mass becomes large.

In order to determine the exact value of the constant appearing in (8.26),

---

[1] This transformation would be of the type

$$t' = \psi(t),$$

whence we have

$$(dt')^2 = \left(\frac{d\psi}{dt}\right)^2 (dt)^2.$$

Now suppose that the function ψ is so chosen that

$$\left(\frac{d\psi}{dt}\right)^2 \equiv e^{f(t)}.$$

Then, the fourth term in the expression (8.8) of the metric would become

$$-c^2 \frac{e^{\nu}}{(d\psi/dt)^2} (dt')^2 \equiv -c^2 e^{\nu - f(t)} (dt')^2,$$

which shows that the coefficient of $(dt')^2$ would now be time-independent. It may also be noted here that the transformation $t \to t'$ does not affect any of the other three terms in the metric.

[2] Actually, by (8.25), ν *may* differ from $-\lambda$ by a constant term; however, in the correspondence limit, both ν and λ approach zero. Hence, we must have: $\nu = -\lambda$.

we note that in the limit of a weak field, see (7.12),

$$g_{44} \simeq -1 -2 \frac{\varphi}{c^2}, \tag{8.27}$$

where φ is the Newtonian potential, given by

$$\varphi = -kM/l, \tag{8.28}$$

k being the constant of gravitation, M the mass of the central body and l the distance of the point of observation from the origin. For weak fields (and especially at large distances from the origin) $l \simeq r$; hence, for large r we must have

$$g_{44} \simeq -1 + 2 \frac{kM}{rc^2}. \tag{8.29}$$

However, in the case under study

$$g_{44} = -e^{\nu} = -1 - \frac{\text{const}}{r}, \tag{8.30}$$

for *all* r. Therefore, in order to have the correct correspondence between (8.29) and (8.30), we must take the constant in (8.30) to be equal to $-2kM/c^2$. The expression (8.8) then becomes

$$ds^2 = \frac{dr^2}{\left(1 - \frac{2kM}{c^2 r}\right)} + r^2 (d\theta^2 + \sin^2 \theta \, d\varphi^2) - \left(1 - \frac{2kM}{c^2 r}\right) c^2 \, dt^2, \tag{8.31}$$

which is the desired expression for the *Schwarzschild metric*, valid in the exterior of the given central mass.

We find that the spatial metric, at any instant of time, is given by

$$dl^2 = \frac{dr^2}{\left(1 - \frac{2kM}{c^2 r}\right)} + r^2 (d\theta^2 + \sin^2 \theta \, d\varphi^2), \tag{8.32}$$

whence it follows that the radial distance from the origin to the point (R, θ, φ) would be *formally* given by

$$l = \int_0^R \frac{dr}{\left(1 - \frac{2kM}{c^2 r}\right)^{1/2}} \neq R. \tag{8.33}$$

Thus, the radial coordinate R is not equal to the radial distance l; it is, of course, a monotonic measure of l. However, the circumference of a circle passing through the point (R, θ, φ), with centre at the origin, is indeed $2\pi R$, for the relevant part of the metric is of the Euclidean type. Consequently, the ratio of the circumference of this circle to its radius would be

$$\frac{2\pi R}{l} \neq 2\pi ; \tag{8.34}$$

cf. a similar situation in (6.4).

Finally, we note that, by virtue of the condition (6.18), we must have:

$$\left(1 - \frac{2kM}{c^2 r} \right) > 0. \tag{8.35}$$

Hence, the radial coordinate r must be greater than the limiting value r_0, given by

$$r_0 = \frac{2kM}{c^2}. \tag{8.36}$$

The characteristic radius r_0 is known as the *Schwarzschild radius* of the object under study. At the surface $r = r_0$, g_{44} vanishes and along with it g_{11} diverges. It also turns out that the geometrical properties of the region $r < r_0$ are characteristically different from those of the region $r > r_0$. Normally, a material body of mass M extends over a region of space much larger than a sphere of perimeter $2\pi r_0$; for instance, the value of r_0 in the case of the Sun is about 3 km, which is much smaller than the actual radius of the Sun ($\sim 7 \times 10^5$ km). No peculiarities arise in cases such as this one.

The situation is, however, different if the object under study is confined entirely to the region $r < r_0$. Such objects, popularly known as *black holes*, possess rather peculiar properties and have lately drawn considerable attention. We propose to discuss them in some detail in Sec. 8.8 ; see also Sec. 9.6.

8.3.   Precession of the planetary orbits

We shall now consider the motion of a 'free' particle in the space-time continuum characterized by the SCHWARZSCHILD metric

$$ds^2 = g_{ik}\, dx^i\, dx^k, \tag{8.11}$$

with $(x^i) \equiv (r, \theta, \varphi, ct)$ and

$$g_{11} = e^\lambda = \frac{1}{1 - \dfrac{2kM}{c^2 r}}, \quad g_{22} = r^2, \quad g_{33} = r^2 \sin^2 \theta,$$

$$g_{44} = -e^\nu = -\left(1 - \frac{2kM}{c^2 r}\right), \quad g_{ik} = 0 \quad \text{for} \quad i \neq k. \tag{8.37}$$

We assume that the mass of the 'free' particle is negligible in comparison with the mass of the central body to which the field is due, so that its presence does not modify the metric to any appreciable extent. The time track of the particle would be a *geodesic*, determined by the differential equations

$$\frac{d^2 x^i}{ds^2} + \Gamma_{kl}^i \frac{dx^k}{ds} \frac{dx^l}{ds} = 0, \quad i = 1, 2, 3, 4. \tag{6.74}$$

The quantities Γ_{kl}^i are given by (8.17); one simply has to substitute for λ and ν from (8.37). Equations (6.74) then become

$$\frac{d^2 r}{ds^2} + \frac{1}{2} \frac{d\lambda}{dr} \left(\frac{dr}{ds}\right)^2 - r\, e^{-\lambda} \left(\frac{d\theta}{ds}\right)^2 - r \sin^2 \theta\, e^{-\lambda} \left(\frac{d\varphi}{ds}\right)^2$$

$$+ \frac{1}{2} e^{\nu-\lambda} \frac{d\nu}{dr} \left(c \frac{dt}{ds}\right)^2 = 0, \tag{8.38}$$

$$\frac{d^2 \theta}{ds^2} + \frac{2}{r} \frac{dr}{ds} \frac{d\theta}{ds} - \sin \theta \cos \theta \left(\frac{d\varphi}{ds}\right)^2 = 0, \tag{8.39}$$

$$\frac{d^2 \varphi}{ds^2} + \frac{2}{r} \frac{dr}{ds} \frac{d\varphi}{ds} + 2 \cot \theta \frac{d\theta}{ds} \frac{d\varphi}{ds} = 0, \tag{8.40}$$

and

$$\frac{d^2 t}{ds^2} + \frac{d\nu}{dr} \frac{dr}{ds} \frac{dt}{ds} = 0. \tag{8.41}$$

The plane in which the initial velocity vector and the initial position vector of the particle lie may be taken as the equatorial plane ($\theta = \pi/2$). Of course, initially $d\theta/ds = 0$; then, by (8.39), $d^2\theta/ds^2$ is also zero. Consequently, θ remains constant throughout the motion, and the orbit is permanently confined to the equatorial plane. Other equations also get simplified ; remembering that $\lambda = -\nu$, we have

$$\frac{d^2r}{ds^2} - \frac{1}{2}\frac{d\nu}{dr}\left(\frac{dr}{ds}\right)^2 - re^{\nu}\left(\frac{d\varphi}{ds}\right)^2 + \frac{1}{2}e^{2\nu}\frac{d\nu}{dr}\left(c\frac{dt}{ds}\right)^2 = 0, \quad (8.42)$$

$$\frac{d^2\varphi}{ds^2} + \frac{2}{r}\frac{dr}{ds}\frac{d\varphi}{ds} = 0, \quad (8.43)$$

and

$$\frac{d^2t}{ds^2} + \frac{d\nu}{ds}\frac{dt}{ds} = 0, \quad (8.44)$$

where

$$e^{\nu} = 1 - \frac{2kM}{c^2 r}. \quad (8.45)$$

Now, Eq. (8.43) can be written as

$$\frac{1}{r^2}\frac{d}{ds}\left(r^2\frac{d\varphi}{ds}\right) = 0,$$

whence it follows that

$$r^2\frac{d\varphi}{ds} = \text{const} = \frac{A}{ic}, \quad \text{say}; \quad (8.46)$$

this result is the *relativistic* analogue of KEPLER'S second law of planetary motion, the constant A being a measure of the angular momentum of the motion concerned. Next, integrating (8.44) once, we get

$$\frac{dt}{ds} = \frac{B}{ic}e^{-\nu}, \quad (8.47)$$

B being the relevant constant of integration. For the r-equation we may use, instead of (8.42), a simpler equation directly obtainable from the SCHWARZS-CHILD metric (with $\theta = \pi/2$ and $d\theta = 0$), viz.

$$1 = e^{-\nu}\left(\frac{dr}{ds}\right)^2 + r^2\left(\frac{d\varphi}{ds}\right)^2 - e^{\nu}\left(c\frac{dt}{ds}\right)^2. \quad (8.48)$$

Eliminating ds and dt with the help of Eqs. (8.46) and (8.47), we get

$$1 = -e^{-\nu}\frac{A^2}{c^2 r^4}\left(\frac{dr}{d\varphi}\right)^2 - \frac{A^2}{c^2 r^2} + B^2 e^{-\nu}. \quad (8.49)$$

Substituting for e^ν from (8.45) and changing from the variable r to its reciprocal u, we obtain

$$\left(\frac{du}{d\varphi}\right)^2 + u^2 = \frac{c^2}{A^2}.(B^2 - 1) + \frac{2km}{A^2}\,u + \frac{2km}{c^2}\,u^3. \qquad (8.50)$$

Differentiating (8.50) with respect to φ, we finally obtain

$$\frac{d^2u}{d\varphi^2} + u = \frac{kM}{A^2} + \frac{3kM}{c^2}\,u^2, \qquad (8.51)$$

which is the desired differential equation of the orbit of the particle. It differs from its Newtonian counterpart through the presence of the second term on the right-hand side ; as expected, this term vanishes if $c \to \infty$.

The solution of (8.51), in the absence of the perturbation term $3kMu^2/c^2$, is well known :[1]

$$u = \frac{kM}{A^2}\,(1 + \varepsilon \cos \varphi); \qquad (8.52)$$

clearly, u is a periodic function of φ, with period 2π. The moving particle is nearest to the central body, corresponding to the position known as *perihelion*, when u is maximum, i.e. when φ is an even multiple of π ; on the other hand, it is farthest, corresponding to the position known as *aphelion*, when u is minimum, i.e. when φ is an odd multiple of π. Thus, we have a *stable* elliptic orbit of eccentricity ε and semi-latus rectum $A^2/(kM)$, the semi-major axis of the ellipse being $a = A^2/\{kM(1 - \varepsilon^2)\}$.

The perturbed solution, on the assumption that the perturbation is small (which is generally true), may be written as

$$u = \frac{kM}{A^2}\,[1 + \varepsilon \cos (\eta\varphi)]. \qquad (8.53)$$

Substituting (8.53) into (8.51) and equating the coefficients of $\cos (\eta\varphi)$ on the two sides of the equation, we get

$$1 - \eta^2 = 6\,\frac{k^2M^2}{c^2A^2}$$

---

[1] There is, of course, a variety of solutions of the differential equation in question ; however, in this section we are interested only in a *periodic* solution,

or, since η is expected to be close to unity,

$$\eta \simeq 1 - 3 \frac{k^2 M^2}{c^2 A^2}. \tag{8.54}$$

The period of the motion is now $2\pi/\eta$ instead of 2π; consequently, in going from one perihelion (or aphelion) to the next, the radius vector sweeps an angle

$$\frac{2\pi}{\eta} \simeq 2\pi + 6\pi \frac{k^2 M^2}{c^2 A^2}. \tag{8.55}$$

In other words, the apses of the orbit advance at the rate of δ radians per revolution of the orbit, where

$$\delta = + 6\pi \frac{k^2 M^2}{c^2 A^2}; \tag{8.56}$$

the plus sign implies that the resulting precession of the orbit is in the *same* sense as the revolution of the particle in the orbit (see Fig. 8.1).[1]   In terms

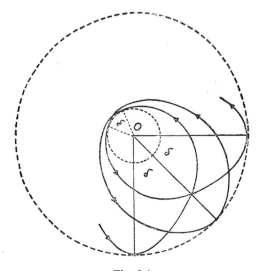

Fig. 8.1.

[1] It may be mentioned here that a similar effect, though one-sixth in magnitude, results from a calculation within the frmework of the special theory of relativity, where the only new feature of the calculation arises from the velocity dependence of the mass of the particle. This was first demonstrated by SOMMERFELD, in 1915-16, in connection with his investigations of the elliptic orbits of an electron around a nucleus; see Problem 3.3.

of a, the semi-major axis of the orbit, and T, the time period of revolution, the expression for δ becomes

$$\delta = +6\pi \frac{kM}{c^2a(1 - \varepsilon^2)} \tag{8.57}$$

$$= +24\pi^3 \frac{a^2}{c^2T^2(1 - \varepsilon^2)}; \tag{8.58}$$

here, use has been made of KEPLER's formula

$$T^2 = 4\pi^2a^3/(kM).$$

The foregoing results can be subjected to decisive observational test by applying them to the motion of a planet around the Sun. Here, the Sun would serve as the central body to which the SCHWARZSCHILD metric is due, while the planet would serve as the 'free' particle; the assumption that the mass of the moving particle is negligible in comparison with the mass of the central body, so that the field of force (that is, the metric of the space-time) is not appreciably affected, holds very well in this case. The most suitable planet in this connection is Mercury which, being nearest to the Sun, would be expected to show the largest effect. Actually, it has been known since the time of LEVERRIER[1] that the advance of the perihelion of Mercury's orbit shows a remainder which cannot be explained by perturbations due to other planets; according to a calculation by NEWCOMB, towards the end of the last century, the value of this remainder was $41''.24 \pm 2''.09$ per century[2]. The more recent estimate, however, is $42''.56 \pm 0''.94$ per century[3].

The corresponding theoretical expectation, as obtained from the formulae derived above, comes out to be $43''.03 \pm 0''.03$ per century. Agreement with the observed value is, clearly, very good[4].

In the case of the planets Venus, Earth, Mars and Jupiter, the theoretical expectations of the relativistic precession are $8''.63$, $3''.84$, $1''.35$ and $0''.06$ per

---

[1] U. J. LEVERRIER, *Ann. Obs. Paris* **5** (1859).

[2] S. NEWCOMB, *Astron. papers*, Washington **6**, 108 (1898). EINSTEIN, who was the first to obtain formulae (8.57—8.58) in 1915 (see the relevant issue of the *Berlin Sitzungsberichte*, p. 831) quoted the observed value as $45'' \pm 5''$ per century.

The word 'century', of course, means the *terrestrial* century.

[3] See G. M. CLEMENCE, *Revs. Mod. Phys.* **19**, 361 (1947); also F. R. TANGHERLINI, *Nuovo Cimento*, Supp. to vol. **20**, 1 (1961), especially p. 59.

[4] For the most recent evidence, made possible by measuring echo delays of radar signals transmitted from Earth to Mercury, see I. I. SHAPIRO *et. al.*, *Phys. Rev. Lett.* **28**, 1594 (1972).

century, respectively. The corresponding observational position is as follows.

In the case of the Earth, the observed value is much less accurate than in the case of Mercury; it is $4''.6 \pm 2''.7$ per century.[1] Agreement with the theoretical expectation is not unreasonable.

In the case of Venus, the magnitude of the precession is larger than in the case of the Earth; however, due to the excessively low eccentricity of its orbit, it is exceedingly difficult to measure the orbital precession with good accuracy[2] and, then, to separate out the part due to the relativistic effect. Actually, a better representation of this effect, from the point of view of observation, is provided by the product $\varepsilon\delta$ which, in this case, is equal to 0.059 (in the same units as before), in comparison with 0.064 in the case of the Earth and 8.85 in the case of Mercury. Nevertheless, the motion of Venus has been investigated in detail by DUNCOMBE[3] who finds that the remainder precession in this case is such that the product $\varepsilon\delta$ is 0.057 ± 0.033; agreement with the theoretical value is as good as in the case of the Earth.

There are no observational data as yet to compare with theory in the case of planets further from the Sun than the Earth is. It is hoped that with improved observations and calculations (say, by a factor of ten) the relativistic precession of the orbit of Mars, for which $\varepsilon\delta = 0.126$, could also be checked with theory; at the same time, the state of comparison in the case of Venus and Earth could be made more reliable than at present.

Corroborative evidence has emerged from optical observations of the minor planet Icarus for which $a \simeq 1.08$ A.U., $T \simeq 409$ days and $\varepsilon \simeq 0.83$, so that the expected value of the relativistic precession δ is about $10''.05$ per century. Comparing the product $\varepsilon\delta$, which is about 8.31 in this case, with the same parameter for the major planets, one finds that Icarus is a very promising object for verifying this particular prediction of the general theory of relativity. In fact, it has three distinct advantages over the planet Mercury: (i) it approaches the Earth more closely than Mercury does, so observations can be more accurate, (ii) it is almost a point object and (iii) it can be seen at night as well. Observations extending only over two decades have already shown[4] that, to an accuracy of about 20%, the perihelion of Icarus does advance by the amount expected on the basis of this theory.

---

[1] See H. R. MORGAN, *Astron. J.* **51**, 127 (1945); also G. M. CLEMENCE, *loc. cit.*

[2] Note that in the limit $\varepsilon \to 0$, which corresponds to a circular orbit, the precession is altogether unobservable.

[3] See V. L. GINZBURG, 'Experimental Verifications of the General Theory of Relativity' in *Recent Developments in General Relativity* (Pergamon Press, London, 1962), pp. 57-71; Sec. 1

[4] I. I. SHAPIRO, M. E. ASH and W. B. SMITH, *Phys. Rev. Lett.* **20**, 1517 (1968).

SCHIFF[1] has suggested that a crucial test of the general theory of relativity would be to observe the precession of a gyroscope mounted on the Earth or on an artificial satellite. This involves three independent contributions:

(i)   the *geodesic precession* which arises from the "parallel transport" of the axis of the gyroscope around the Earth,

(ii) the *Lense-Thirring precession* which arises from the influence of the rotation of the Earth on its otherwise SCHWARZSCHILD metric, and

(iii) the *Thomas precession*, a special-relativity effect, which arises if a non-gravitational force **F** is acting on the gyroscope. There is no such force if the gyroscope is mounted on a satellite; if mounted on the Earth, a nonzero **F** would be needed to support the weight of the gyroscope.

All these contributions are of the order of a few seconds of arc per century. Experimental prospects of observing this effect have been discussed by EVERITT, FAIRBANK and HAMILTON.[2]

GINZBURG,[3] on the other hand, has suggested extended observation of the orbits of artifical planets as well as of artifical satellites to obtain further evidence for the precession predicted by the general theory of relativity.

8.4. Deflection of a ray of light

In this section we propose to trace the course of a ray of light passing through a gravitational field. We again consider the SCHWARZSCHILD metric, characterized by the elements (8.37) of the fundamental tensor (g_{ik}). The world line corresponding to the propagation of light would be a *null* geodesic, defined by the equations

$$ds = 0 \qquad\qquad (8.59)$$

and

$$\frac{d^2x^i}{d\lambda^2} + \Gamma^i_{kl}\frac{dx^k}{d\lambda}\frac{dx^l}{d\lambda} = 0, \qquad\qquad (8.60)$$

where λ is some *nonvanishing* invariant parameter of the world line; see Sec. 6.5.

---

[1] L. I. SCHIFF, *Phys. Rev. Lett.* 4, 215 (1960); *Am. J. Phys.* **28**, 340 (1960); *Proc. Nat. Acad. Sci.*, USA **46**, 871 (1960).

[2] C. W. F. EVERITT, W. M. FAIRBANK and W. O. HAMILTON, in *Relativity*, ed. M. CARMELI, S. I. FICKLER and L. WITTEN (Plenum Press, New York, 1970), p.145.

[3] V. L. GINZBURG, *loc. cit.* See also his popular article in *Scientific American* **200**, 149 (May, 1959).

Proceeding in the same manner as in Sec. 8.3, we obtain, instead of (8.46) and (8.47),

$$r^2 \frac{d\varphi}{d\lambda} = A', \qquad (8.61)$$

and

$$\frac{dt}{d\lambda} = B' e^{-\nu}, \qquad (8.62)$$

where A' and B' are the relevant constants of integration. Further, we have from Eqs. (8.11), (8.37) and (8.59), for the propagation of light in the plane $\theta = \pi/2$ (with $d\theta=0$),

$$0 = e^{-\nu} \left(\frac{dr}{d\lambda} \right)^2 + r^2 \left(\frac{d\varphi}{d\lambda} \right)^2 - e^{\nu} \left(c \frac{dt}{d\lambda} \right)^2. \qquad (8.63)$$

Eliminating $d\lambda$ and dt from Eqs. (8.61)–(8.63), substituting for e^{ν} from (8.45), changing over to the variable $u = 1/r$, and finally differentiating with respect to φ, we get

$$\frac{d^2 u}{d\varphi^2} + u = \frac{3kM}{c^2} u^2. \qquad (8.64)$$

Comparing (8.64) with (8.51), we note that passage from material motion to light propagation is primarily characterized by the vanishing of the interval ds, which implies, by (8.46), that $A \to \infty$; hence the disappearance of the first term on the right-hand side of (8.51).

The corresponding Newtonian equation is

$$\frac{d^2 u}{d\varphi^2} + u = 0, \qquad (8.65)$$

whose solution can be written as

$$u = \frac{1}{R} \sin \varphi \quad \text{or} \quad r = \frac{R}{\sin \varphi}, \qquad (8.66)$$

which corresponds to a *straight* path, with R as the shortest distance from the gravitating centre. We note that

$$\frac{du}{d\varphi} = \frac{1}{R} \cos \varphi = \begin{cases} + \dfrac{1}{R} & \text{for} \quad \varphi = 0, & (8.67) \\[2mm] 0 & \text{for} \quad \varphi = \pi/2, & (8.68) \\[2mm] - \dfrac{1}{R} & \text{for} \quad \varphi = \pi ; & (8.69) \end{cases}$$

these cases correspond, respectively, to (i) the infinitely distant *initial* point, (ii) the point of closest approach and (iii) the infinitely distant *final* point of

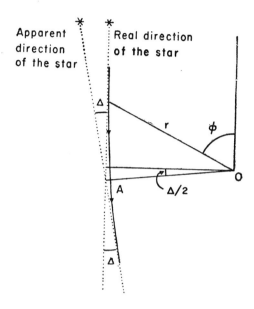

Fig. 8.2.

the path ; see the dotted vertical line in Fig. 8.2. Further, we note that since the perturbation term in (8.64) vanishes for $u = 0$ (i.e. for $r = \infty$, corresponding to which $\varphi = 0$ or π), the values (8.67) and (8.69) of $du/d\varphi$ would hold in the perturbed case as well.

Equation (8.64) will obviously not lead to a straight path ; a bending will inevitably result. In order to calculate the amount of bending, we must integrate the perturbed differential equation ; we may, however, assume that the perturbation is small, which *is* generally the case.[1] Multiplying (8.64) by $2du/d\varphi$ and integrating once, we get

$$\left(\frac{du}{d\varphi}\right)^2 + u^2 = \frac{2kM}{c^2} u^3 + \frac{1}{R^2} ; \qquad (8.70)$$

---

[1] It may be noted that there is a large variety of solutions to Eq. (8.64), a very special one being $u = c^2/3kM = $ const., which corresponds to a *circular* path of 'radius' $3kM/c^2$. We are, however, interested in that path which differs *only slightly* from the Newtonian straight path.

the constant $1/R^2$ has been chosen so that for $u = 0$, $du/d\varphi = \pm 1/R$ [see Eqs. (8.67) and (8.69)]. Thus,

$$\frac{du}{d\varphi} = \pm \frac{1}{R} (1 - \sigma^2)^{1/2}, \tag{8.71}$$

where

$$\sigma^2 = R^2 u^2 \left(1 - \frac{2kM}{c^2} u \right); \tag{8.72}$$

note that $\sigma = 0$ at infinity and $\sigma = 1$ at the point of closest approach. In view of the smallness of the perturbation, we can write from (8.72)

$$u \simeq \frac{\sigma}{R} \left(1 + \frac{kM}{c^2} \frac{\sigma}{R} \right) \tag{8.73}$$

and, hence,

$$\frac{du}{d\sigma} \simeq \frac{1}{R} \left(1 + \frac{2kM}{c^2} \frac{\sigma}{R} \right). \tag{8.74}$$

We then obtain for any point on the first half of the path [see Eqs. (8.71) and (8.74)],

$$\varphi = \int_0^\sigma \frac{d\varphi}{du} \frac{du}{d\sigma} d\sigma \simeq \int_0^\sigma \frac{1 + \dfrac{2kM}{c^2} \dfrac{\sigma}{R}}{\sqrt{1 - \sigma^2}} d\sigma$$

$$= \sin^{-1} \sigma - \frac{2kM}{c^2 R} \sqrt{1 - \sigma^2} \Big|_0^\sigma, \qquad \sigma < 1. \tag{8.75}$$

Now, the apse A of the path corresponds to $du/d\varphi = 0$ and, hence, to $\sigma = 1$; see (8.71). Consequently, we have, from (8.75), over the first half of the path

$$\varphi = \frac{\pi}{2} + \frac{2kM}{c^2 R}; \tag{8.76}$$

the second term here gives the amount of bending caused up to the point A. In the second half of the path, φ will further increase by an amount equal to (8.76); see Fig. 8.2. It may be noted that in the latter case, $d\varphi/du$ would be negative but the limits of σ would also be reversed; so, the final answer would be the same as that given by (8.76). Thus, the net angle between the initial and final directions of the path would be $\pi + 4kM/(c^2 R)$, the net bending, Δ, being

$$\Delta = \frac{4kM}{c^2 R} = \frac{2r_0}{R}, \tag{8.77}$$

where r_0 is the SCHWARZSCHILD radius of the central body ; see (8.36).

For a ray of light which almost grazes past the limb of the Sun, the foregoing result becomes

$$\Delta_S = \frac{4k}{c^2} \frac{\text{solar mass}}{\text{solar radius}} = 1''.75. \tag{8.78}$$

This conclusion of the general theory of relativity has also been subjected to observational tests, and the findings provide a considerable scope for discussion ; however, before we enter into that it appears worthwhile to make some remarks on the emergence of the formula (8.77).

First of all, it may be noted that the derivation of the formula (8.77), as given here, is essentially due to FLAMM.[1] EINSTEIN's own derivation[2] was based on an application of the HUYGENS principle of secondary wavelets, which may also be understood in terms of the FERMAT principle of least time. According to the latter, the *actual* path of light would be determined by the variational condition

$$\delta \int dt = \delta \int \frac{dl}{c'} = 0, \tag{8.79}$$

where dt is the time taken by light in traversing the spatial distance dl, c' being the *local* speed of propagation. Now, if we write the metric of the four-dimensional continuum as

$$ds^2 = dl^2 - \left(1 - \frac{2kM}{c^2 r}\right) c^2 \, dt^2, \tag{8.80}$$

assuming dl^2 to be Euclidean, then the condition (8.59), which is characteristic of the propagation of light, gives

$$c' = \frac{dl}{dt} = c \left(1 - \frac{2kM}{c^2 r}\right)^{1/2} \tag{8.81}$$

$$\simeq c \left(1 + \frac{\varphi}{c^2}\right) ; \tag{8.82}$$

thus, the speed of light c' is a space variable ! Consequently, the propagation

---

[1] L. FLAMM, *Phys. ZS.* **17**, 448(1916).

[2] A. EINSTEIN, *Berlin Sitzungsberichte* (1915), p. 831 and *Ann. der Phys.* **49**, 769(1916), Sec. 22. English translation of the latter paper is available in *The Principle of Relativity*, loc. cit.

of light in a gravitational field corresponds to propagation in a refracting medium whose refractive index changes continuously along the path. The path would, therefore, be a curved one and may be traced quantitatively with the help of HUYGENS' principle. However, the spatial line element dl, which appears in the integrand of the variational integral (8.79), is itself non-Euclidean ; see Eq. (8.31). Therefore, in the application of HUYGENS' principle this must also be taken into account. When this is done, the net bending obtained agrees completely with the one derived above.

It must be mentioned here that EINSTEIN, in an earlier investigation[1], considered this problem from the point of view of the principle of equivalence alone. This principle also leads to the formula (8.82) for the variation of the speed of light with the field potential, and an application of HUYGENS' principle then leads to a finite bending of the path of light. However, the spatial geometry employed in that investigation was Euclidean ; thus, only one aspect of the space-time metric was actually accounted for. The result obtained thereby was exactly one half of that given by (8.77).[2]

SCHIFF[3] has demonstrated that the full result (8.77) can also be obtained

---

A. EINSTEIN, *Ann. der Phys.* **35**, 898(1911); English translation available in *The Principle of Relativity, loc. cit.*

[2] It appears quite obvious that if we regard the light beam as a stream of corpuscles, then even the Newtonian theory of gravitation would lead to a finite bending of the path of the beam. In this connection, we must remember that the resulting curvature, by the very nature of the gravitational field, would be independent of the mass of the corpuscles; it would, of course, depend on the initial velocity of the corpuscles, which is c, and on the apsidal distance R. The factor (kM) would enter through the strength of the field. The amount of bending would, then, be given by the inclination of the asymptotes of the hyperbolic path pursued by the corpuscles. A straightforward inverse-square calculation leads to a result exactly equal to the one obtained by EINSTEIN in 1911, i.e. one half of that given by (8.77). See Problem 8.1.

This argument appears to have been put forward first by J. SOLDNER, in 1801, who also carried out the calculation indicated above: *Berliner Astr. Jahrb.* (1804), p. 161; reprinted in *Ann. der Phys.* **65**, 593(1921).

One important difference must be noted here, namely that in the Newtonian calculation the velocity of a corpuscle would increase as it enters regions of stronger potential, while formula (8.81) leads to a corresponding decrease in c'; this difference is, however, well known to be characteristic of the two different aspects, corpuscular and undulatory, of the propagation of light. The sign of the deflection is, however, the same in the two cases.

[3] L. I. SCHIFF, *Am. J. Phys.* **28**, 340(1960). This paper also discusses the important question of the extent to which the three 'crucial' tests, dealt with in Secs. 8.3—8.5 of this text, can be considered as evidence for the full structure of the general theory of relativity. See also R. H. DICKE, *ibid.* **28**, 344(1960).

without going into the detailed structure of the general theory of relativity. According to him, the principle of equivalence alone is sufficient for this purpose; one has only to take into account the *first order* changes in the lengths of identically constructed rods as well as those in the periods of identically constructed clocks. In a way, this amounts to taking into account the non-Euclidean character of the spatial geometry. The calculation, however, is needed only in the *first* approximation and, hence, the principle of equivalence suffices.

We shall now discuss the findings of the various investigations carried out to verify the result (8.78). The procedure normally adopted for this purpose is to photograph the position of a group of stars in the (angular) vicinity of the Sun during the time of a total solar eclipse. The geometrical arrangement of the various positions on the photographic plate is then compared with the one expected in the absence of any influencing field, such as the gravitational field of the Sun, in the path of the light of these stars. The latter arrangement can, in principle, be obtained from catalogued star positions; however, the accuracy of these may not, in general, be sufficient for the purpose of the delicate determination under consideration. It is, therefore, advisable to take a similar photograph of the same star field *under the night sky*, i.e. six months before or after the eclipse. A comparison of the two photographs should then reveal a systematic displacement of the star positions, *away* from the central position (see Fig. 8.2), in accordance with the hyperbolic law

$$\Delta = 1''.75/x, \tag{8.83}$$

where x is the apsidal distance, measured in units of the solar radius.

Data from six total eclipses during the period 1919-52 have been analyzed in detail by MIKHAILOV[1], taking into account (i) the probable errors declared by the original investigators, (ii) the ones arrived at by later analysts, and (iii) the ones requiring the so-called *scale correction* (which is necessitated by slightly different focal settings, temperature effects, etc.). The final results of his analysis are summarized in Table 8.1.[2] MIKHAILOV thus arrived at the weighted mean 1''.93, to an accuracy of about 10%. Agree-

---

[1] A. A. MIKHAILOV, *Mon. Not. Roy. Astron. Soc.* **119**, 593(1959). See also H. V. KLÜBER, in *Vistas in Astronomy*, Vol. 3, ed. A. BEER (Pergamon Press, Oxford, 1960).

[2] Entries in this table have been extracted from a table prepared by MIKHAILOV, *loc. cit.*, and reproduced by M. G. ADAM in her paper: *Proc. Roy. Soc. London* A **270**, 297 (1962).

ment with the theoretical value is not perfect; however, there is every reason to believe that the effect does exist, is in the right direction and is not inconsistent with the theory. MIKHAILOV's analysis further shows that the hyperbolic dependence of Δ on x, as given by Eq. (8.83), though not in disagreement with observation, does require an unambiguous verification as yet.

TABLE 8.1 — DEFLECTION OF A RAY OF LIGHT TRAVERSING THE SOLAR FIELD

Eclipse	Number of stars observed	Limits of x	Observed deflection (against the theoretical value of 1".75)	
			Observer's mean value	New reduction
1919, May 29	7	2.0— 5.4	1".98[1]	2".07 ± 0".09
1922, Sep. 21	71	2.1 —13.0	1".72[2]	1".83 ± 0".11
1929, May 9	18	1.5— 7.5	2".24[3]	1".96 ± 0".08
1936, June 19	29	2.0— 7.2	2".70[4]	2".68 ± 0".37
1947, May 20	51	3.3 —10.2	2".01[5]	2".20 ± 0".38
1952, Feb. 25	11	2.1— 8.9	1".70[6]	1".43 ± 0".18

[1] F. W. DYSON, A. S. EDDINGTON and C. DAVIDSON, *Phil. Trans. Roy. Soc. London* A, **220**, 291 (1920).

[2] See R. J. TRUMPLER, Proceedings of the Congress : 'Jubilee of Theory of Relativity', Berne, 1955 ; reported in the *Helv. Phys. Acta*, Supp. IV, 106 (1956). A summary of TRUMPLER's report also appears in F. R. TANGHERLINI, *loc. cit.*, p. 63.

[3] E. FREUNDLICH, H. v. KLÜBER and A. V. BRUNN, *Zeits. für Astroph.* **3**, 171 (1931); *ibid.* **6**, 218 (1933) ; *ibid.* **14**, 242 (1937).

[4] See A. A. MIKHAILOV, *Astron. Z.* **33**, 912 (1956); this entry does not appear in TRUMPLER's report.

[5] G. VAN BIESBROEK, *Astron. J.* **55**, 49 (1950).

[6] G. VAN BIESBROEK, *Astron. J.* **58**, 57 (1953).

Corroborative evidence has come recently from observations of two radio sources, 3C273 and 3C279, during their occultation by the Sun in October 1969. Measuring the deflection of the 9.602-GHz radiation from the radio source 3C279, SEIELSTAD, SRAMEK and WEILER[1] obtained: $\Delta = 1".77 \pm$

[1] G. A. SEIELSTAD, R. A. SRAMEK and K. W. WEILER, *Phys. Rev. Lett.* **24**, 1373 (1970).

0".20. Studying phase differences between the two sources, using 2.388-GHz radiation, MUHLEMAN, EKERS and FOMALONT[1] obtained: $\Delta = 1".82 + 0".20$.

A phenomenon closely related to the one discussed in this section is the so-called fourth test of the general theory of relativity. We propose to discuss that in Sec. 8.8.

8.5. Rates of clocks in a gravitational field. Shift in the spectral lines

We shall now consider another test of the theory, which concerns with the retardation of clocks in a gravitational field. In Sec. 6.1, we came across a similar effect, arising in a *noninertial* system of reference; from there, one could arrive, through an application of the principle of equivalence, at the corresponding effect arising in a gravitational field.[2] We, however, prefer to base our considerations on the space-time metric

$$ds^2 = g_{ik} \, dx^i \, dx^k. \tag{8.84}$$

As usual, the first three coordinates of the continuum determine the location of an event, while the fourth one refers to the time of its occurrence; $(x^i) \equiv (x^\alpha, ct)$. It may be specifically mentioned that the allotment of coordinates to the various events is understood to have been carried out in a manner which, though arbitrary, is unambiguous and satisfies the conditions of uniqueness and continuity, characteristic of the Gaussian coordinates.

Now, consider a standard clock C_1 stationed at the space point A_1 (x_1^α). Its line element would be given by, see (8.84),

$$ds_1^2 = (g_{44})_{A_1} (c \, dt)^2 \,; \tag{8.85}$$

hence, the corresponding 'proper' time interval $d\tau_1$ recorded by the clock would be, see Sec. 2.4,

$$d\tau_1 = \frac{ds_1}{ic} = \sqrt{-(g_{44})_{A_1}} \, dt. \tag{8.86}$$

---

[1] D. O. MUHLEMAN, R. D. EKERS and E. B. FOMALONT, *Phys. Rev. Lett.* 24, 1377 (1970).

[2] See, for instance, W. PAULI, *loc. cit.*, Sec. 53 (β).

Next, consider a similar clock C_2 stationed at the space point A_2 (x_2^α). Its line element would be given by

$$ds_2^2 = (g_{44})_{A_2} \ (c \ dt)^2 ; \tag{8.87}$$

hence, the corresponding 'proper' time interval $d\tau_2$ recorded by this clock would be

$$d\tau_2 = \frac{ds_2}{ic} = \sqrt{-(g_{44})_{A_2}} \ dt . \tag{8.88}$$

Thus, to *one and the same* interval dt of the coordinate time t there correspond *different* 'proper' time intervals recorded by clocks stationed at space points with different values of the element g_{44} ; we have

$$\frac{d\tau_1}{d\tau_2} = \frac{\sqrt{-(g_{44})_{A_1}}}{\sqrt{-(g_{44})_{A_2}}}. \tag{8.89}$$

Further, let us consider an atomic process, such as the emission of light, taking place at the point A_1, the 'proper' frequency associated with the process being ν_1. Then, during an interval $d\tau_1$ of the 'proper' time relevant to the point A_1, $(\nu_1 \ d\tau_1)$ pulses will be produced. These pulses will be received at the point A_2 during the corresponding interval $d\tau_2$ of the 'proper' time relevant to the point A_2. Accordingly, the frequency of the pulses received at the point A_2 will be recorded as

$$\nu_2 = \frac{\nu_1 \ d\tau_1}{d\tau_2} = \nu_1 \frac{\sqrt{(-g_{44})_{A_1}}}{\sqrt{(-g_{44})_{A_2}}}. \tag{8.90}$$

However, an *identical* atomic process taking place at A_2 should have, in terms of the proper time τ_2, the exact frequency ν_1. Hence, the observed frequency ν_2 of the light emitted *elsewhere* differs from the frequency ν_1 of the same light emitted *locally*, according to the relation

$$\frac{\nu_2}{\nu_1} = \frac{\sqrt{(-g_{44})_{A_1}}}{\sqrt{(-g_{44})_{A_2}}} . \tag{8.91}$$

In the correspondence limit, we can write these results as, see (7.12),

$$\frac{\nu_2}{\nu_1} = \frac{d\tau_1}{d\tau_2} \simeq 1 + \frac{\varphi_1 - \varphi_2}{c^2}, \tag{8.92}$$

so that

$$\frac{\Delta \nu}{\nu_1} = \frac{\nu_2}{\nu_1} - 1 \simeq \frac{\varphi_1 - \varphi_2}{c^2}; \qquad (8.93)$$

here, φ_1 and φ_2 are the Newtonian gravitational potentials at the points A_1 and A_2, respectively, while $\Delta \nu$ is the *Einstein shift* of the spectral line observed. It is clear that if the field at the point of emission is stronger than the field at the point of reception, i.e. if φ_1 is more negative than φ_2, then $d\tau_1 < d\tau_2$ and $\nu_2 < \nu_1$. Thus, the presence of a gravitational field (in comparison with the case when it is absent) causes a *retardation* of the clocks and a consequent *reddening* of the spectral lines. The EINSTEIN shift is, therefore, commonly referred to as the *gravitationl red shift*. Of course, a spectral line emitted at a place of weaker potential would, if observed at a place of stronger potential, show a shift towards the *blue*; it is, therefore, advisable to refer to this shift as simply the *gravitational shift*.

As already remarked, the foregoing results can also be obtained by considering the phenomenon first in an accelerated system of reference and then translating the resulting formulae, with the help of the principle of equivalence, to the case of the corresponding gravitational field; it was precisely in this manner that EINSTEIN, starting with a uniformly accelerated system of reference, obtained these results in 1911. It must be noted, however, that such a derivation would never give exact formulae, such as (8.89) and (8.91); one would obtain only approximate results, such as (8.92) and (8.93). The reason for this lies in the fact that the validity of the application of the principle of equivalence is rather limited; the *local* character of this validity, in the case of an *inhomogeneous* gravitational field (see Sec. 7.2), restricts the application of this principle to the lowest approximation alone, in which approximation the results (8.92—8.93) hold.

It is interesting to note that the first-order result (8.93) can be obtained, rather straightforwardly, by considering the energy content of a photon moving in a gravitational field. The photon carries an inertial, and hence also a gravitational, mass given by the expression $h\nu/c^2$, ν being the associated frequency. Consequently, its passage from a point where the gravitational potential is φ_1 to a point where the potential is φ_2 will entail an expenditure of work given by $h\nu/c^2$ times the potential difference $(\varphi_2 - \varphi_1)$. This will result in a decrease in the energy content of the photon and, hence, a decrease in its frequency. The latter will be given by

$$\nu_1 - \nu_2 = \frac{E_1 - E_2}{h} = \nu \, \frac{\varphi_2 - \varphi_1}{c^2}, \qquad (8.94)$$

in agreement with (8.93) ; here, ν may be regarded as the average frequency of the photon.

The foregoing considerations make it rather questionable whether the observational (or experimental) verification of the above formula for the frequency shift in the spectral lines could be regarded as an evidence in favour of the basic formalism of the *general* theory of relativity!

Any way, we shall now discuss the results of the various investigations that have been carried out in order to verify these results. To start with, let us have an estimate of the magnitude of the expected effect. For the light emitted at the surface of the Sun and observed at the surface of the Earth,

$$\frac{\Delta\nu}{\nu} \simeq \frac{\varphi_S}{c^2} = -2.12 \times 10^{-6} ; \qquad (8.95)$$

here, φ_S denotes the potential at the surface of the Sun while φ_E , the potential at the surface of the Earth, has been neglected in comparison. It is customary to express this shift in terms of an *effective* velocity which would cause an equivalent Doppler shift ; in the present case, this turns out to be

$$v_{\text{eff}} \simeq c \frac{\Delta\nu}{\nu} \simeq -0.635 \text{ km/sec.} \qquad (8.96)$$

It is evident that mass motions in the solar atmosphere, which are, in general, of the same order of magnitude as (8.96), will complicate the issue of interpreting the observed shifts properly. Also, there are other disturbing effects, such as the pressure effects,[1] the radial outflow,[2] the line asymmetries[3] (which are probably of a chromospheric origin), etc., with the result that the relativistic contribution to the observed shift of the spectral lines is very difficult to sort out. It has, therefore, been suggested[4] that simple wavelength and shift measurements be replaced by detailed contour determinations of the lines so that one may be able to interpret the observational data discriminately.

Nevertheless, BLAUMONT and RODDIER[5] have reported a determination of the solar red shift in strontium with considerable accuracy ; they employed

---

[1] L. E. JEWELL, *Astroph. J.* 3, 89 (1896).

[2] C. E. ST. JOHN, *Astroph. J.* 67, 195 (1928).

[3] L. A. HIGGS, *Mon. Not. Roy. Astron. Soc.* 121, 421 (1960) ; 124, 51 (1962).

[4] M. G. ADAM, *loc. cit.*

[5] J. E. BLAUMONT and F. RODDIER, *Phys. Rev. Letters* 7, 437 (1961).

sunlight to excite secondary emission from an atomic beam of strontium. Within the limits of experimental error, their results appear to be in agreement with the theoretical ones. Moreover, at the limb of the Sun they found a shift rather larger than the theoretical value ; following previous authors, they attribute the excess to a pressure shift.

In regard to stellar lines, greater effect is expected in the case of stars with higher M/R ratio. As first pointed out by EDDINGTON, this is to be found in the case of white dwarfs, the most promising case being that of Sirius B, the companion of the bright star Sirius A. Here, the expected shift is about 130 times the solar one, corresponding to an effective Doppler velocity of about 83 km/sec. Earliest investigations in this connection were carried out by ADAMS.[1] whose results are often quoted as having verified the prediction satisfactorily. However, detailed analyses have tended to show that those tests on Sirius B were rather inconclusive ;[2] the major trouble here arises because of the proximity of the very bright star Sirius A, whose spectrum overlies that of the star under examination. Recently, GREENSTEIN, OKE and SHIPMAN[3] have reported a very accurate determination of the gravitational red shift of the $H\alpha$ and $H\gamma$ line profiles for Sirius B. They find an effective Doppler velocity of 89 ± 16 km/sec, in fair agreement with the value predicted by the theory.

An investigation, similar to the one discussed in the preceding paragraph, was initiated by POPPER[4] who carried out measurements on the star 40 Eridani B, the second brightest white dwarf ; the expected result in this case is about 17 km/sec. POPPER estimated that the ratio of the observed shift to the expected one lay in the interval 1/2 to 2, which could not be regarded as a satisfactory verification of the relativity formula. In this case again we have a recent report, of WIESE and KELLEHER,[5] according to which the profiles of the Stark-broadened Balmer lines from 40 Eridani B show a systematic shift towards the red ; the gravitational part of this shift corresponds to an effective Doppler velocity of 18 ± 4 km/sec.

GREENSTEIN and TRIMBLE[6] have carried out a detailed investigation of the gravitational red shift for a group of about forty white dwarfs, whose mean

---

[1] W. S. ADAMS, *Proc. Nat. Acad. Sci.*, Washington **11**, 382 (1925).

[2] See M. G. ADAM, *loc. cit.*

[3] J. L. GREENSTEIN, J. B. OKE and H. L. SHIPMAN, *Ap. J.* **169**, 563 (1971).

[4] D. M. POPPER, *Astroph. J.* **120**, 316 (1954).

[5] W. L. WIESE and D. E. KELLEHER, *Ap. J.* **166**, L59 (1971).

[6] J. L. GREENSTEIN and V. L. TRIMBLE, *Ap. J.* **149**, 283 (1967) ; see also *ibid.* **177**, 441 (1972).

radius is about 0.01 $R_\odot$ and mean mass about 1 $M_\odot$; the expected shift, therefore, is about 100 times larger than the solar shift, which corresponds to an effective Doppler velocity of about 63.5 km/sec. The range observed by GREENSTEIN and TRIMBLE is 55—65 km/sec.

TRUMPLER, in his report, *loc. cit.*, refers to a series of red-shift measurements carried out on a number of stars of spectral class 0 ($M \approx 100$ $M_\odot$, $R \approx 5$ $R_\odot$) and concludes that on an average the data thus obtained are in reasonable agreement with the theory.[1] It must, however, be said that, apart from the large experimental errors declared, the question of physical interpretation of the observed shifts is, in general, not so straightforward; the evidence, therefore, remains somewhat ambiguous.

8.6.   Terrestrial measurements of the Einstein shift

It is obvious that the frequency shifts, in the spectral lines, caused by the gravitational field of the Earth would be far less in magnitude than those caused by stellar fields. For instance, a level difference of h metres near the surface of the Earth would result in a fractional shift given by, see (8.93),

$$\frac{\Delta\nu}{\nu} \simeq \frac{gh}{c^2} = 1.09 \times 10^{-16} h; \qquad (8.97)$$

thus, for a level difference of 50 m we expect a fractional shift of 5.45×10^{-15}, which is only a few billionths of the corresponding shift expected in the case of spectral lines emitted at the surface of the Sun and observed at the surface of the Earth. Accordingly, the question of measuring EINSTEIN shift *terrestrially* never arose until a remarkable discovery by MÖSSBAUER, known as the MÖSSBAUER effect, made it possible to use certain radioactive nuclei as extremely accurate 'clocks', thus shifting the relevant field of activity from (observational) optical spectroscopy to (experimental) nuclear spectroscopy. Within a year or two of this discovery, a number of experiments were performed which showed, rather conclusively, that the EINSTEIN shift did exist and was in agreement with the theoretical prediction (8.97), both in sign and magnitude.

MÖSSBAUER[2] discovered in 1958 that some of the low-energy gamma radia-

---

[1] A summary of these results also appears in F. R. TANGHERLINI, *loc. cit.*, p. 21.

[2] R. L. MÖSSBAUER, *Z. Physik* **151**, 124 (1958) ; *Naturwissenschaften* **45**, 538 (1958) ; *Z. Naturforsch.* **14a**, 211 (1959).

For an account of the MÖSSBAUER effect, especially from the point of view of its relevance to the EINSTEIN shift, see O. R. FRISCH, *Contemporary Physics* **3**, 194 (1961-62).

tion (in the range of about 10^4 eV) emitted by long-lived 'isomeric' states of nuclei, with life times of the order of $10^{-7}-10^{-8}$ sec, were practically *recoil-free*; the recoil momentum in their case was taken up by the crystal *as a whole*, with the result that there was no significant Doppler shift and hence no consequent line broadening.[1] One could thus obtain in the laboratory *extremely sharp* gamma rays—in fact, so sharp that their resonance on absorption by another piece of the same substance could be destroyed, as was demonstrated by MÖSSBAUER himself, by a relative speed of the order of 1 cm/sec between the source and the absorber, i.e. by the introduction of a fractional frequency displacement of the order of 10^{-10}, or even 10^{-11}.

Soon after MÖSSBAUER'S discovery, a number of authors, especially POUND and REBKA[2] and CRANSHAW,[3] suggested that the recoil-less gamma radiation may be employed to measure the gravitational shift *terrestrially*. Subsequently, POUND and REBKA[4] investigated quantitatively the resonant absorption of the 14.4 keV gamma radiation emitted by 10^{-7} sec Fe^{57}, with a view to assessing their suitability for the purpose of the proposed measurements. They found, by studying the variation of the intensity of absorption as a function of the relative velocity between the source and the absorber, that the resonant absorption of these radiation is halved by a Doppler speed of about 0.017 cm/sec, i.e. by the introduction of a fractional frequency displacement of about 5.7×10^{-13}! Now, the terrestrial gravitational shift arising from a level difference of 50 m has already been seen to be about 5.45×10^{-15}, which is almost one per cent of the half line width of the radiation considered. In view of the accuracy obtainable in nuclear spectroscopy, the proposed experiments seemed quite feasible, even promising.

The first practical report in this connection came from CRANSHAW,

---

[1] It may be noted that the recoil momentum may either excite lattice vibrations or be transferred as linear momentum to the entire crystal; in the former case, there will be a Doppler broadening because of the recoil of the emitting nucleus. The criterion that lattice vibrations are not excited and, hence, the recoil momentum is taken by the crystal as a whole may be stated as

$$p^2/(2m) \ll k\theta,$$

where $p(=h\nu/c)$ is the magnitude of the recoil momentum, m the mass of the recoiling nucleus, k the BOLTZMANN constant and θ the DEBYE temperature of the crystal. For a proof of this criterion, see J. WEBER, *General Relativity and Gravitational Waves* (Interscience, New York, 1961), p. 62.

[2] R. V. POUND and G. A. REBKA, JR., *Phys. Rev. Letters* 3, 439 (1959).

[3] See J. P. SCHIFFER and W. MARSHALL, *Phys. Rev. Letters* 3, 556 (1959).

[4] R. V. POUND and G. A. REBKA, JR., *Phys. Rev. Letters* 3, 554 (1959).

SCHIFFER and WHITEHEAD[1] who employed the same radiation as discussed above and worked with a height difference of 12.5 m ; the expected fractional shift was now 1.36×10^{-15}. Computing the relative frequency shifts, between the emission and absorption lines, from the observed counting rate differences, these authors concluded that the ratio of the observed (gravitational) shift to the expected one was $+0.96 \pm 0.45$. Doubts were cast on this result when POUND and REBKA[2] pointed out that a shift of the same order of magnitude could also be produced by a temperature difference of less than 1°C, actually about 0.6°C, between the source and the absorber ; that it could be so was pointed out independently by JOSEPHSON[3] as well.

These authors[2,3] noted that the afore-mentioned temperature effect is a consequence of the relativistic time dilation caused by the motion of the nuclei (in the crystal lattice) due to thermal vibrations. If v_0 is the mean frequency of the gamma ray when the emitting nucleus is at rest in an inertial system of reference while Δv_T is the shift in the frequency of the ray when the emitting nucleus is in thermodynamic equilibrium with the lattice at $T°K$, then

$$\frac{\Delta v_T}{v_0} = \left(1 - \frac{\langle v^2 \rangle}{c^2}\right)^{1/2} - 1$$

$$\simeq -\frac{1}{2}\frac{\langle v^2 \rangle}{c^2} \simeq -\frac{3}{2}\frac{kT}{mc^2};\qquad (8.98)$$

The approximations are valid only if $\langle v^2 \rangle$, the mean square velocity of a nucleus (of mass m) due to thermal vibrations, is much smaller than c^2 and, at the same time, the temperature of the crystal is high enough for the equipartition law to hold. Accordingly, the fractional shift in the resonant frequency of absorption would be

$$\frac{\Delta v}{v_0} \simeq -\frac{3}{2}\frac{k\,(T_s - T_a)}{mc^2}$$

$$= -2.44 \times 10^{-15}\,\Delta T,\qquad (8.99)$$

where $\Delta T(= T_s - T_a)$ is the difference between the temperatures, in K, of

[1] T. E. CRANSHAW, J. P. SCHIFFER and A. B. WHITEHEAD, *Phys. Rev. Letters* **4**, 163 (1960).

[2] R. V. POUND and G. A. REBKA, JR., *Phys. Rev. Letters* **4**, 274 (1960).

[3] B. D. JOSEPHSON, *Phys. Rev. Letters* **4**, 341 (1960).

the source and the absorber while for m we have substituted the mass of an Fe^{57} nucleus.

POUND and REBKA, however, indicated that for the case under study, the DEBYE temperature being 420 °K, the equipartition law would not apply at room temperatures; actually, the numerical coefficient in (8.99) would be smaller by a factor of about 1.1, leading to the theoretical result

$$\frac{\Delta v}{v_0} = -2.21 \times 10^{-15} \Delta T; \qquad (8.100)$$

cf. the corresponding fractional shift sought by CRANSHAW, SCHIFFER and WHITEHEAD. Experimental observations gave the value $(-2.09 \pm 0.24) \times 10^{-15}$ per K for the foregoing coefficient at room temperature. Thus, the second-order Doppler shift provides a satisfactory explanation of the temperature dependence of the mean frequency of the Fe^{57} recoil-less gamma resonance absorption; in other words, this experiment provides another convincing evidence for the phenomenon of relativistic time dilation.

Having studied the temperature effect thoroughly, POUND and REBKA returned to the original problem of measuring the gravitational shift;[1] obviously, they were now in a position to take into account the 'spurious' shift caused by the temperature difference between the source and the absorber, thus taking a lead over the earlier findings of CRANSHAW, SCHIFFER and WHITEHEAD. Moreover, they took into account the frequency shift inherent in a particular source-absorber combination; this was accomplished by repeating various sets of observations with the positions of the source and the absorber interchanged, and then combining the results of all these sets. With a level difference of 74 ft, the expected fractional shift, for a two-way passage, was about 4.91×10^{-15}. Experimental data yielded a net fractional shift of $(5.13 \pm 0.51) \times 10^{-15}$, i.e. about (1.04 ± 0.10) times the expected value.

In the subsequent years, both groups of workers have made significant improvements on their previous results. While CRANSHAW and SCHIFFER[2] have reported an observed shift of 0.859 ± 0.085 times the value predicted by the theory of relativity, POUND and SNIDER[3] have reported an observed shift of 0.999 ± 0.008 times the predicted value. Thus, the EINSTEIN shift of

[1] R. V. POUND and G. A. REBKA, JR., *Phys. Rev. Letters* **4**, 337(1960).

[2] T. E. CRANSHAW and J. P. SCHIFFER, *Proc. Phys. Soc. London* **84**, 245(1964).

[3] R. V. POUND and J. L. SNIDER, *Phys. Rev.* **140**, B788(1965).

(nuclear) spectral lines, caused by the gravitational field of the Earth, has been verified to a very high degree of accuracy.

8.7.   The clock paradox

In his very first paper on relativity, EINSTEIN[1] pointed out a rather puzzling consequence of the phenomenon of time dilation which led to an apparent paradox, known as the *clock paradox*.   Suppose there are two identical standard clocks, A and B, such that the former is permanently at rest in an inertial system of reference S, while the latter undertakes a *round* trip, with a time-dependent velocity $\mathbf{v}(t)$, the initial and final positions of the trip being coincident with the fixed position of the former.   Now, if the rest clock, A, indicates a time interval $t\,(= t_2 - t_1)$ for the duration of the foregoing trip, the travelling clock, B, would be expected to indicate, for the same duration, a shorter time interval t', as given by the relativistic formula

$$t' = \int_{t_1}^{t_2} \sqrt{1 - \{v(t)/c\}^2}\, dt. \tag{8.101}$$

The primary reason for the predicted asymmetry between the recordings of the two clocks lies in the fact that while the clock A has remained throughout at rest, the clock B has been roving in space (with the result that it has been continually slow in comparison with the clock A).   It should be noted that the fact of the clock B having undergone accelerations during the trip does not show up *quantitatively* in the expression (8.101); it is only the relative speed of the clocks that appears decisive.

Now suppose that the whole situation is viewed from the system of reference S' in which the clock B is permanently at rest while the clock A undertakes the round trip, the two clocks again coming together at the end of the trip.   Then, the 'travelling' clock A should be losing time (in comparison with the 'rest' clock B)!   The situation is clearly paradoxical, for it seems difficult to say from first principles as to which of the two clocks, on reuniting, would *really* indicate lesser time.

Now, the difference in the readings of any two clocks *located at one and the same point in space* is, undoubtedly, an absolute and objective fact; it should depend neither on the system of reference adopted nor on the method of description.   Hence, the question of the relative measures of the duration

---

[1] A. EINSTEIN, *Ann. der Phys.* 17, 891(1905); English translation available in *The Principle of Relativity, loc, cit,*

of a given time interval, as recorded by two clocks in different states of motion (but coincident both initially and finally), must have an unambiguous answer, *independent* of the choice of the reference system.

Considering the matter a little more closely, one finds that the paradox here arises from the fact that we are, quite wrongly, regarding the systems S and S' as *literally* equivalent; actually, whereas the system S is, by assumption, inertial, the system S' is necessarily noninertial.[1] Consequently, the customary time dilation formula—the one on which Eq. (8.101) is based—is not applicable in S'. On the contrary, one has to employ, for a calculation in S', a modified time dilation formula, see (8.102) below, in which the influence, on the rates of clocks, caused by the field of force arising from the accelerated motion of the system is *also* taken into account. When that is done, the conclusion reached in S' is the same as the one reached in S. In fact, one can say, quite generally, that the time gained by the clock A in comparison with the clock B, by virtue of the field of force prevailing in system S', is such that, after compensating for the loss incurred by this clock (because of its relative motion in this system), it still has the predicted lead over the clock B. There is, thus, no paradox in this problem.

The foregoing argument would not cause any surprise or suspicion if we kept in mind that a standard clock always records its 'proper' time τ, which is an *invariant*; consequently, the ratio of the recordings of two standard clocks would be the same in *all* systems of reference. Difference between various systems lies *only* in the manner of computation of these recordings and *not* in the final outcomes! For instance, the 'proper' time interval $d\tau$ associated with the line element (dx^i) is given by

$$d\tau = \frac{ds}{ic} = \frac{1}{ic} (g_{ik} \, dx^i \, dx^k)^{1/2}$$

$$= dt \left[-g_{44} - 2g_{\alpha 4} \left(\frac{u^\alpha}{c} \right) - g_{\alpha\beta} \left(\frac{u^\alpha u^\beta}{c^2} \right) \right]^{1/2}; \qquad (8.102)$$

here, $u^\alpha \, (= dx^\alpha / dt : \alpha = 1, 2, 3)$ are the velocity components of the motion concerned while g_{ik} are the components of the metric tensor appropriate to the system of reference adopted. In an inertial system, g_{ik} have their normal values (6.12); accordingly, (8.102) reduces to the conventional time dilation formula

$$d\tau = dt \left[1 - \frac{u^2}{c^2} \right]^{1/2}. \qquad (8.103)$$

---

[1] This was specifically emphasized by EINSTEIN in *Naturwissenschaften* **6**, 697(1918).

In a noninertial system of reference, the g_{ik} are expected to be different from the normal values ; it is, then, necessary to employ the full formula (8.102), and not the ordinary one (8.103). The final result for $d\tau$ would be the same in each case.

Turning to *finite* time intervals, we note that the measure τ associated with two arbitrary world points 1 and 2 depends critically on the choice of the

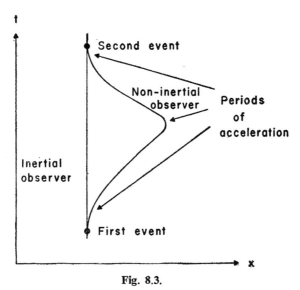

Fig. 8.3.

world line joining the two points ; of course, *it would still be independent of the choice of the system of reference.* Figure 8.3 depicts the situation obtaining in this problem. The world points here are those corresponding to the *initial* and *final* instants of the trip ; the world line of the clock A is parallel to the t-axis of system S, indicating a state of relative rest, while that of the clock B is curved, indicating a state of relative motion involving periods of acceleration during the trip. Now, we know, see Sec. 2.4, that the 'proper' time interval associated with a linear track joining two given world points 1 and 2 is greater than that for any other track joining these points ; accordingly, the recording of the clock A would be greater than that of any other clock. And, because of the *invariance* of the measures involved, the result would be valid, quantitatively as well as qualitatively, for *any* observer whatsoever.[1]

---

[1] A typical case has been worked out in detail by C. MOLLER, *loc. cit.*, Sec. 98. An instructive, first-order calculation can be found in O. R. FRISCH, *loc. cit.*, pp. 200-201,

We thus arrive at the conclusion that it is the state of motion (of a clock) with respect to an *inertial* system of reference that decides its *real* rate (of showing time) unambiguously, though the computation thereof may be carried out in any system of reference. Accordingly, it can be stated, with IVES,[1] that of several clocks in relative motion the one which runs fastest has the lowest root-mean-square velocity with respect to an *inertial* system! This, then, provides a novel method of determining the average speed of a clock, or of any other suitable object, undergoing a round trip, viz. by comparing the temporal recording of the given clock with the one which, with respect to an *inertial* system of reference, has remained stationary during this period.[2] It is remarkable that such a determination does not involve any distance measurements relating to the path of the travelling clock (nor any kinetic energy measurements or mass measurements)!

We note that one of the most striking features of the foregoing results is the inherent *asymmetry* in the temporal measures of various clocks undergoing *round* trips. By contrast, for clocks undergoing *uniform translational* motion, thus ruling out the possibility of a face-to-face comparison of the time intervals shown by different clocks, the time rates are, no doubt, symmetrical, for observers in *any* system of reference find the clocks of the other systems to be running slower. To some authors, notably DINGLE,[3] the foregoing symmetry appears to be more appealing than anything else, with the result that they expect absolutely symmetrical results—irrespective of the nature of the motions involved. Thus, in the case of the clock problem under discussion here, these authors expect that, on re-uniting, the two clocks would not display any discordance in their recordings of the duration of the trip. In other words, they wish to resolve the paradox by asserting that the two clocks would indicate the same time interval.

However, this position, which arises from a literal belief in what may be called 'absolute relativism', does not appear to be tenable. Some of the recent experiments have yielded results which can be interpreted to have verified that a *real asymmetry in the behaviour of clocks undergoing round trips does indeed exist*! For instance, the results obtained by POUND and

---

[1] H. E. IVES, *J. Opt. Soc. America* **27**, 305(1937).

[2] See, e.g., C. W. SHERWIN, *Phys. Rev.* **120**, 17(1960).

[3] H. DINGLE, *Nature* **144**, 888 (1939); *Proc. Phys. Soc., London* A **69**, 925 (1956); *Nature* **180**, 1275 (1957). M. SACHS, *Physics Today* **24**, No. 9, 23 (1971); see also the ensuing correspondence in *Physics Today* **25**, No. 1, 9 (1972).

REBKA,[1] on the temperature dependence of the frequency shift for the recoil-less gamma radiation, show that clocks moving with a larger mean square velocity run slower than the ones moving with a smaller mean square velocity. Now, since during thermal motions the mean distance between the emitting and absorbing nuclei remains unchanged, the situation essentially corresponds to the one visualized in the clock problem; of course, here both clocks A and B undertake a large number of round trips.[2]

Mention may also be made of a similar experiment conducted by HAY, SCHIFFER, CRANSHAW and EGELSTAFF[3] who attached the source and the absorber of the gamma radiation to a *rotating* wheel, the former near the centre and the latter around the periphery. In this case again, the distance between the source and the absorber remained fixed during the experiment; however, they undertook a large number of round trips (with different mean square velocities). A frequency shift, in consonance with the theoretical expectation, was observed.

A dramatic confirmation of the foregoing results has been reported by HAFELE and KEATING who, during October 1971, flew four cesium beam clocks on regularly scheduled commercial jet flights around the world—once eastward and once westward. For the actual flight paths of each trip, the theory predicted that the flying clocks, as compared with the reference clocks at the U. S. Naval Observatory (which remained stationary with respect to the *rotating* frame of the Earth), would lose 40 ± 23 nanoseconds during the eastward trip and would gain 275 ± 21 nanoseconds during the westward trip.[4] Carrying out the desired comparison, after the flying clocks had completed their round trips and were re-united with the reference clocks, HAFELE and KEATING found[5] that the flying clocks indeed lost 59 ± 10 nanoseconds in the eastward trip and gained 273 ± 7 nanoseconds in the westward trip.

---

[1] R. V. POUND and G. A. REBKA, JR., *Phys. Rev. Letters* **4**, 274 (1960): discussed, in greater detail, in the preceding section.

[2] Compare this experiment with those of IVES and STILWELL (mentioned in Sec. 4.2), which, no doubt, were concerned with a measurement of the second-order Doppler effect but in which a continuous separation of the light source and the analyzer took place.

[3] J. J. HAY, J. P. SCHIFFER, T. E. CRANSHAW and P. A. EGELSTAFF, *Phys. Rev. Letters* **4**, 165 (1960). For a more precise verification, see W. KÜNDIG, *Phys. Rev.* **129**, 2371 (1963).

[4] J. C. HAFELE and R. E. KEATING, *Science* **177**, 166 (1972). For details of the theoretical derivation, see J. C. HAFELE, *Nature* **227**, 270 (1970); *Nature Phys. Sci.* **229**, 238 (1971); *Am. J. Phys.* **40**, 81 (1972). See also R. K. PATHRIA, *Nature* **241**, 263 (1973) and R. SCHLEGEL, *Nature* **242**, 180 (1973).

[5] J. C. HAFELE and R. E. KEATING, *Science* **177**, 168 (1972).

Clearly, these results provide an unambiguous empirical resolution of the clock "paradox" using *macroscopic* clocks.

One important aspect of the clock problem appears worthy of a special emphasis, i.e. the magnitude of the asymmetry among the clocks *A* and *B* is determined *solely* by the mean square velocity during the trip and not by the accelerations involved.[1] In other words, the loss of time incurred by the roving clock, though arising from the fact that this clock is the one that undergoes 'real' accelerations during the trip, is not quantitatively dependent on the strengths and durations of these accelerations; it is determined solely by the (average) speed of motion. Thus, mere relative motion of two clocks is not sufficient for determining their relative time-reckoning behavior; it is of primary importance to know their *individual* states of motion with respect to an *inertial* system of reference or, in other words, with respect to the rest of the universe.[2]

Finally, we touch upon a question which has been the centre of much of the recent controversy on the clock paradox, viz. the extent to which the formalism of the general theory of relativity is essential for the resolution of the paradox. This question has been discussed in a large number of papers, many of which appeared during the years 1957-60. Clearly, it is not possible here to dwell on the considerations of the various authors one by one; it appears sufficient to summarize the final position regarding this issue and refer the more interested reader to the relevant literature.[3]

Now, the very fact that the problem does not involve *permanent* gravitational fields, as produced by massive bodies, implies that the full-fledged formalism of general relativity is apparently unnecessary. Of course, the introduction of a non-Euclidean metric, appropriate to the noninertial system *S'*, does make the computation of the various time intervals elegant and, so to say, straightforward, with the result that most of the authors, especially text-book writers, prefer to handle the problem this way. In

---

[1] This position is strikingly upheld by experiments in which the nuclear clocks move under high, incessant accelerations but their timings remain apparently unaffected by these; for details, see C. W. SHERWIN, *loc. cit.*

[2] This point has been discussed at length by G. BUILDER, *Australian J. Phys.* **11**, 279 (1958). See also H. E. IVES, *loc. cit.*

[3] A useful reference, along with that of C. W. SHERWIN, *loc. cit.*, is the one due to J. TERRELL: *Nuovo Cimento* **16**, 457 (1960). These two papers, together with L. MARDER, *Time and the Space Traveller* (George Allen and Unwin, London, 1971), provide a fairly complete covering of the literature on the clock paradox.

principle, however, a computation based on the special theory of relativity (including, of course, the Doppler effect) is indeed possible.[1]

8.8. Recent developments

A. *The fourth test of general relativity*

In 1964 SHAPIRO[2] suggested a new test of the general theory of relativity which consists in comparing the time it takes for a radar signal to make a round trip to an inner planet, Mercury or Venus, when the signal on its way passes close to the Sun, with the time it takes when the signal is practically unaffected by the solar field. The excess time taken in the former case is due to the delay caused by (i) the slowing down of standard clocks in a gravitational field, and (ii) the modified space geometry near the central mass. The net delay in the reception of an echo turns out to be, see Problem 8.13,

$$\Delta t \simeq (4kM/c^3) \ln [(r_e + r_p + x)/(r_e + r_p - x)], \qquad (8.104)$$

where r_e is the distance between the Earth and the Sun, r_p is the distance between the planet and the Sun while x is the distance between the Earth and the planet ; see Fig. 8.4. The factor $(4kM/c^3)$, in the case of the solar field, is about 20 μs. The time delay is *minimum* at 'inferior conjunction' when the planet is between the Earth and the Sun (i.e. $x = r_e - r_p$) ; one then obtains

$$\Delta t \simeq (4kM/c^3) \ln (r_e/r_p), \qquad (8.105)$$

which is of the order of 10 μs. The delay is maximum when the signal grazes past the limb of the Sun (i.e. $x \simeq \sqrt{r_p^2 - R^2} + \sqrt{r_e^2 - R^2}$, R being the radius of the Sun) ; one then obtains

$$\Delta t \simeq (4kM/c^3) \ln (4r_e r_p/R^2), \qquad (8.106)$$

---

[1] For details of such an approach, see G. BUILDER, *Australian J. Phys.* **10**, 226, 424 (1957) ; *Phil. Sci.* **26**, 135 (1959). See also J. TERRELL, *loc. cit.*, and A. P. FRENCH, *Special Relativity* (Norton, New York, 1968), pp. 154-159.

[2] L. I. SHAPIRO, *Phys. Rev. Lett.* **13**, 789 (1964) ; *Phys. Rev.* **141**, 1219 (1966) ; *Phys. Rev.* **145**, 1005 (1966).

which is of the order of 200 μs.

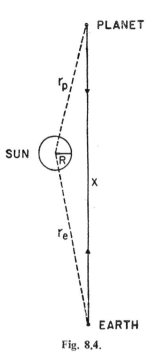

Fig. 8.4.

SHAPIRO and coworkers have carried out this test using both Mercury and Venus as their target. Preliminary results,[1] reported in 1968, verified the prediction of the general theory of relativity with an experimental uncertainty of about 20%. Newer observations[2] have reduced the level of uncertainty to about 2%.

HOFFMANN[3] and RICHARD[4] have proposed yet another test which arises from the increase, in the optical path length of radiation, caused by the gravitational field of an intervening body such as the Sun. The proposal consists in measuring the shift of pulsar frequencies in comparison with the frequency of an earthbound clock. The fractional variation in the frequency of the pulses arriving at the Earth after passing through the solar field is expected to be of the order of 5×10^{-10}.

[1] I. I. SHAPIRO et al., Phys. Rev. Lett. **20**, 1265 (1968).
[2] I. I. SHAPIRO et al., Phys. Rev. Lett. **26**, 1132 (1971).
[3] B. HOFFMANN, Nature **218**, 667 (1968).
[4] J.-P. RICHARD, Phys. Rev. Lett. **21**, 1483 (1968).

B.   Gravitational waves

Soon after the development of the general theory of relativity, EINSTEIN[1] examined the *weak-field* solutions of his field equations and discovered the possibility of gravitational waves.   To see this, we put

$$g_{ik} = g_{ik}^{(0)} + h_{ik},\tag{8.107}$$

where $g_{ik}^{(0)}$ denote the normal unperturbed values, (6.12), of the g_{ik} while h_{ik} denote *first-order* departures from these values.   The curvature tensor R_{iklm}, given by Eqs. (7.30), may then be approximated as

$$R_{iklm} \simeq \frac{1}{2}\left[\frac{\partial^2 h_{im}}{\partial x^k \partial x^l} + \frac{\partial^2 h_{kl}}{\partial x^i \partial x^m} - \frac{\partial^2 h_{il}}{\partial x^k \partial x^m} - \frac{\partial^2 h_{km}}{\partial x^i \partial x^l}\right].\tag{8.108}$$

The RICCI tensor R_{km}, given by Eqs. (7.34), may in turn be approximated as

$$
\begin{aligned}
R_{km} &= g^{il}R_{iklm}\\
&\simeq \frac{1}{2}\left[\frac{\partial^2 h_m^l}{\partial x^k \partial x^l} + \frac{\partial^2 h_k^i}{\partial x^i \partial x^m} - \frac{\partial^2 h}{\partial x^k \partial x^m} - \Box\, h_{km}\right]\\
&= \frac{1}{2}\left[\frac{\partial \Upsilon_m}{\partial x^k} + \frac{\partial \Upsilon_k}{\partial x^m} - \Box\, h_{km}\right],
\end{aligned}\tag{8.109}
$$

where

$$\Upsilon_m = \frac{\partial}{\partial x^l}\left(h_m^l - \frac{1}{2}h\,\delta_m^l\right) = \frac{\partial}{\partial x^l}\,\varphi_m^l, \quad \text{say,}\tag{8.110}$$

$\Box$ is the D'Alembertian operator, while

$$h = h_i^i \simeq g^{(0)in}\, h_{ni}.\tag{8.111}$$

We now make an infinitesimal coordinate transformation, from (x^i) to (x'^i), such that in the new system of coordinates the functions Υ_m are identically zero.   This can be accomplished by writing

$$x'^i = x^i + f^i,\tag{8.112}$$

[1] A. EINSTEIN, *Sitzber. Preuss. Akad. Wiss.* (1916), 688; (1918), 154.

where f^i are certain functions of the coordinates. Then,

$$g_{ik} = \frac{\partial x'^l}{\partial x^i} \frac{\partial x'^m}{\partial x^k} g'_{lm} = \left(\delta^l_i + \frac{\partial f^l}{\partial x^i} \right) \left(\delta^m_k + \frac{\partial f^m}{\partial x^k} \right) g'_{lm}$$

$$\simeq g'_{ik} + \frac{\partial f^m}{\partial x^k} g'_{im} + \frac{\partial f^l}{\partial x^i} g'_{ik}$$

$$\simeq g'_{ik} + \frac{\partial f_i}{\partial x^k} + \frac{\partial f_k}{\partial x^i} ; \tag{8.113}$$

consequently,

$$h'_{ik} = g'_{ik} - g^{(0)}_{ik} = h_{ik} - \frac{\partial f_i}{\partial x^k} - \frac{\partial f_k}{\partial x^i} \tag{8.114}$$

and hence

$$h'^i_k = h^i_k - \frac{\partial f^i}{\partial x^k} - g^{ir} \frac{\partial f^k}{\partial x^r} ; \qquad h' = h - 2 \frac{\partial f^i}{\partial x^i}. \tag{8.115}$$

Substituting these results into the formulae

$$Y'_m = \frac{\partial}{\partial x'^l} \varphi'^l_m \qquad \text{and} \qquad Y_m = \frac{\partial}{\partial x^l} \varphi^l_m$$

where

$$\varphi'^l_m = h'^l_m - \frac{1}{2} h' \delta^l_m \qquad \text{and} \qquad \varphi^l_m = h^l_m - \frac{1}{2} h \delta^l_m,$$

we obtain

$$Y'_m = Y_m - \Box f_m. \tag{8.116}$$

Choosing f_m such that $\Box f_m = Y_m$, we make Y'_m vanish. Equation (8.109) then reduces to

$$R'_{km} = -\frac{1}{2} \Box' h'_{km} ; \tag{8.117}$$

at the same time, our h'_{km} satisfy the conditions

$$\frac{\partial}{\partial x'^l} \left(h'^l_m - \frac{1}{2} h' \delta^l_m \right) = 0. \tag{8.118}$$

In the same approximation, the EINSTEIN tensor G'_{km} is given by

$$G'_{km} = R'_{km} - \frac{1}{2} g'_{km} R' \simeq -\frac{1}{2} \Box' \varphi'_{km}, \qquad (8.119)$$

where φ'_{km} satisfy the conditions

$$\frac{\partial}{\partial x'^l} \varphi'^l_m = 0. \qquad (8.120)$$

The field equations then reduce to, omitting the primes,

$$\Box \varphi^l_m \simeq -2 G^l_m = -\frac{16\pi k}{c^4} T^l_m, \qquad (8.121)$$

where (T^l_m) denotes the energy-momentum tensor; conditions (8.120) may now be written as

$$\frac{\partial}{\partial x^l} \varphi^l_m = 0. \qquad (8.122)$$

In empty space, Eqs. (8.121) become

$$\Box \varphi^l_m = 0. \qquad (8.123)$$

Equations (8.121) and (8.123) represent, respectively, a first-order improvement on the classical POISSON and LAPLACE equations for a gravitational field. Major difference between the new equations and the classical ones lies in the replacement of the three-dimensional Laplacian operator ∇^2 by the four-dimensional D'Alembertian operator $\left(\nabla^2 - \frac{1}{c^2} \frac{\partial^2}{\partial t^2} \right)$. Consequently, according to EINSTEIN's theory of gravitation, gravitational influences propagate through space with a *finite* speed equal to the speed of light—and not as an *instant action-at-a-distance*! The possibility of gravitational waves is manifest in these equations.

Total power P radiated by a gravitating source through gravitational radiation can be calculated in the same way as in the case of electromagnetic radiation, i.e. by calculating the total energy flux in the wave zone; see Problem 5.14. For a spinning rod, EINSTEIN showed that

$$P = \frac{32k}{5c^5} I^2 \omega^6, \qquad (8.124)$$

where I is the moment of inertia of the rod and ω the angular frequency of rotation ; the frequency of the radiation emitted is, for reasons of symmetry, *twice* the frequency of rotation. For a typical iron rod available in the laboratory, rotating at a speed close to the breaking point, $P = O(10^{-30})$ erg/s, which is ridiculously low for detection.[1] From a pulsar, such as the one in the Crab Nebula, the total power radiated through gravitational radiation may be $O(10^{38})$ erg/s, leading to a fluctuation of the *spatial* curvature, at the Earth, with amplitude $O(10^{-41})$ cm^{-2}; compare this with (i) the overall curvature of the universe, which is $O(10^{-56})$ cm^{-2}, (ii) the curvature, at the Earth, due to the solar field, which is $O(10^{-34})$ cm^{-2}, and (iii) the curvature, at the surface of the Earth, due to its own field, which is $O(10^{-26})$ cm^{-2}. Further, a stellar system undergoing an asymmetric gravitational collapse may emit a burst of gravitational radiation, with total energy output $O(10^{52})$ ergs in about a second's duration. This appears to be an excellent source of these radiations.

We are now faced with the question of detection of these waves, if at all they exist. The basic principle involved here is quite simple, i.e. a varying gravitational field induces relative displacements between test particles.[2] If, for simplicity, we consider two test particles, coupled by a spring, immersed in a gravitational field that fluctuates due to the passage of gravitational waves, the test system will undergo an oscillatory motion which, at resonance, may absorb considerable energy from the waves. From this simpleminded system we may pass on to an elastic solid — to improve the efficiency of detection ; of course, to attain resonance, one of the normal modes of the solid should be as close as possible to the (suspected) frequency of the waves. Piezoelectric crystals bonded to the ends of the solid would convert the mechanical energy of oscillation into electrical energy ; the resulting pulses can be amplified and detected.

The pioneering work in this direction was undertaken by WEBER who, after several years of experimentation, reported in 1967 that his detectors (huge aluminum cylinders, about 153 cm long, 66 cm in diameter and weighing about a ton each), had recorded *isolated* events which could possibly be attri-

---

[1] The weakness of these radiation, as compared with electromagnetic radiation studied in Problem 5.14, is related to the fact that in most cases $(km^2/e^2) \ll 1$. It is further weakened by the fact that, because of momentum conservation, *dipole* gravitational radiation are completely absent ; the lowest order must, therefore, be *quadrupole*, which is weaker by another factor of the order of (v^2/c^2).

[2] For details, see J. WEBER, *loc. cit.*, Chap. 8.

buted to gravitational waves (of unknown origin).[1] In 1969, he reported a number of *coincident* events, within about 0.4-sec resolving time of the apparatus, recorded by his detectors at the University of Maryland and at the Argonne National Laboratory (about 1000 km apart) which, he concluded, were excited by gravitational radiation.[2] Use of directional antennas seemed to suggest that the source of these radiations was a 10^{10} solar-mass object situated at the centre of the galaxy.[3]

Spurred by WEBER's reports, several groups of workers set out to observe gravitational radiation, using similar (though not identical) techniques. Surprisingly, none of them has so far observed anything resembling, in quantity or quality, the findings of WEBER. This has raised the important question whether WEBER's observations have anything at all to do with gravitational radiation. In this connection, we note that a rigorous statistical analysis of WEBER's observations against known activity (of terrestrial, solar or cosmic-ray origin) indicates some positive correlation between the two.[4] It, therefore, appears that at least some of WEBER's events could possibly be caused by processes other than gravitational radiation. However, the implications of these correlations are somewhat uncertain, and the failure of other groups to observe what WEBER thinks he has already observed does not automatically rule out the possibility that gravitational radiation may after all have been detected. Of course, the final word on the subject has not yet been said ; the interested reader may refer to the various progress reports for further information.[5]

C. Gravitational collapse, neutron stars and black holes

In most of the phenomena studied in this chapter we were concerned with small, first- or second-order, departures from the Newtonian theory of gravitation. In this section we shall be concerned with situations where departures

[1] J. WEBER, *Phys. Rev. Lett.* **18**, 498 (1967).

[2] J. WEBER, *Phys. Rev. Lett.* **22**, 1320 (1969); **24**, 276 (1970).

[3] J. WEBER, *Phys. Rev. Lett.* **25**, 180 (1970).

[4] R. A. ADAMYANTS, A. D. ALEKSEEV and N. I. KOSOSNITSYN, *JETP Lett.* **15**, 194 (1972) ; J. A. TYSON, C. G. MACLENNAN and L. J. LANZEROTTI, *Phys. Rev. Lett.* **30**, 1006(1973).

[5] S. W. HAWKING, *Contemp. Phys.* **13**, 273(1972) ; P. S. APLIN, *ibid.* **13**, 283(1972) ; D. W. SCIAMA, *Gen. Rel. and Grav.* **3**, 149(1972) ; W. H. PRESS and K. S. THORNE, *Ann. Rev. Astron. and Astroph.* **10**, 335(1972) ; J. L. LOGAN, *Phys. Today* **26**, No. 3, 44(1973) ; *ibid.* **26**, No. 10, 17(1973).

from the Newtonian theory are so large that curvature in the space-time continuum assumes an overriding importance and the new theory of gravitation, with all its peculiarities, shows up in full glory. This happens when the radius of a given object (assumed spherical, with mass M) is comparable with its *Schwarzschild radius* r_0 ($= 2kM/c^2$), i.e. for a given value of M the density of the object is comparable with ρ_0, where

$$\rho_0 \sim (c^6/k^3M^2). \tag{8.125}$$

For a typical star ($M \sim 10^{33}$ g), this implies a density of the order of 10^{16} g/cm^3 and a radius of the order of 10^{5-6} cm. Such a situation is obtainable in a *neutron star*, which results from the *gravitational collapse* of an ordinary star and represents one of the last stages of its evolutionary career. A complete discussion of this phenomenon would require much more space than can be afforded in this text; we, therefore, give here only a brief outline of the most important aspects of the problem and refer the more interested reader to other sources.[1]

For our purpose the problem of central importance in this study is that of "equilibrium configurations of cold-catalyzed stars,[2] of mass comparable to the solar mass but density much higher than the solar density". For such a star, the equilibrium configuration is solely determined by a competition between the zero-point motion of the particles and the gravitational pull of matter towards the centre of the star; for the former we need a suitable equation of state, $P = P(\rho)$, for the system while for the latter we require an appropriate generalization of the classical equation, $(dP/dr) = -g\rho$, of hydrodynamic equilibrium. The end result of such a calculation is shown

---

[1] B. K. HARRISON, K. S. THORNE, M. WAKANO and J. A. WHEELER, *Gravitation Theory and Gravitational Collapse* (University of Chicago Press, Chicago, 1965). For simpler accounts, see J. A. WHEELER, in *Gravitation and Relativity*, ed. H.-Y. CHIU and W. F. HOFFMANN (Benjamin, New York, 1964), Chap. 10; K. S. THORNE, *Science* **150**, 1671 (1965) and *Scientific American* **217**, 88(Nov. 1967); E. E. SALPETER, in *Relativity Theory and Astrophysics: Stellar Structure*, Vol. 10 of Lectures in Applied Mathematics (American Mathematical Society, Providence, R. I., 1967).

[2] *Cold* in the sense that the temperature T of the system under study is much smaller than the FERMI temperature T_F, so that the system may be regarded as completely degenerate; see R. K. PATHRIA, *Statistical Mechanics* (Pergamon, Oxford, 1972), Sec. 8.1. Accordingly, the temperature of the system becomes a 'dead parameter' of the problem.

Catalyzed in the sense that the system has reached the endpoint of thermonuclear evolution, so that there is no other source of kinetic energy left except the zero-point motion of the particles,

in Fig. 8.5, where all possible configurations of a cold-catalyzed star are shown in terms of a mass-density relationship represented by the solid curve ; the mass of the star is expressed in units of solar mass $M_\odot$ and the density refers to the centre of the star. The lower dashed curve was derived

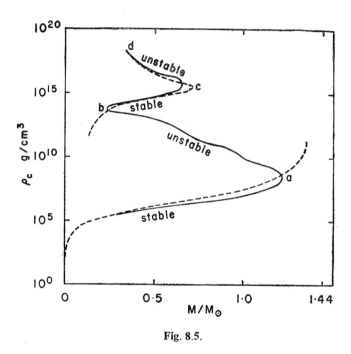

Fig. 8.5.

by CHANDRASEKHAR[1] for white dwarf stars, without the use of general relativity and without allowing for the crushing of electrons onto the nuclei ; the upper dashed curve was deduced by OPPENHEIMER and VOLKOFF[2] for an idealized model of a neutron star. The full curve is due to HARRISON, WAKANO and WHEELER ; it contains a lower stable portion (applicable to white dwarf stars), an upper stable portion (applicable to neutron stars) and two unstable portions which do not evoke much interest.[3]   We are now in a

---

[1] S. CHANDRASEKHAR, *Mon. Not. Roy. Astr. Soc.* 95, 207(1935) ; *Introduction to the Study of Stellar Structure* (University of Chicago Press, Chicago, 1939). For a simpler account, see R. K. PATHRIA, *loc. cit.*, Sec. 8.4.

[2] J. R. OPPENHEIMER and G. VOLKOFF, *Phys. Rev.* 55, 374(1939).

[3] If, by chance, a star could be created in such an equilibrium configuration, the slightest disturbance would set it off—either into implosion or explosion.

position to discuss the fate of any given star when it has reached the end-point of its thermonuclear evolution. Two possibilities arise here:

(i) If the mass of the star is less than 1.2 $M_\odot$, corresponding to point a in the figure, then, after consuming its nuclear fuel, the star settles down in a white-dwarf configuration. It then radiates at a very slow rate, drawing energy from a quasi-static gravitational contraction.

(ii) If, on the other hand, the mass of the star is greater than 1.2 $M_\odot$ it *cannot* attain a white-dwarf configuration. The gravitational contraction goes on compressing matter in the central regions of the star to higher and higher densities, until it reaches values $O(10^{11})$ g/cm³. At that stage, the electrons—to remain free—require too much zero-point energy; instead, they find it energetically favorable to coalesce with the nuclei present in the system and produce neutron-rich nuclei, e.g. Mn^{56} instead of Fe^{56}, etc. Disappearance of electrons removes the major source of internal pressure that held the star against the gravitational pull towards the centre; consequently, the star begins to collapse catastrophically! The subsequent picture is qualitatively as follows.[1]

Within a fraction of a second after the initiation of collapse, nearly all the electrons and Fe^{56} nuclei inside the star's core are transformed into highly neutron-rich nuclei and free neutrons, and the core is in *free* fall. In about two seconds the core acquires a kinetic energy (of collapse) equivalent to a sizeable fraction of its rest-mass energy, and the neutrons in the core are compressed into regions of the order of their Compton wavelength ($\rho_c \sim 10^{14}$ g/cm³). At this point the zero-point energy of the neutrons, and with it their zero-point pressure, begins to rise rapidly. The collapsing core is suddenly faced with a huge central pressure which calls its collapse to a halt and sends a shock wave propagating outward through it. In this core shock front, the huge kinetic energy of collapse is converted into heat and temperatures of over 10^{10} K are reached. At such high temperatures and densities, elementary particle transformations proceed at a rapid rate and the heat produced in the core shock front is converted into high-energy neutrinos. The mean free path of neutrinos is less than 100 m under these extreme conditions. Hence, instead of escaping freely from the star, they diffuse outward, depositing the energy released by the collapse of the core in the envelope of the star and thereby raising the envelope to temperatures as high as

[1] K. S. THORNE, *Science* **150**, 1671 (1965).

10^{11} K.   At these enormous temperatures explosive nuclear burning is initiated in the envelope, with a consequent release of additional thermal energy.
Because of the huge thermal energies generated by neutrino deposition and
nuclear burning, the envelope of the star becomes gravitationally unbound.
An exploding shock wave forms and blows the envelope away from the core
with speeds approaching the speed of light and the huge thermal energies of
the expanding envelope are converted into radiation so intense that the
luminosity of the exploding star approaches that of a galaxy.   The observers
at the Earth refer to this spectacle as a *supernova explosion.*

What happens to the collapsed core left behind in the process depends
essentially on the mass of the core.   As one possibility, it may attain a
neutron star configuration, corresponding to the portion *bc* of the solid curve
in Fig. 8.5.[1]   As another, particularly if the core is too massive, it may not
settle down into any cold, dead state ; rather, after it has sufficiently cooled,
it may undergo another catastrophic collapse—this time to zero volume—
leading to a *black hole.*   It may be mentioned here that a less massive supernova core, or a cold star with mass less than the CHANDRASEKHAR limit of
1.2 $M_\odot$, can also be induced to collapse to zero volume if it is compressed
sufficiently.

A black hole—an object confined to the interior of its SCHWARZSCHILD surface ($r=r_0=2kM/c^2$)—is one of the strangest objects of Nature.   Its spacetime
is so strongly curved that no light can come out of it, no matter can be
ejected from it and no measuring rod can survive being put into it.[2]   Any
kind of object that falls into it loses its separate identity ; it is remembered
only through the mass, the charge, the angular momentum and the linear
momentum it contributed to the hole.   How the physics of a black hole looks
depends essentially on the choice of the observer.   Suppose that the observer
follows the collapsing matter through its collapse down into the black hole.
Then he will see it crushed to indefinitely high density, and he himself will be
torn apart by indefinitely increasing tidal forces.   No restraining force has
the power to hold him away from this catastrophe, once he has crossed the
critical SCHWARZSCHILD surface.   The final collapse occurs a *finite* time after
the passage of this surface, but it is inevitable.   Time and space are inter-

---

[1] The precise value of the upper limit on the mass of a neutron star, which is about
0.7 $M_\odot$ in Fig. 8.5, is in some doubt because of uncertainties in the equation of state of cold
matter at densities $0(10^{15})$ g/cm^3.   The true upper limit may be as large as 2 $M_\odot$.

[2] R. RUFFINI and J. A. WHEELER, *Physics Today* 24, No. 1, 30(1971) ; R. PENROSE,
Nature 236, 377(1972) and *Scientific American* 226, No. 5, 38(1972); S. CHANDRASEKHAR,
Contemp. Phys. 15, 1 (1974).

changed inside a black hole in an unusual manner ;[1] the direction of increasing proper time for the observer is now the direction of decreasing values of the radial coordinate r. The observer has no more power to return to a larger value of r (or at least stay where he is) than we have power to turn back (or at least stop) the hands on the clock of life.

The situation is different for an observer who watches the collapse from far away. As a price for his safety, he is deprived of any chance to see more than mere initial steps on the way to collapse. All signals, with accompanying information, from the later phases of collapse are held back by the space-time metric of the collapsing system. More specifically, the collapsing system *appears* to slow down as the SCHWARZSCHILD surface is approached ; light emitted by the system becomes more and more red-shifted and clocks attached to the system appear to run more and more slowly. It takes an *infinite* time for the collapsing system to reach the SCHWARZSCHILD surface and, for the far-out observer, it never gets beyond there.[2]

For observational evidence on neutron stars and black holes, reference may be made to other literature.[3]

Problems

8.1. Assuming that the light corpuscles of NEWTON are susceptible to gravitation in the same way as normal material particles are, examine the Newtonian orbit of such a corpuscle (coming from infinity and going to infinity) past a massive object such as the Sun. Show that the path of the corpuscle undergoes a bending Δ, given by $2kM/c^2R$; *cf.* expression (8.77) and recall the footnote on p. 230.

8.2. (i) Show that, for a mass-particle moving in the SCHWARZSCHILD geometry, closed orbits with perihelion closer than $2r_0$ are not possible.

---

[1] This is related to the fact that the elements g_{11} and g_{44} of the SCHWARZSCHILD metric change sign when r passes through the value $r_0(=2kM/c^2)$; see Eq. (8.31).

[2] This is related to the fact that $g_{44} \to 0$ as $r \to r_0$; now see Eqs. (8.89) and (8.91).

[3] For neutron stars, see A. HEWISH *et al.*, *Nature* **217**, 709 (1968) ; T. GOLD, *Nature* **218**, 731 (1968) ; A. J. R. PRENTICE, *Nature* **225**, 438 (1970) ; F. PACINI and M. J. REES, *Nature* **226**, 622 (1970) ; P. A. STURROCK, *Nature* **227**, 465 (1970) ; D. PINES and J. SHAHAM, *Nature Physical Science* **235**, 43 (1972).

For black holes, see A. G. W. CAMERON, *Nature* **229**, 178 (1971) ; R. STOTHERS, *Nature* **229**, 180 (1971) ; P. DEMARQUE and S. C. MORRIS, *Nature* **230**, 516 (1971) ; G. W. GIBBONS and S. W. HAWKING, *Nature* **232**, 465 (1971) ; E. J. DEVINNEY, JR., *Nature* **233**, 110 (1971) ; R. E. WILSON, *Nature* **234**, 406 (1971).

See also S. SOFIA, *Nature Physical Science* **234**, 155 (1971) ; P. J. E. PEEBLES, *Gen. Rel. and Grav.* **3**, 63 (1972).

(ii) If a mass-particle coming from infinity has an impact parameter less than $(3/2)\, r_0$, even an open orbit is not possible. Instead, the particle is captured by the field.

(iii) Light can move perpetually in a circle of "radius" $(3/2)\, r_0$.

(iv) If a ray of light coming from infinity has an impact parameter less than $(3\sqrt{3}\,/2)\, r_0$, it cannot get out of the gravitational field ever again.

The symbol r_0 in this problem denotes the SCHWARZSCHILD radius of the central body, viz. $2kM/c^2$.

8.3. Discuss the significance of the parameter B in Eq. (8.47) and of parameter B' in Eq. (8.62).

8.4. (i) Show that the transformation

$$r' = \frac{1}{2} \left\{ \left(r^2 - \frac{2kM}{c^2}\, r \right)^{1/2} + \left(r - \frac{kM}{c^2} \right) \right\},$$

which implies that

$$r = r' \left(1 + \frac{kM}{2c^2 r'} \right)^2,$$

puts the SCHWARZSCHILD metric (8.31) into the *isotropic* form

$$ds^2 = \left(1 + \frac{kM}{2c^2 r'} \right)^4 \{dr'^2 + r'^2\, (d\theta^2 + \sin^2\theta\; d\varphi^2)\} - \left\{ \frac{1 - (kM/2c^2 r')}{1 + (kM/2c^2 r')} \right\}^2 c^2\, dt^2 .$$

(ii) For $r' \ll (kM/2c^2)$, one may write

$$ds^2 = \left(1 + \frac{2kM}{c^2 r'} + \ldots \right) \{dr'^2 + r'^2\, (d\theta^2 + \sin^2\theta\; d\varphi^2)\}$$

$$ - \left(1 - \frac{2kM}{c^2 r'} + \frac{2k^2 M^2}{c^4 r'^2} + \ldots \right) c^2\, dt^2 ,$$

which may be generalized to the form

$$ds^2 = \left(1 + \gamma\, \frac{2kM}{c^2 r'} + \ldots \right) \{dr'^2 + r'^2\, (d\theta^2 + \sin^2\theta\; d\varphi^2)\}$$

$$ - \left(1 - \alpha\, \frac{2kM}{c^2 r'} + \beta\, \frac{2k^2 M^2}{c^4 r'^2} + \ldots \right) c^2\, dt^2 ,$$

where α, β and γ are the EDDINGTON-ROBERTSON parameters (which may be different from unity). Show that in the generalized metric the bending of light rays and the precession of planetary orbits increase by a factor of $\frac{1}{2}\, (\alpha + \gamma)$ and $\frac{1}{3}\, [2\alpha\, (\alpha + \gamma) - \beta]$, respectively.

8.5. According to the generalized metric of the preceding problem, the "effective" optical path length for light propagating in a weak gravitational field is determined by the factor $[1 + (\alpha + \gamma)\, kM/c^2 r']$. This is equivalent to propagation in a refracting medium whose refractive index is space-dependent. Using this concept, study the deflection of light rays in a weak gravitational field and show that

$$\Delta = (\alpha + \gamma)\,(2kM/c^2R)\;,$$

instead of the standard result $(4kM/c^2R)$.

8.6. Show that the proper acceleration of a slow-moving particle in the SCHWARZSCHILD field of an object (of mass M) is given by

$$g = -\frac{kM}{r^2}\left(1 - \frac{2kM}{c^2r}\right)^{-1/2},$$

where r is the radial coordinate of the instantaneous position of the particle.

8.7. (i) Show that, in the SCHWARZSCHILD matric, circular orbits are possible with any radius $r > 3kM/c^2$. Derive expressions for $(d\varphi/d\tau)$ and $(d\varphi/dt)$ for such an orbit, and compare the latter with the corresponding Newtonian result.

(ii) What is the total proper time for such an orbit? What is the total coordinate time? Compare the two and discuss what happens as $r \to 3kM/c^2$.

8.8. A satellite-borne clock C is compared with an earth-bound clock C_0. Show that

$$\frac{d\tau}{d\tau_0} \simeq 1 + \frac{kM}{c^2R} - \frac{3}{2}\frac{kM}{c^2r},$$

where M is the mass of the earth, R the radius of the earth and r the radius of the satellite's orbit (assumed circular). Study the variation of this factor as r changes from the value R to values much greater than R.

8.9. Linearize the SCHWARZSCHILD line element and show that the motion of a particle in the resulting metric corresponds to the case of a uniform acceleration field.

8.10. Proceeding on the basis of the principle of equivalence, derive Eq. (8.93) for the gravitational shift of a spectral line.

8.11. SIR CHARLES DARWIN[1] cites the following delightful example of the *twin paradox*—a biological version of the clock paradox : 'On New Year's Day 1984 an astronaut, *B,* sets out from earth at a speed of 0.8 c and travels to the nearest star, α-Centauri, which is just about 4 light-years away as measured in the earth's frame of reference. Having reached the star he immediately turns around and returns to earth at the same speed, arriving home on New Year's Day 1994 by earth's time. The astronaut has a brother, A, who remained on earth and they agreed to send each other greetings by radar-telephone on every New Year's day until the traveler gets back.'

Verify that while A sends 10 messages during the intervening time, B sends only 6 (including the one sent on the last day of his trip).

8.12. Show that the following statement, if valid in one inertial frame of reference, is valid in *all* inertial frames of reference : 'Of several standard clocks undergoing a round trip, the one for which the mean value of the quantity $(1 - u^2/c^2)^{1/2}$ during the trip turned out to be the largest is also the one that, on the whole, recorded time at the fastest rate; here, u is the (instantaneous) speed of the clock with respect to the inertial observer.'

---

[1] C. G. DARWIN, *Nature* **180**, 976 (1957).

8.13. (i)   Show that, in the SCHWARZSCHILD metric, the coordinate time for the travel of a light signal, starting and ending at the radial coordinate r_1, is given by

$$t \simeq \frac{2}{c} \int_p^{r_1} \{e^{-2\nu} + e^{-\nu} p^2 (r^2 - p^2)^{-1}\}^{1/2} \, dr$$

$$\simeq 2 \frac{x_1}{c} \left(1 - \frac{r_0}{2r_1} \right) + 2 \frac{r_0}{c} \ln \left(\frac{r_1 + x_1}{p} \right); \qquad x_1 = (r_1^2 - p^2)^{1/2} \, .$$

Here, $e^\nu = 1 - (r_0/r)$, where $r_0 = (2kM/c^2)$, while p is the radial coordinate corresponding to the point of closest approach to the gravitating object.   Verify that this value of t always exceeds the Newtonian value $2x_1/c$.

(ii)   Show that the expression for the *excess* in t, in the case of a light signal bouncing back from an inner planet, can be written in the form (8.104).

CHAPTER 9

RELATIVISTIC COSMOLOGY

9.1. The cosmological principle and Weyl's postulate

In the preceding chapter we applied the formalism of general relativity to the motion of material particles and to the propagation of light in the gravitational field of a single massive body. The results of that study were applied to phenomena taking place in the solar system, and an appreciable degree of agreement was observed. This raises the hope of understanding relativistically the phenomena taking place on a scale even larger than the one encountered in the solar system, e.g. the phenomena associated with the space-time of the universe itself. Studies in this direction were initiated by EINSTEIN[1] in 1917; since then, considerable attention has been paid to this problem by a number of eminent cosmologists.

As a result of these investigations, we now have a number of rival theories of cosmology, which differ from one another more on attitude than on outcome[2]. And the observational data are not yet in a position to make a final choice among the various theories. In our study we shall be concerned exclusively with the *relativistic* cosmology; it has the distinction of being not only the first systematic theory in this field but also the most simple and straightforward of all the different cosmological disciplines.

---

[1] A. EINSTEIN, *Berlin Sitzungsberichte* (1917), p. 142; English translation available in *The Principle of Relativity, loc. cit.*

[2] For a discussion of these, see H. BONDI, *Cosmology* (University Press, Cambridge, 1960); H. P. ROBERTSON and T. W. NOONAN, *Relativity and Cosmology* (W. B. Saunders, Philadelphia, 1968); D. W. SCIAMA, *Modern Cosmology* (University Press, Cambridge, 1971); S. WEINBERG, *Gravitation and Cosmology* (John Wiley, New York. 1972).

As our starting point we adopt, as is done (with amazing unanimity) in all theoretical approaches to this problem, the so-called *cosmologica i principle*. This principle is based on the assumption that the universe as a whole, when looked at from a large-scale point of view (i.e. with stellar and nebular condensations virtually smeared out), is spatially homogeneous and isotropic. This assumption is amply justified empirically, for observations deep into the universe indicate that, to a fairly good degree of accuracy, the nebulae and stellar systems appear, on the whole, to be distributed uniformly *throughout*, and in *all* directions of, the cosmic region probed so far.[1] Consequently, it would be expected that, at any given instant of time, observers (like ourselves) stationed *anywhere* in the universe would see essentially the same picture of the universe as we do; this statement constitutes the cosmological principle. Accordingly, *any* point in the universe, e.g. our own location, may be taken as the origin of the spatial coordinates. Further, for *all* observers in the universe a *common* 'cosmical' time may be adopted, with the result that, at any given instant of this (public) time, the metric of the space-time continuum representing the development of our 'smeared-out' universe is everywhere the same.[2]

The cosmological principle, however, asserts the (over-all) homogeneity and isotropy of the universe only with respect to space, not with respect to time. Thus, whereas any arbitrary point in space can be taken as the origin of the spatial coordinates, any arbitrary instant of the cosmical time cannot be taken as the origin of the time coordinate; in fact, as is commonly held, the universe has been, and is expected to keep on, going through a well-defined pattern of evolution in time. Accordingly, our cosmological principle is a rather narrow principle, in that it discriminates sharply between space and time; it is a postulate of simplicity in respect of space but not so in respect of time.

In 1948 BONDI and GOLD[3] suggested a generalization of the cosmological principle with a view to removing the aforementioned asymmetry. Accord-

---

[1] A. SANDAGE, G. A. TAMMANN and E. HARDY, *Astroph. J.* **172**, 253 (1972). Following WERTZ, if a systematic deviation from large-scale homogeneity is characterized by a function proportional to $r^{-\theta}$, then the limit on θ turns out to be -0.115 ± 0.030, which does not differ significantly from $\theta = 0$.

[2] The adoption of a *common* 'cosmical' time, which reminds one of the *absolute* time of NEWTON, is made possible by the fact that a spatially homogeneous universe would have a uniform gravitational potential and hence a uniform g_{44} (which, by a suitable choice of the time coordinate, can be made equal to -1 *throughout* the universe).

H. BONDI and T. GOLD, *Mon. Not. Roy. Astron. Soc.* **108**, 252 (1948).

ing to the generalized principle, the universe would present the same picture to all observers carrying out observations not only *from anywhere in space* but also *at any instant of time.* They called this principle the *perfect cosmological principle*, in contrast with the 'restricted' one discussed above. In view of the perfect principle, the question of a temporal evolution of the universe would not even arise ; in fact, the very idea of defining a cosmical time *with a fixed origin* would become superfluous. The associated theory of cosmology is referred to as the *steady-state* theory[1] (it is, of course, *nonstatic* all right). However, with one apparently satisfactory feature incorporated into the formalism, the theory has otherwise to face difficulties of a rather peculiar nature, the most crucial one being the necessity of a *continual* creation of matter in order to keep the mean density in the expanding universe constant. Of late, this theory has received serious setbacks on the observational front as well.

In relativistic cosmology, however, we work with the restricted principle alone ; consequently, we have ample scope for discussing the evolution of the universe, as a whole, with the passage of time. The smeared-out structure, which is throughout homogeneous and isotropic, provides the best workable model of the universe ; we call it the *substratum.* If the actual universe is nonstatic (which *is* really the case), then the substratum is to be regarded as endowed with a systematic motion of the kind we have in an ordinary fluid. This fluid, the smeared-out matter of the universe, may be characterized by a number of physical parameters, e.g. the density, the pressure, etc. (which are independent of the position in the substratum) and the flow velocity (which would be a function of the position).[2]

The nebulae and other stellar systems, which may be regarded as the 'particles' of the substratum (or of the *cosmic fluid*) will partake of this general motion in the substratum. They will also possess motions over and above this general one—the so-called random motions ; however, these may be regarded as negligible, or at least much less significant, in comparison with the overall motion. The resulting situation is normally summed up as

---

[1] See H. BONDI and T. GOLD, *loc. cit.*; F. HOYLE, *ibid.* **108**, 372 (1948) and **109**, 365 (1949). See also H. BONDI, *Cosmology, loc. cit.*, Chap. XII.

W. H. McCREA, *Proc. Roy. Soc. London* A **206**, 562 (1951), discusses a way of basing the steady-state theory on general relativity.

[2] The density of the fluid will arise from the matter density and the radiation (and thermal) energy density in the universe, while the pressure will arise from the random motions of the nebulae and the stars, the thermal motion of the molecules, the radiation, etc.

WEYL's postulate :[1]

> *The particles of the substratum (representing the nebulae) lie, in the space-time of the cosmos, on a bundle of geodesics diverging from a point in the (finite or infinite) past.*

The most important implication of this postulate is that since the cosmical geodesics do not intersect anywhere except at the singular point in the past (and possibly a similar singular point in the future), we have not more than one geodesic passing through each point of the space-time ; thus, the velocity of the general cosmical motion at any given world point is a *unique* function of the coordinates of that point.   The analogy of the streamlined hydrodynamic flow to the motion of the substratum is, therefore, appropriate.

It can be readily seen that a bundle of geodesics satisfying WEYL's postulate must possess a set of (three-dimensional) hypersurfaces, which are orthogonal to the geodesics.   It then appears natural to choose these hypersurfaces as the (purely spatial) surfaces $t =$ const., t being the cosmical time, and allot the spatial coordinates x^α ($\alpha = 1, 2, 3$), *in a unique and continuous manner*, to the various geodesics in the bundle.   Thus, the spatial coordinates of each particle of the substratum are, *once and for all*, fixed !   At any particular instant of time, the various particles occupy uniquely defined positions in space (as given by the instantaneous tips of the respective geodesics) ; at a later instant, the actual positions of the particles, and the distances between them, would be different.   Note, however, that the spatial coordinates, like labels, remain unchanged during the process of evolution, and the variation in the positions and distances of the particles arises from a change in the *scale factor* $R(t)$ of the substratum itself ; see details in the following sections.

The fact that in this choice of the coordinate system the spatial coordinates of the various particles of the substratum remain constant in time implies that our system of observation is itself partaking of the motion of the substratum.   In other words, we are employing a system which is moving along with the local smeared-out matter ; such a system is referred to as a *co-moving system*.

9.2.   The cosmical line element

We shall now examine the space-time metric of the universe in the light of the discussion of the preceding section   First of all we make the observation

---

[1] H. WEYL, *Phys. ZS.* **24**, 230 (1923).

that in a co-moving system of reference the orthogonality of the hypersurfaces $t=$ const. and the geodesics $x^\alpha =$ const. implies that the cosmical metric can be written in the form

$$ds^2 = \gamma_{\alpha\beta} \, dx^\alpha \, dx^\beta - c^2 \, dt^2 \qquad (\alpha, \beta = 1, 2, 3), \qquad (9.1)$$

where the quantities $\gamma_{\alpha\beta}$ are certain *universal* functions of the spatial coordinates x^α and the time coordinate t. In view of the over-all homogeneity and isotropy of the physical space,[1] the dependence of $\gamma_{\alpha\beta}$ on t must be such that it makes no direct reference to the coordinates x^α; in other words, t must enter into $\gamma_{\alpha\beta}$ only through a common universal factor, say $R(t)$, which is independent of x^α. Thus, we may write

$$\gamma_{\alpha\beta} = [R(t)]^2 \, l_{\alpha\beta}, \qquad (9.2)$$

where $l_{\alpha\beta}$ depend only on x^α.

The foregoing result can also be understood in the following way. Imagine, at any given instant of time t_1, a geometrical figure formed by some of the particles of the substratum (say, a polygon with certain nebulae as its vertices). Then, at a later instant of time t_2, the nebulae will have occupied new positions in space (though the spatial coordinates attached to them remain unchanged) and our polygon will have undergone an appropriate modification. Now, all directions in space being equally good, the resulting polygon must be geometrically similar to the original one; in other words, all its characteristic lengths must have undergone the same magnification. Moreover, because of spatial homogeneity, the linear dimensions of *any* geometrical figure located *anywhere* in space would undergo the same magnification. Now, since

$$\frac{dl(t_2)}{dl(t_1)} = \frac{\sqrt{\gamma_{\alpha\beta}(t_2) \, dx^\alpha \, dx^\beta}}{\sqrt{\gamma_{\alpha\beta}(t_1) \, dx^\alpha \, dx^\beta}}, \qquad (9.3)$$

the only way to satisfy this requirement *universally* is to have $\gamma_{\alpha\beta}$ in the form

---

[1] It will be noted that the cosmic space presents itself as homogeneous and isotropic *only* to co-moving observers; the situation is, therefore, even more restricted than the one obtaining in the special theory of relativity. In fact, the various coordinate transformations that appear in the sequel are mere analytical manipulations within the framework of the co-moving systems and do not, in any way, imply the introduction of new systems having motions over and above that of the substratum; moreover, for *all* the infinitely many observers in space there seems to exist a unique (practically absolute) time t.

(9.2); the universal magnification factor would then be

$$R(t_1, t_2) = \frac{R(t_2)}{R(t_1)}.$$ (9.4)

The variable $R(t)$ can, therefore, be looked upon as the *cosmical scale factor* (also referred to in the preceding section) which, at any given time t, determines the instantaneous spatial distances in the universe.

We shall now examine the details of the three-space which is characterized by the time-independent spatial metric $(l_{\alpha\beta}\, dx^\alpha\, dx^\beta)$ and is known to be homogeneous and isotropic. By a well known theorem of differential geometry (the so-called SCHUR's theorem), this space must be one of *constant curvature*;[1] of course, the curvature may be (i) positive (the corresponding space being spherical or elliptic), (ii) zero (the corresponding space being Euclidean or 'parabolic') or (iii) negative (the corresponding space being hyperbolic).

The metric for a three-space of *constant positive* curvature may be written as, see Eqs. (33) and (34) of the Appendix,

$$dl^2 = \frac{(dl^2)_{\text{Euclidean}}}{\left(1 + \dfrac{r^2}{4a^2}\right)^2} = \frac{1}{\left(1 + \dfrac{r^2}{4a^2}\right)^2}[dr^2 + r^2 d\theta^2 + r^2 \sin^2\theta d\varphi^2],$$ (9.5)

where a is the *radius of curvature* of the space. The transformation

$$r' = \frac{r}{\left(1 + \dfrac{r^2}{4a^2}\right)}$$ (9.6)

puts the metric (9.5) into the form

$$dl^2 = \frac{dr'^2}{(1 - r'^2/a^2)} + r'^2 d\theta^2 + r'^2 \sin^2\theta\, d\varphi^2,$$ (9.7)

which defines the so-called EINSTEIN model of the universe. Sometimes one introduces, in place of the radial coordinate r', an 'angle' χ such that

$$r' = a \sin \chi \quad (0 \leqslant \chi \leqslant \pi);$$ (9.8)

[1] See, for instance, T. LEVI-CIVITA, *The Absolute Differential Calculus* (Blackie and Son, London, 1929), Secs. 8.5 and 8.6.

the line element (9.7) then takes the form

$$dl^2 = a^2[d\chi^2 + \sin^2\chi \, (d\theta^2 + \sin^2\theta \, d\varphi^2)]. \tag{9.9}$$

The space under consideration now appears as the three-dimensional 'surface' of a 'sphere' of radius a in a fictitious four-dimensional space with 'spherical' coordinates $(\rho, \chi, \theta, \varphi)$; see also Problems 9.1 and 9.2.

Likewise we can obtain expressions for the line element of a three-space of *constant negative* curvature; this would correspond to the case

$$a^2 < 0, \tag{9.10}$$

so that a would now be imaginary. The desired expressions follow directly from the ones appropriate to the case of positive curvature by replacing the quantity a by ib; the actual curvature of the space would now be determined by b, which may accordingly be regarded as the radius of curvature of the space. We thus have:

$$dl^2 = \frac{1}{\left(1 - \dfrac{r^2}{4b^2}\right)^2} [dr^2 + r^2 d\theta^2 + r^2 \sin^2\theta \, d\varphi^2] \tag{9.11}$$

$$= \frac{dr'^2}{(1 + r'^2/b^2)} + r'^2 d\theta^2 + r'^2 \sin^2\theta \, d\varphi^2 \tag{9.12}$$

$$= b^2[d\chi^2 + \sinh^2\chi \, (d\theta^2 + \sin^2\theta \, d\varphi^2)], \tag{9.13}$$

the relevant substitutions being

$$r' = \frac{r}{\left(1 - \dfrac{r^2}{4b^2}\right)} \tag{9.6'}$$

and

$$r' = b \sinh \chi \qquad (0 \leqslant \chi < \infty). \tag{9.8'}$$

In the limits $a \to \infty$ and $b \to \infty$, the curved spaces considered above reduce to a flat (Euclidean) space.

We can now write down the space-time metric of the universe [see (9.1), (9.2), (9.5) and (9.11)]:

$$ds^2 = \{R(t)\}^2 \frac{\sum\limits_{\alpha} (dx^\alpha dx^\alpha)}{\left[1 + \dfrac{\kappa}{4} \sum\limits_{\alpha} (x^\alpha x^\alpha)\right]^2} - c^2 dt^2. \tag{9.14}$$

It will be noted that the coordinates x^α appearing here are the Cartesian counterpart of the coordinates (r, θ, φ) and are expressed in terms of the radius of curvature of the space; the function $R(t)$ in (9.14) thus includes the radius of curvature of the space (a or b, as the case may be) and, consequently, is of the dimensions of a length.[1] Further, the constant κ is a parameter which is $+1, 0$ or -1 according as the space is of positive curvature, flat or of negative curvature.

Expression (9.14) is known as the ROBERTSON-WALKER line element and is common to all theories postulating a homogeneous isotropic substratum.[2]

Thus, we finally have two quantities, $R(t)$ and κ, which are to be determined in accordance with the observed physical properties of the actual universe. For this, we must solve the relevant field equations for the metric (9.14). As a result, we expect a large variety of possible solutions, each representing a particular model of the universe.

9.3.   Emergence of the various models of the universe

Before entering into the proposed discussion we consider a modification of the field equations of Sec. 7.5, which was suggested by EINSTEIN himself in 1917. Looking at the *smeared-out* universe from the Newtonian point of view, EINSTEIN noticed a serious shortcoming of the POISSON equation,

$$\nabla^2 \varphi = 4\pi k \mu, \tag{7.38}$$

viz. its incompatibility with the obvious solution

$$\varphi = \text{const.} \quad (\mu \neq 0); \tag{9.15}$$

in other words, one could not have, according to the Newtonian point of view, a *homogeneous* universe with a *finite* density of matter. This defect could be easily remedied by modifying the POISSON equation to

$$\nabla^2 \varphi - \lambda \varphi = 4\pi k \mu, \tag{7.38'}$$

where λ is a certain universal constant[3] having the dimensions of (length)$^{-2}$.

---

[1] In the sequel, $R(t)$ denotes the radius of curvature of the cosmic space *at time t*.

[2] H. P. ROBERTSON, *Astroph. J.* **82**, 284(1935) and **83**, 187, 257(1936). A. G. WALKER, *Proc. Lond. Math. Soc.* (2), **42**, 90(1936).

[3] A positive λ implies some sort of a (cosmical) repulsion, superposed on the (conventional) gravitational attraction. The field of a point mass, for instance, is now 'screened' over distances of the order of $(\lambda)^{-1/2}$, which would be of the order of the radius of curvature of the three-space. For smaller distances, as encountered within a galaxy, the aforementioned modification would be unimportant.

Now one indeed obtained a solution for the *smeared-out* universe, viz.

$$\varphi = \text{const.} = -4\pi k\mu/\lambda. \tag{9.16}$$

EINSTEIN thereupon felt that a similar modification may be necessary in the field equations of general relativity as well. Accordingly, he introduced the so-called *cosmological term* into his field equations whereby they became, see the relevant equations of Sec. 7.5,

$$G_{ik} - \Lambda g_{ik} \equiv R_{ik} - \left(\Lambda + \frac{1}{2}R\right)g_{ik} = 8\pi\frac{k}{c^4}T_{ik}, \tag{9.17}$$

that is,

$$G_k^i - \Lambda\delta_k^i \equiv R_k^i - \left(\Lambda + \frac{1}{2}R\right)\delta_k^i = 8\pi\frac{k}{c^4}T_k^i; \tag{9.18}$$

the constant Λ here is referred to as the *cosmological constant* [which, again, has the dimensions of (length)$^{-2}$]. Contracting the last result, we get

$$R - \left(\Lambda + \frac{1}{2}R\right)4 = 8\pi\frac{k}{c^4}T, \tag{9.19}$$

so that

$$R = -8\pi\frac{k}{c^4}T - 4\Lambda. \tag{9.20}$$

Substituting this into (9.18), we finally obtain

$$R_k^i + \Lambda\delta_k^i = 8\pi\frac{k}{c^4}\left(T_k^i - \frac{1}{2}T\delta_k^i\right); \tag{9.21}$$

cf. Eq. (7.67).

Next, we need an appropriate expression for the energy-momentum tensor characterizing the cosmic fluid. We adopt, for this purpose, the most natural model, viz. that of a perfect fluid. Now, for *incoherent* matter

$$T^{ik} = \varepsilon^0 u^i u^k, \tag{9.22}$$

where $\varepsilon^0(=\mu^0c^2)$ is the 'proper' energy density (as measured in the local rest system of the fluid) while (u^i) is the unit velocity four-vector. We also know (see Sec. 3.6) that the purely spatial part of this tensor, $(T^{\alpha\beta})$, represents the momentum current density; for instance, the component $T^{\alpha\beta}$, in particular, stands for the current density, in the β-direction, of the α-component of the momentum. In the local rest system of the incoherent matter, this part would vanish. However, in the case of a fluid, even in the local rest system, a momentum current arising from the fluid pressure p^0 would exist.

In view of the fact that our fluid is perfect (in the sense that it cannot withstand shear), the pressure on a surface imagined in the fluid would act *normally* to the surface; the spatial tensor $(T^{\alpha\beta})$ would, therefore, be diagonal. Further, since the pressure is transmitted equally in all directions (PASCAL's law), we must have (in the local rest system)

$$T^{\alpha\beta} = p^0 \, \delta^{\alpha\beta}, \tag{9.23}$$

where

$$\delta^{\alpha\beta} = \begin{cases} 1 & \text{for } \alpha = \beta, \\ 0 & \text{for } \alpha \neq \beta. \end{cases} \tag{9.24}$$

Of course, the components T^{x4} and $T^{4\beta}$, which stand for the energy flow, must be identically zero in the system chosen; T^{44} is, as usual, $-\varepsilon^0$. The relevant tensor components in the rest system are, then,

$$(T^{ik})^0 = \begin{bmatrix} p^0 & 0 & 0 & 0 \\ 0 & p^0 & 0 & 0 \\ 0 & 0 & p^0 & 0 \\ 0 & 0 & 0 & -\varepsilon^0 \end{bmatrix}. \tag{9.25}$$

In an arbitrary system of reference, the energy-momentum tensor must be such that for $p^0 = 0$ it leads to (9.22) and for $u = 0$ to (9.25). Written in the mixed form, we have

$$T^i_k = (\varepsilon^0 + p^0)\, u^i\, u_k + p^0\, \delta^i_k; \tag{9.26}$$

here, (δ^i_k), unlike the $(\delta^{\alpha\beta})$ of (9.24), is a tensor.

We substitute (9.26) into the right-hand side of the cosmical field equations (9.21), and use (9.14) with the relevant formulae of tensor calculus to obtain the left-hand side. As a result, we obtain the following explicit relations:

$$\frac{3}{R^2}\left[\kappa + \frac{1}{c^2}\left(\frac{dR}{dt}\right)^2\right] = \Lambda + 8\pi\, \frac{k}{c^4}\, \varepsilon, \tag{9.27}$$

and

$$\frac{1}{R^2}\left[\kappa + \frac{1}{c^2}\left\{ \left(\frac{dR}{dt}\right)^2 + 2R\,\frac{d^2R}{dt^2} \right\} \right] = \Lambda - 8\pi\,\frac{k}{c^4}\,p\,;\qquad (9.28)$$

we omit superscripts from ε and p and, henceforth, understand them to be the (averaged) quantities measured by *co-moving* observers.

Equations (9.27) and (9.28) imply that

$$\frac{d\varepsilon/dt}{(\varepsilon + p)} = -\frac{3}{R}\,\frac{dR}{dt}\,;\qquad (9.29)$$

thus, if the equation of state, i.e. the ε-p relation, of the cosmic fluid is known, the functional dependence of ε on R (or, what is the same thing, of p on R) can be determined straightaway.[1] It is then sufficient to consider only one of the two equations, say (9.27), and write

$$t = \int \frac{\frac{1}{c}\,dR}{\sqrt{\left[\Lambda + 8\pi\,\dfrac{k}{c^4}\,\varepsilon\,(R) \right]\dfrac{R^2}{3} - \kappa}} + \text{const.}\qquad (9.30)$$

Equations (9.29) and (9.30) solve the problem of determining the geometrical evolution of the universe.

About the nature of the cosmic fluid, there are two extreme choices. Firstly, one may assume that most of the energy density in the universe arises from radiation; in that case, the fluid would behave, more or less, in the same way as radiation. We would then have: $p \simeq \varepsilon/3$; consequently, Eq. (9.29) would lead to the relation

$$\varepsilon \propto R^{-4},\qquad (9.31)$$

which can be understood directly in terms of the WIEN's displacement law. It is conceivable that the assumption leading to (9.31) was fairly valid in the early stages of the cosmic evolution.

Secondly, one may assume that most of the energy density in the universe exists in the form of condensed matter; in that case, our fluid would behave

---

[1] It may be mentioned here that the relation (9.29) can be obtained directly by considering the thermodynamic equilibrium of the cosmic fluid; see, for instance, L. D. LANDAU and E. M. LIFSHITZ, *Classical Theory of Fields*, loc. cit., Sec. 107.

practically like a gas, in which the bulk of the energy appears as the rest energy of the particles and only a small fraction as the energy of thermal agitation. The pressure of the fluid would then be much smaller than the energy density: $p \ll \varepsilon$; consequently, Eq. (9.29) would lead to the relation

$$\varepsilon \propto R^{-3} . \tag{9.32}$$

Now, the actual situation in the universe at the present epoch must be somewhere between the two aforementioned extremes; observationally, however, it is known[1] that the pressure (i.e. the energy density associated with all types of random motion) is much smaller, by a factor of about $10^5 - 10^6$, than the energy density due to condensed matter. Energy density due to radiation is also estimated to be much less than that due to condensed matter—by a factor of about 10^3. Consequently, the second of the two choices seems much closer to reality than the first. Accordingly, we adopt relation (9.32) for our considerations in the sequel. Equation (9.30) then becomes

$$t = \int \frac{\frac{1}{c} \, dR}{\sqrt{\frac{1}{3} \Lambda R^2 + \frac{C}{R} - \kappa}} + \text{const.,} \tag{9.33}$$

where

$$C = \frac{8\pi k}{3c^4} (\varepsilon R^3) \tag{9.34}$$

—a *positive constant* which measures the total energy content of the universe (and has the dimensions of a length). Customarily, (9.33) is written in the form [see also (9.27)]

$$\left(\frac{dR}{dt} \right)^2 = c^2 \left[\frac{1}{3} \Lambda R^2 + \frac{C}{R} - \kappa \right] . \tag{9.35}$$

Different choices of the parameters Λ and κ lead to a large variety of cosmological models; a detailed analysis of the distinctive features of these models, particularly their relation with the known behavior of the actual universe, is carried out in the next section.

It appears worthwhile to mention here that MILNE and McCREA, in 1934, showed that the differential equation (9.35) could also be obtained within

---

[1] See H. P. ROBERTSON and T. W. NOONAN, *loc. cit.*, Secs. 17.1 and 18.4,

the framework of the Newtonian theory.[1] This is not surprising because in the relativistic treatment too we have employed a *universal* time t, independent of all observers, and have introduced the spatial line element independently of the temporal part; see Eq. (9.1). The physical interpretation of the various terms appearing in Eq. (9.35) also turns out to be similar, except that the constant κ is no longer a representative of the curvature of the three-space; it is rather a measure of the total energy of the particles in the universe, while C becomes a measure of the total matter content. The physical significance of this similarity is debatable.

9.4. Distinctive features of the various models

Before we discuss the evolutionary aspects of the problem, it would be worthwhile to have a feeling for the *instantaneous* geometry of the physical space; this presents the following three possibilities (see also Sec. 9.2) :

(i) $\kappa = 0$; correspondingly, the space is Euclidean and the surface area of a sphere of radius r in it is $4\pi r^2$. The model is 'flat' and 'open'; the former signifies the absence of curvature, the latter implies that there is no upper limit to the size of a sphere that can be drawn in this space.

(ii) $\kappa = +1$; correspondingly, the space is spherical. The radial distance L of the point (r', θ, φ) from the origin is $a\chi = a \sin^{-1} (r'/a)$; see (9.8) and (9.9). The distance L is invariably larger than the radial coordinate r'; it is only for $r' \ll a$ that $L \simeq r'$. The deviation becomes considerable when r' is comparable to a. The surface area of a sphere of 'radius' r' is $4\pi r'^2 = 4\pi a^2 \sin^2 \chi$; being equal to $4\pi L^2 \left(\dfrac{\sin \chi}{\chi} \right)^2$, it is invariably smaller than $4\pi L^2$. As we move out from the origin the surface area of our sphere first increases, reaches its maximum value, $4\pi a^2$, at a distance $\pi a/2$ (corresponding to $\chi = \pi/2$); thereafter it starts decreasing and finally reduces to zero, corresponding to a point-surface (at the 'opposite pole' of the space), distant πa (for $\chi = \pi$). πa is, therefore, the largest direct distance that can exist in such a space. At the same time we note that the coordinate value r' is at most equal to a [see also (9.7)], and is equal

[1] E. A. MILNE, *Quart. J. Math.* (Oxford Ser.) **5**, 64 (1934); W. H. McCREA and E. A. MILNE, *ibid.* **5**, 73 (1934). See also W. H. McCREA, in *Radio Astronomy Today* (Manchester University Press, 1963), p. 206,

to zero at the 'opposite pole' as well as at the origin. The total volume of this space is equal to, see (9.9),

$$V = \int_{\varphi=0}^{2\pi} \int_{\theta=0}^{\pi} \int_{\chi=0}^{\pi} a^3 \sin^2 \chi \sin \theta \, d\chi \, d\theta \, d\varphi$$

$$= 2\pi^2 a^3, \tag{9.36}$$

a, of course, being the radius of curvature of the space; see also Eq. (A.36) which follows directly from the metric (9.5). Thus, a space of positive curvature is closed on itself; its volume, at any instant of time, is finite though it has no boundaries as such.

(iii) $\kappa = -1$; correspondingly, the space is hyperbolic. The radial distance L of the point (r', θ, φ) from the origin is $b\chi = b \sinh^{-1} (r'/b)$; see (9.8') and (9.13). This distance is invariably smaller than the radial coordinate r'; again, it is only for $r' \ll b$ that $L \simeq r'$. The surface area of a sphere of radius r' in this space is $4\pi r'^2 = 4\pi b^2 \sinh^2 \chi$; being equal to $4\pi L^2 \left(\dfrac{\sinh \chi}{\chi} \right)^2$, this is invariably larger than $4\pi L^2$. Clearly, the surface area of our sphere in this case is a monotonically increasing function of the radial distance and, in principle, there is no upper limit to their values. A space of negative curvature is, thus, an open one whose total volume, at any instant of time, is infinite.

We shall now consider the variety of solutions afforded by the cosmological differential equation

$$\left(\frac{dR}{dt} \right)^2 = c^2 \left[\frac{1}{3} \Lambda R^2 + \frac{C}{R} - \kappa \right], \tag{9.35}$$

which enable us to visualize the geometrical evolution of the universe.[1] To recapitulate, the function $R(t)$ here stands for the radius of curvature of the universe,[2] Λ is the cosmological constant whose sign and magnitude

---

[1] A detailed, systematic study of this problem was initiated by A. FRIEDMANN, *Zeits. f. Phys.* 10, 377 (1922); 21, 326 (1924). The resulting solutions were applied to the actual universe by G. LEMAITRE, *Ann. Soc. Sci. Bruxelles* A 47, 49 (1927).

For a summary of the work done on static and nonstatic models till the end of 1932, see H. P. ROBERTSON, *Revs. Mod. Phys.* 5, 62 (1933).

[2] For $\kappa = +1$, $R(t)$ is also a measure of the actual size of the universe. However, for $\kappa = 0$, and for Newtonian cosmology, $R(t)$ is no more than a scale factor for measuring distances in the (nonstatic) universe.

are both as yet unknown, C is a positive constant characteristic of the universe,

$$C = \frac{8\pi k}{3c^4}\,(\varepsilon R^3), \tag{9.34}$$

while κ is the index of curvature of the space, which may be $+1$, 0 or -1.

We start with the consideration of *closed* models: $\kappa = +1$; this case affords the largest variety of models (see the first row in Fig. 9.1).

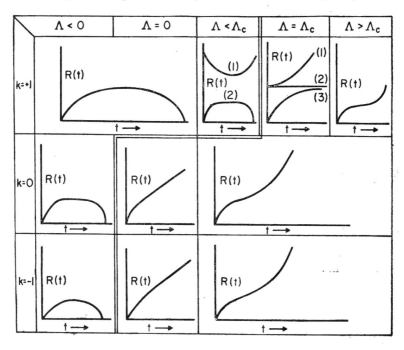

Fig. 9.1.

For $\Lambda \leqslant 0$, we have an oscillating model;[1] the function $R(t)$ first increases with t but the rate of increase goes on decreasing until $R(t)$ attains a maximum value, which is a solution of the cubic equation

$$\frac{1}{3}\,\Lambda R^3 - R + C = 0, \tag{9.37}$$

namely

$$R_{\max} = \frac{2}{|\Lambda|^{1/2}}\,\sinh\left\{\frac{1}{3}\,\sinh^{-1}\left(\frac{3}{2}\,C\,|\Lambda|^{1/2}\right)\right\}.$$

---

[1] The case $\Lambda = 0$ affords a parametric solution for the function $R(t)$; it turns out to be a cycloid of height C and period $\pi C/c$; see Eqs. (9.46). Considered first by FRIEDMANN, this model was later advocated strongly by EINSTEIN, *Berlin Sitzungsberichte* (1931), p. 235.

After this, a phase of contraction sets in and the model passes through all its previous stages of evolution in the reverse order; this continues till R becomes very small, eventually producing densities where present understanding of physics breaks down (with, perhaps, a further expansion from the 'primaeval' matter and so on, in an endless cycle of pulsations). Of course, our analytical expressions would probably cease to be applicable when R is too small; it seems more plausible that, as R becomes small enough, the resulting high density of matter (and radiation) would start off some unknown mechanism which brings about a rather early beginning of the next phase of expansion.

For $0 < \Lambda < \Lambda_c$, where $\Lambda_c = 4/(9C^2)$, we have two solutions:

(1) A solution characterized by the passage of $R(t)$ through a minimum value,

$$R_{\min} = \frac{2}{\Lambda^{1/2}} \cos \left\{ \frac{\pi}{6} + \frac{1}{3} \sin^{-1} \left(\frac{\Lambda}{\Lambda_c} \right)^{1/2} \right\},$$

along with the extreme behavior

$$R(t) \propto \exp \left[(\Lambda/3)^{1/2} c \, |t| \right], \tag{9.38}$$

both in the infinite past and the infinite future; its first phase, therefore, corresponds to contraction and the second one to expansion.

(2) An oscillating solution, formally similar to the one for $\Lambda \leqslant 0$, with

$$R_{\max} = \frac{2}{\Lambda^{1/2}} \sin \left\{ \frac{1}{3} \sin^{-1} \left(\frac{\Lambda}{\Lambda_c} \right)^{1/2} \right\}.$$

For $\Lambda = \Lambda_c$, we have three solutions, of which one is purely *static* [labelled (2) in the figure]; this corresponds to a *constant* value of R, given by

$$R_c = (\Lambda_c)^{-1/2} = \frac{3}{2} C. \tag{9.39}$$

This is the famous EINSTEIN model, discovered in 1917;[1] the very essential role played by the cosmological constant Λ in this case is obvious. Another solution [labelled (1)] is the so-called EDDINGTON-LEMAITRE model;[2] this is

[1] A. EINSTEIN, *Berlin Sitzungsberichte* (1917), p. 142; English translation available in *The Principle of Relativity, loc. cit.*

[2] A. S. EDDINGTON, *Mon. Not. Roy. Astr. Soc.* 90, 668(1930); G. LEMAITRE, *ibid.* 91, 490 (1931).

an ever-expanding model, such that, as $t \to -\infty$, R approaches R_c asymptotically from above. It may, therefore, be looked upon as originally conforming to an EINSTEIN type of model which, at some instant of time in the distant past and for some physically inevitable reason, left the EINSTEIN state and started expanding—first gradually, then faster and finally in the manner of (9.38). The third solution in this case corresponds to a model which starts according to the pattern determined by[1]

$$R(t) = (9C/4)^{1/3} (ct)^{2/3}, \tag{9.40}$$

but the expansion gradually slows down, with R approaching $\tilde{R}_c$ asymptotically from below. This represents the limiting case of an oscillating model, with *infinite* time period.

For $\Lambda > \Lambda_c$, we have the LEMAITRE model,[2] which starts in accordance with (9.40) and indeed the expansion first slows down; however, this slowing down continues only upto a particular stage of the evolution, after which the rate of expansion goes on increasing. The model ends up, as $t \to \infty$, in the manner of (9.38).

We shall now consider *flat* models : $\kappa = 0$; see the second row in Fig. 9.1. Here, for $\Lambda < 0$ we have the explicit solution

$$R = \left[\frac{3}{2} \frac{C}{|\Lambda|} (1 - \cos \{ct\, 3\, |\Lambda|\,)^{1/2}\}) \right]^{1/3}, \tag{9.41}$$

which corresponds to an oscillating model. For $\Lambda = 0$, the solution is exactly (9.40), valid for *all* t. This yields a monotonically expanding model, in which the rate of expansion gradually slows down with time; $R(t)$, however, goes on increasing indefinitely.[3] For $\Lambda > 0$, we again have an explicit solution,

$$R = \left[\frac{3}{2} \frac{C}{\Lambda} (\cosh \{ct(3\Lambda)^{1/2}\} - 1) \right]^{1/3}, \tag{9.42}$$

which corresponds to an ever-expanding universe; qualitatively, this model is similar to the LEMAITRE model discussed above.

---

[1] Quite generally, when R is much smaller than C and $\Lambda^{-1/2}$, Eq. (9.35) reduces to $(dR/dt) \simeq c(C/R)^{1/2}$ which leads to (9.40).

[2] This model was first discussed by G. LEMAITRE in 1927, *loc. cit.*

[3] This model was first proposed by A. EINSTEIN and W. DE SITTER in 1932; see *Proc. Nat. Acad. Sci.* **18**, 213.

Next, we have the *open curved* models : $\kappa = -1$; see the third row in Fig. 9.1. Here again we have three different possibilities, corresponding to the choices $\Lambda < 0$, $\Lambda = 0$ and $\Lambda > 0$. Of course, in these cases we do not obtain explicit solutions which would hold for all t ;[1] nevertheless, the solutions here possess the same qualitative features as in the corresponding cases of flat models.

A special model which could not be included in the foregoing study is the one corresponding to a *flat* space, which is (practically) *devoid of matter or radiation*, i.e. a (practically) empty Euclidean space.   In this case, we obtain directly from Eqs. (9.27) and (9.28), with substitutions $\kappa = 0$ and $\varepsilon = p = 0$,

$$R(t) \propto \exp\left[(\Lambda/3)^{1/2}\, ct\right] \quad \text{for all } t\,; \tag{9.43}$$

see also Eq. (9.35).   Clearly, in this case,

$$\frac{1}{R}\frac{dR}{dt} = \text{const.} \left[= \left(\frac{\Lambda}{3}\right)^{1/2} c\right]; \tag{9.44}$$

thus, the model possesses the remarkable property of being *stationary*, i.e. presenting the same picture at all times, with no preferred choice of the origin $t = 0$.   It is, thus, physically akin to the model envisaged in the steady-state theory of cosmology.[2]   However, in relativistic cosmology this model does not evoke much interest, for it suffers from the undesirable defect of being empty.   Nonetheless, it does represent the common limiting case to which all indefinitely expanding models of the universe, with $\Lambda > 0$, tend as $t \to \infty$ (i.e. as $R \to \infty$ and, hence, $\varepsilon \to 0$).

It may be mentioned here that the foregoing model was shown by LEMAITRE,[3] and independently by ROBERTSON,[4] to follow directly from the famous DE SITTER model, discovered as early as 1917,[5] by means of a suitable space-time transformation.

We thus obtain a large variety of models for the geometrical evolution of

---

[1] The case $\Lambda = 0$ does afford a parametric solution for the function $R(t)$; see Eqs. (9.47).

[2] In the steady-state theory, the parameter ε stays constant (with the help of a continuous creation of matter in the expanding universe) ; accordingly, $C \propto R^3$. The terms $(1/3)\Lambda R^2$ and C/R can then be combined into a single term, $(1/3)\Lambda' R^3$ say, with $\Lambda' = \Lambda + 3C/R^3 = \Lambda + 8\pi k\varepsilon/c^4$. This leads naturally to Eqs. (9.43) and (9.44), with Λ replaced by Λ'.

[3] G. LEMAITRE, *J. Math. and Phys.* (M.I.T.) **4**, 188 (1925).

[4] H. P. ROBERTSON, *Phil. Mag.* **5**, 835 (1928).

[5] W. DE SITTER, *Proc. Akad. Wetensch. Amsterdam* **19**, 1217 (1917).

the universe. To a considerable extent, this variety arises from the latitude provided by the uncertainty in the value, even the sign, of the cosmological constant Λ. Unfortunately, there is no final opinion yet regarding the necessity, or otherwise, of the introduction of this constant into the cosmological theory.

9.5. Red shift in the spectral lines of distant galaxies

In this section we propose to discuss the important problem of the observed red shift in the spectral lines of distant galaxies. This observation seems to imply a universal recession of galaxies from one another, leading thereby to the conclusion that the universe, at the present epoch, happens to be in a phase of expansion—hence the term *expanding universe*. In order to make a quantitative analysis of this problem we have to consider the temporal behavior of the function $R(t)$, for it determines the manner in which the distance scales in the universe change with time. It is, therefore, advisable to write down, if possible, analytical expressions for this function for all those models that seem worthy of consideration.

For simplicity, we set $\Lambda = 0$. The cosmological equation (9.35) then simplifies to the following:

$$\left(\frac{dR}{dt}\right)^2 = c^2 \left(\frac{C}{R} - \kappa\right), \quad C = \frac{8\pi k}{3c^4} (\varepsilon R^3). \tag{9.45}$$

In turn, we obtain three different solutions; see the respective (R, t)-plots in the second column of Fig. 9.1.

In the case of a *closed* model ($\kappa = +1$), we have the parametric solution

$$R = \frac{C}{2} (1 - \cos \eta), \quad t = \frac{C}{2c} (\eta - \sin \eta), \tag{9.46}$$

which corresponds to a cycloidal variation of R, starting from the (catastrophic!) value $R = 0$ at $t = 0$, passing through the maximum value $R = C$ at $t = \pi C/(2c)$ and again returning to the (catastrophic!) value $R = 0$ at $t = \pi C/c$. The parameter η covers the range $(0, 2\pi)$; the phase of expansion corresponds to the range $(0, \pi)$ while the phase of contraction corresponds to the range $(\pi, 2\pi)$. This refers to a single cycle of evolution; in reality, we face the possibility of a succession of such cycles—even an endless succession!

In the case of an *open curved* model ($\kappa = -1$), we again have a parameteric solution, viz.

$$R = \frac{C}{2}\,(\cosh \eta - 1), \quad t = \frac{C}{2c}\,(\sinh \eta - \eta)\,; \tag{9.47}$$

here, for $0 < \eta < \infty$, the function R increases monotonically with t. Of course, for $-\infty < \eta < 0$, R decreases monotonically with t; however, in the face of actual observations, this possibility does not arouse any interest.

In the case of a *flat* model ($\kappa = 0$), we have the explicit solution

$$R = (9C/4)^{1/3}\,(ct)^{2/3}, \tag{9.48}$$

which corresponds to the EINSTEIN-DE SITTER model discussed earlier; this too is monotonically expanding.

We now take up the question of the observed red shift of the spectral lines from distant galaxies. The most plausible explanation of this phenomenon lies in the time-variation of the scale factor $R(t)$ of the universe. A light wave received on the Earth at time t_1 would have a wavelength proportional to the value $R(t_1)$ of the scale factor R. When it left its source, at time t_2, its wavelength was proportional to the value $R(t_2)$ of the scale factor at *that* time. Accordingly,

$$\frac{\lambda_{\text{received}}}{\lambda_{\text{emitted}}} \equiv 1 + z = \frac{R(t_1)}{R(t_2)}, \tag{9.49}$$

as if the wave got stretched along with the space itself! For $z \ll 1$, we can write

$$z \simeq (\dot{R}/R)\,(t_1 - t_2) = (H/c)\,\delta l, \tag{9.50}$$

where $\delta l\ \{ = c(t_1 - t_2)\}$ is the distance of the source from the observer while H is the *Hubble parameter* of the universe:

$$H(t) = (\dot{R}/R). \tag{9.51}$$

Equation (9.50) embodies the so-called *Hubble law* of galactic red shifts.[1]

It is tempting to view this phenomenon as a (cosmological) Doppler effect, arising from the 'running away' of the cosmical bodies from one another.

---

[1] E. HUBBLE, *Astroph. J.* **84**, 158, 270, 517(1936); *Mon. Not. Roy. Astr. Soc.* **113**, 658 (1953). See also M. HUMASON, N. U. MAYALL and A. R. SANDAGE, *Astron. J.* **61**, 97 (1956).

To the *first order of approximation*, the fractional shift in the frequency of a light wave, observed in a co-moving system, would be given by 'the ratio of the velocity of recession v of the source (with respect to the system of observation) to the velocity of light' ; thus

$$\frac{\Delta \nu}{\nu} \simeq -\frac{v}{c}. \tag{9.52}$$

Comparing (9.50) and (9.52), we conclude that

$$v \simeq H \, \delta l, \tag{9.53}$$

which is another version of the HUBBLE law. Relation (9.53) enables us to estimate the linear dimensions of the 'observable' universe, i.e. of the order of (c/H).

In view of (9.45), we can write

$$H = c \left[\frac{8\pi k}{3c^2} \mu - \frac{\kappa}{R^2} \right]^{1/2}, \tag{9.54}$$

where $\mu (= \varepsilon/c^2)$ is the average mass density in the universe. Thus, if $\Lambda = 0$, our universe will correspond to a *closed* model, a *flat* model or an *open curved* model according to whether the observed value of the HUBBLE parameter is less than, equal to or greater than $\sqrt{(8\pi k \mu/3)}$, i.e. whether the average mass density in the universe is greater than, equal to or less than the critical value μ_c, where

$$\mu_c = \frac{3H^2}{8\pi k}. \tag{9.55}$$

Now, the most recent estimate of the HUBBLE parameter is[1] (55 ± 7) km sec^{-1} Mpc^{-1}, i.e. about $(1.78 \pm 0.23) \times 10^{-18}$ sec^{-1}. Accordingly, the critical mass density μ_c is about 0.57×10^{-29} g/cc. Unfortunately, the observational information on the average mass density in the universe is not good enough for judging whether $\mu > \mu_c$, $\mu = \mu_c$ or $\mu < \mu_c$; in fact, according to the cosmological data to date, μ can be placed anywhere around 10^{-29} g/cc within a factor of 10 or so. Hence, the question of the geometrical nature of the space around us remains unsettled for lack of sufficiently accurate data.

A knowledge of the Hubble parameter H can also be utilised for estimating the cosmical time t corresponding to the present epoch—in other words,

[1] A. SANDAGE, *Astroph. J.* **178**, 1(1972).

the *age of the universe*. Assuming, first of all, the *flat* model, we obtain from (9.48) and (9.51)

$$H = \frac{2}{3t} \quad \text{or} \quad t = \frac{2}{3H}. \tag{9.56}$$

Substituting the observed value of H, this gives

$$t \sim 3.7 \times 10^{17} \text{ sec} = 1.2 \times 10^{10} \text{ yr}, \tag{9.57}$$

which is comparable with the estimated upper limit for the age of the oldest galaxy.[1]

Next, assuming the *closed* model, we have from (9.46) and (9.51)

$$Ht = \frac{\sin \eta \, (\eta - \sin \eta)}{(1 - \cos \eta)^2}. \tag{9.58}$$

Now, the right-hand side, for $0 < \eta < \pi$ (which is necessary for *expansion*), lies in the range $\left(0, \frac{2}{3}\right)$. Accordingly, the age of the universe in this case would be less than the estimate (9.57).

Finally, assuming the *open curved* model, we have from (9.47) and (9.51)

$$Ht = \frac{\sinh \eta \, (\sinh \eta - \eta)}{(\cosh \eta - 1)^2}, \tag{9.59}$$

which, for $0 < \eta < \infty$, lies in the range $\left(\frac{2}{3}, 1\right)$. Accordingly, the age of the universe in this case would be greater than the estimate (9.57) but less than about 1.8×10^{10} yr.

It is again not possible to decide on the basis of these estimates whether the actual universe conforms to one type of model or the other. Obviously, for settling this issue, one needs considerably more and refined data about the universe than are available at present.

While empirical evidence remains impartial in regard to different models, theoretical opinion is now shifting in favor of the 'oscillating' models.[2]

---

[1] A. SANDAGE, *Astroph. J.* **183**, 711(1973).

[2] See, for instance, M. J. REES, *Observatory* **89**, 193(1969); A. R. SANDAGE, *Physics Today*, **23**, no. 2, 34(1970); E. R. HARRISON, *Physics Today* **25**, no. 12, 30(1972); R. K. PATHRIA, *Nature* **240**, 298(1972).

According to these, the red shift we have been discussing here would eventually become *blue shift* (when the phase of contraction sets in) ; ultimately, there would be a 'reuniting' of the galaxies into a practically singular state—just like the one obtaining initially. However, this ultimate situation may not be taken too seriously because one doesn't really expect it to be a *true* singularity, but only a 'small' space-time region for which our theoretical approximations would presumably break down ; then, by means of some unknown mechanism, expansion may restart, setting the universe on another cycle of evolution.

Apart from the welcome feature that such a universe is a closed one, with no boundaries to bother us about the boundary conditions to be satisfied by the solutions of the relevant field equations, there is another interesting point about it, i.e. a universe like this, in a certain sense, satisfies the perfect cosmological principle as well (see Sec. 9.1). Averaged over a large number of oscillations, it would present practically the same appearance *throughout* time while, at any given instant of time, it would be homogeneous and isotropic *throughout* space.

9.6. Cosmological parameters for the zero-pressure Friedmann models

A comparative study of the various cosmological models can be carried out in terms of the dimensionless parameters q and σ, where

$$q \equiv - \ddot{R}R/\dot{R}^2, \tag{9.60}$$

and

$$\sigma \equiv 4\pi k\mu/(3H^2) = c^2C/(2H^2R^3) ; \tag{9.61}$$

obviously, q is referred to as the *deceleration parameter* of the universe and σ the *density parameter*. Since the HUBBLE parameter $H = \dot{R}/R$, Eq. (9.35) which governs the evolution of all zero-pressure FRIEDMANN models of the universe leads to the relationships

$$\Lambda = 3(H^2/c^2) (\sigma - q) \tag{9.62}$$

and

$$\kappa/R^2 = (H^2/c^2) (3\sigma - q - 1) ; \tag{9.63}$$

moreover, the critical parameter Λ_c takes the form

$$\Lambda_c \equiv 4/(9C^2) = (H^2/c^2) (3\sigma - q - 1)^3/(9\sigma^2). \tag{9.64}$$

For the age of the universe, we obtain (the suffix 0 denotes the present epoch)

$$t_0 = \int^{R_0} \frac{dR}{c \left(\frac{1}{3} \Lambda R^2 + \frac{C}{R} - \kappa \right)^{1/2}} + \text{const.} \tag{9.33'}$$

$$= \frac{1}{H_0} \int^1 \frac{dy}{\{(\sigma_0 - q_0) \, y^2 + 2\sigma_0 y^{-1} - (3\sigma_0 - q_0 - 1)\}^{1/2}} + \text{const.} \tag{9.65}$$

For *big-bang* models, the singular state $R = 0$ provides a natural choice for the origin of the time scale t; Eq. (9.65) then becomes

$$t_0 = \frac{1}{H_0} \int_0^1 \frac{dy}{\{(\sigma_0 - q_0) \, y^2 + 2\sigma_0 y^{-1} - (3\sigma_0 - q_0 - 1)\}^{1/2}} . \tag{9.66}$$

The foregoing relations enable us to derive a number of results involving the kinematics of the universe. In this connection, the parameters H_0, q_0 and σ_0 may be regarded as *observables*, from which information regarding Λ, κ, R_0 and t_0 may be obtained. The empirical value of H_0 has already been quoted in Sec. 9.5. For q_0, the painstaking work of SANDAGE and his co-workers[1] suggests a value of 1 ± 1 which implies that, with a high probability, the universe at present is decelerating ($q_0 > 0$); see Eq. (9.60). The empirical value of σ_0 is again of the order of unity, though the accompanying uncertainty, especially on the higher side, is rather large; of course, we must have $\sigma_0 > 0$.

If we assume that $\Lambda = 0$, the following simplifications result:

$$\sigma_0 = q_0, \tag{9.67}$$

$$\kappa/R_0^2 = (H_0^2 / c^2) \, (2q_0 - 1), \tag{9.68}$$

and

$$t_0 = \frac{1}{H_0} \int_0^1 \frac{dy}{\{2q_0 y^{-1} - (2q_0 - 1)\}^{1/2}} . \tag{9.69}$$

We still have three possibilities:

(i) $\sigma_0 = q_0 > \frac{1}{2}$, which implies that $\kappa = +1$ and hence

$$1/R_0^2 = (H_0^2 / c^2) \, (2q_0 - 1) \tag{9.70}$$

---

[1] A. SANDAGE and E. HARDY, *Astroph. J.* **183**, 743(1973). These findings rule out the value $q = -1$ suggested by the steady-state theory of cosmology; see Eqs. (9.43) and (9.44).

and

$$H_0 t_0 = \frac{2q_0}{(2q_0 - 1)^{3/2}} \cos^{-1} \frac{1}{\sqrt{2q_0}} - \frac{1}{(2q_0 - 1)}, \tag{9.71}$$

which lies between 0 and 2/3.

(ii) $\sigma_0 = q_0 = \frac{1}{2}$, which implies that $\kappa = 0$ and, while R_0 becomes indeterminate, the age of the universe is given by[1]

$$H_0 t_0 = 2/3. \tag{9.72}$$

(iii) $\sigma_0 = q_0 < \frac{1}{2}$, which implies that $\kappa = -1$ and hence

$$1/R_0^2 = (H_0^2 / c^2)(1 - 2q_0) \tag{9.73}$$

and

$$H_0 t_0 = \frac{1}{(1 - 2q_0)} - \frac{2q_0}{(1 - 2q_0)^{3/2}} \cosh^{-1} \frac{1}{\sqrt{2q_0}}, \tag{9.74}$$

which lies between 2/3 and 1.

If, on the other hand, we assume $\kappa = 0$, the following simplifications result:

$$\sigma_0 = \frac{1}{3}(q_0 + 1), \tag{9.75}$$

$$\Lambda = (H_0^2 / c^2)(1 - 2q_0), \tag{9.76}$$

and

$$t_0 = \frac{1}{H_0} \int_0^1 \frac{\sqrt{3}\, dy}{\{(1 - 2q_0)y^2 + 2(q_0 + 1)y^{-1}\}^{1/2}} \cdot \tag{9.77}$$

Again we have three possibilities:

(i) $q_0 > \frac{1}{2}$, which implies that $\sigma_0 > \frac{1}{2}$, $\Lambda < 0$ and

$$H_0 t_0 = 2[3(2q_0 - 1)]^{-1/2} \cos^{-1} \left[\frac{3}{2(q_0 + 1)}\right]^{1/2}. \tag{9.78}$$

(ii) $q_0 = \frac{1}{2}$, which implies that $\sigma_0 = \frac{1}{2}$, $\Lambda = 0$ and

$$H_0 t_0 = 2/3. \tag{9.79}$$

---

[1] This case corresponds to the EINSTEIN-DE SITTER model of the universe for which $\sigma = q = 1/2$ and $Ht = 2/3$ at all times.

(iii) $q_0 < \dfrac{1}{2}$, which implies that $\sigma_0 < \dfrac{1}{2}$, $\Lambda > 0$ and[1]

$$H_0 t_0 = 2[3(1 - 2q_0)]^{-1/2} \cosh^{-1}\left[\frac{3}{2(q_0 + 1)}\right]^{/2}. \qquad (9.80)$$

An instructive study of the big-bang models can be made with the help of the *Robertson diagram* in which one plots $(H_0 t_0)$ as a function of σ_0, for various values of q_0; see Fig. 9.2. Firstly, one notes that, for any given $q_0 < -1$, σ_0 cannot be less than a certain critical value σ^* which is deter-

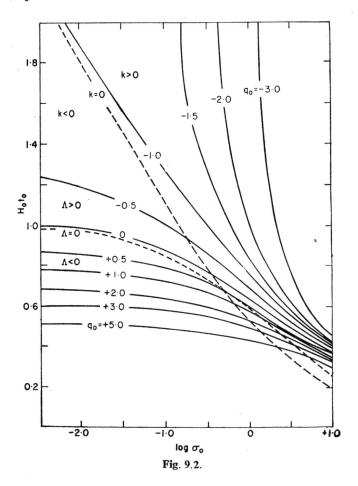

Fig. 9.2.

---

[1] The limiting case $q_0 \to -1$, $\sigma_0 \to 0$ corresponds to the DE-SITTER model of 1917.

mined by the (vertical) asymptote of the corresponding curve in the diagram.[1] The functional dependence of σ^* on q_0 can be determined by noting that the condition $q_0 < -1$ necessarily implies an *accelerating* model with $\kappa = +1$; see Eq. (9.63). Referring to the first row of Fig. 9.1, one further concludes that $\Lambda > \Lambda_c$ (the LEMAITRE model). Accordingly, we have, from Eqs. (9.62)-(9.64),

$$\sigma_0 > \frac{1}{3}(q_0 + 1), \tag{9.81}$$

as well as

$$\sigma_0 > \frac{1}{3}\left\{\frac{(q_0 + 1)^2}{2} + \left[\frac{(q_0 + 1)^4}{4} - \frac{(q_0 + 1)^3}{3}\right]^{1/2}\right\}. \tag{9.82}$$

The former inequality is redundant for $q_0 < -1$ (because $\sigma_0 > 0$ anyway); the latter one establishes the true lower limit for σ_0.

Secondly, one notes that, for any given $q_0 > -1$, $H_0 t_0$ cannot exceed a certain value $(Ht)^*$ which is determined by the (horizontal) asymptote of the corresponding curve in the diagram. By letting $\sigma_0 \to 0$, RINDLER obtained the inequality:

$$H_0 t_0 < \frac{1}{q_0^{1/2}} \sin^{-1}\left(\frac{q_0}{q_0 + 1}\right)^{1/2} \qquad (q_0 > -1); \tag{9.83}$$

for $-1 < q_0 < 0$, (9.83) may preferably be written as

$$H_0 t_0 < \frac{1}{(-q_0)^{1/2}} \sinh^{-1}\left(\frac{-q_0}{1 + q_0}\right)^{1/2}. \tag{9.84}$$

On the other hand, LANDSBERG and BROWN[2] have observed that by replacing the term $(\sigma_0 - q_0) y^2$ in the integrand of (9.66) by $(\sigma_0 - q_0)$ one obtains the following results (*valid for $\sigma_0 > \frac{1}{2}$*):

(i) for $\sigma_0 > q_0$, which implies $\Lambda > 0$,

$$H_0 t_0 > \frac{2\sigma_0}{(2\sigma_0 - 1)^{3/2}} \cos^{-1}\frac{1}{\sqrt{2\sigma_0}} - \frac{1}{(2\sigma_0 - 1)}; \tag{9.85}$$

---

[1] W. RINDLER, *Astroph. J. Letters* **157**, L147 (1969) ; *Essential Relativity* (Van Nostrand Reinhold Co., New York, 1969), Sec. 91.

[2] P. T. LANDSBERG and B. M. BROWN, *Astroph. J.* **182**, 653 (1973).

(ii)   for $\sigma_0 = q_0$, which implies $\Lambda = 0$,

$$H_0 t_0 = \frac{2\sigma_0}{(2\sigma_0 - 1)^{3/2}} \cos^{-1} \frac{1}{\sqrt{2\sigma_0}} - \frac{1}{(2\sigma_0 - 1)}; \qquad (9.86)$$

(iii)   for $\sigma_0 < q_0$, which implies $\Lambda < 0$,

$$H_0 t_0 < \frac{2\sigma_0}{(2\sigma_0 - 1)^{3/2}} \cos^{-1} \frac{1}{\sqrt{2\sigma_0}} - \frac{1}{(2\sigma_0 - 1)}. \qquad (9.87)$$

It can be readily seen that *if* $\sigma_0 < \dfrac{1}{2}$ the foregoing results get replaced by

$$H_0 t_0 \gtrless \frac{1}{(1 - 2\sigma_0)} - \frac{2\sigma_0}{(1 - 2\sigma_0)^{3/2}} \cosh^{-1} \frac{1}{\sqrt{2\sigma_0}}, \qquad (9.88\text{-}9.90)$$

respectively.   In the limit $\sigma_0 \to 0$, one obtains

$$H_0 t_0 \gtrless 1 \qquad (q_0 \lessgtr 0). \qquad (9.91\text{-}9.93)$$

The very last results, though correct as inequalities, are not as useful as RINDLER's inequality (9.83) for $q_0 > -1$. Moreover, for $q_0 < -1$, σ_0 cannot approach zero ; at best it can approach σ^* for which $(H_0 t_0) \to \infty$.

For $\sigma_0 = \frac{1}{2}$, we obtain from either set of results

$$H_0 t_0 \gtrless \frac{2}{3} \qquad \left(q_0 \lessgtr \frac{1}{2}\right). \qquad (9.94\text{-}9.96)$$

In the end we delimit models with $\kappa = +1$ and $\Lambda \leqslant \Lambda_c$; they conform to a black-hole universe.[1]   The density parameter σ_0 is then bounded by the inequalities[2]

$$\frac{1}{3}(q_0 + 1) < \sigma_0 \leqslant \frac{1}{3} \left\{ \frac{(q_0 + 1)^2}{2} + \left[\frac{(q_0 + 1)^4}{4} - \frac{(q_0 + 1)^3}{3} \right]^{1/2} \right\},$$

$$(9.97)$$

which imply that

$$q_0 > \frac{1}{2}, \qquad \sigma_0 > \frac{1}{2}. \qquad (9.98)$$

[1] R. K. PATHRIA, *Nature* **240**, 298 (1972).
[2] P. T. LANDSBERG and R. K. PATHRIA, *Astroph. J.* (in press).

Equations (9.62) - (9.64) then give

$$-\frac{H_0^2}{c^2}(2q_0 - 1) < \Lambda \leqslant \frac{H_0^2}{c^2}\left\{\left[\frac{(q_0 + 1)^2}{2} - 3q_0\right]\right.$$

$$\left. + \left[\frac{(q_0 + 1)^4}{4} - \frac{(q_0 + 1)^3}{3}\right]^{1/2}\right\}, \qquad (9.99)$$

and

$$\frac{1}{R_0^2} \leqslant \frac{H_0^2}{c^2}\left\{\left[\frac{(q_0 + 1)^2}{2} - (q_0 + 1)\right] + \left[\frac{(q_0 + 1)^4}{4} - \frac{(q_0 + 1)^3}{3}\right]^{1/2}\right\}.$$

$$(9.100)$$

The relevant bounds for $H_0 t_0$ are provided by the inequalities (9.85) - (9.87). Adding to the foregoing algebraic inequalities a reasonable upper limit to the empirical value of q_0, say $q_0 \leqslant 2$, we obtain

$$\frac{1}{2} < \sigma_0 \leqslant 2.62$$

$$-3\,(H_0^2 \,/\, c^2) < \Lambda \leqslant 1.85\,(H_0^2 \,/\, c^2)$$

$$\frac{1}{R_0^2} \leqslant 4.85\,(H_0^2 \,/\, c^2)$$

$$0.436 < H_0 t_0 < 0.667.$$

$$\left.\right\} \quad (9.101\text{-}9.104)$$

Further, adopting reasonable limits for the empirical value of H_0, say 40 km sec^{-1} Mpc$^{-1} \leqslant H_0 \leqslant 75$ km sec^{-1} Mpc^{-1}, we finally obtain

$$0.301 \times 10^{-29} < \mu_0 \leqslant 5.54 \times 10^{-29}\,(\text{g cm}^{-3})$$

$$-1.97 \times 10^{-56} < \Lambda \leqslant 1.22 \times 10^{-56}\,(\text{cm}^{-2})$$

$$5.60 \times 10^{27} \leqslant R_0\,(\text{cm})$$

$$0.57 \times 10^{10} < t_0 < 1.63 \times 10^{10}\,(\text{years}).$$

$$\left.\right\} \quad (9.105\text{-}9.108)$$

The upper limit for the age of the universe is consistent with the upper limit, of 1.3×10^{10} years, set by SANDAGE[1] for the age of the oldest galaxy.

---

[1] A. SANDAGE, *Astroph. J.* **183**, 711 (1973).

Problems

9.1. The metric of a four-dimensional Euclidean space is given by

$$dl^2 = \sum_{\alpha=1}^{3} (dx^\alpha \, dx^\alpha) + dx^4 \, dx^4,$$

and the surface of a sphere of radius R in this space is determined by the condition

$$\sum_{\alpha=1}^{3} (x^\alpha \, x^\alpha) + x^4 \, x^4 = R^2.$$

Eliminating x^4 from these equations, we obtain (for the metric on the surface of the sphere)

$$dl^2 = \sum_{\alpha=1}^{3} (dx^\alpha \, dx^\alpha) + \frac{\left(\sum\limits_{\alpha=1}^{3} x^\alpha \, dx^\alpha \right)^2}{R^2 - \sum\limits_{\alpha=1}^{3} (x^\alpha \, x^\alpha)}.$$

Show that this metric is equivalent to the metric (9.7) for a three-dimensional space of constant positive curvature. [Note that the fictitious space (x^α, x^4) is *not* intended to have anything to do with the space-time continuum.]

9.2. (i) Consider an n-dimensional space of constant curvature K. Its metric in the Cartesian coordinates $(x^1, x^2, \ldots, x^n)$ can be written in the form, see Eq. (A. 34),

$$ds^2 = \frac{dx^i \, dx^i}{\left(1 + \dfrac{K}{4} \, x^i x^i \right)^2}.$$

Show that in the polar coordinates $(r, \theta_2, \theta_3, \ldots, \theta_n)$, where

$$x^n = r \cos \theta_2$$
$$x^{n-1} = r \sin \theta_2 \cos \theta_3$$
$$x^{n-2} = r \sin \theta_2 \sin \theta_3 \cos \theta_4, \text{ etc.}$$

$$\begin{Bmatrix} 0 \leqslant \theta_\alpha \leqslant \pi & (\alpha = 2, 3, \ldots, n-1) \\ 0 \leqslant \theta_n \leqslant 2\pi \end{Bmatrix},$$

the metric takes the form

$$ds^2 = \frac{dr^2 + r^2 \{ d\theta_2^2 + \sin^2 \theta_2 \, [d\theta_3^2 + \ldots + \sin^2 \theta_{n-1} \, d\theta_n^2] \}}{\left(1 + \dfrac{K}{4} \, r^2 \right)^2}.$$

(ii) If $K > 0$, define $R = K^{-1/2}$ and substitute $r = 2R \cot \left(\dfrac{1}{2} \theta_1 \right)$ to show that the geometry in this case is the same as for the n-dimensional surface of a sphere of radius R in an $(n + 1)$-dimensional Euclidean space.

9.3. Investigate the influence of the cosmological constant Λ on the SCHWARZSCHILD metric of a point mass.

9.4. Show that for an oscillating universe with $\kappa = +1$ and $\Lambda = 0$

$$\frac{R_0}{R_{\max}} = \frac{2q_0-1}{2q_0} \quad \text{and} \quad \frac{t_0}{t_{\max}} = \frac{2q_0 \cos^{-1}\{1/(2q_0)^{1/2}\} - (2q_0-1)^{1/2}}{\pi q_0},$$

where $R_{\max}$ and $t_{\max}$ correspond to the maximum of the $R(t)$ curve. Use these results to verify that for $R_0 \ll R_{\max}$ (and $t_0 \ll t_{\max}$), the relation (9.40) is obeyed.

9.5. Derive Eq. (9.40) as a limiting case of Eqs. (9.41) and (9.42).

9.6. Derive Eq. (9.48) as a limiting case of Eqs. (9.46) and (9.47). Discuss the implication of this limiting process.

9.7. Observations on the quasar 3C-9 suggest that when it emitted the light that has just reached the earth it was receding from the earth at a speed of $0.8c$. One of the spectral lines identified in its spectrum has a wavelength of about 1200 Å when emitted from a stationary source. At what wavelength would this line have appeared in the observed spectrum of the quasar?

9.8. According to PLANCK's law, the energy density distribution among the various frequencies of black-body radiation is given by[1]

$$du = 8\pi h \, (\nu/c)^3 \, (e^{h\nu/kT} - 1)^{-1} \, d\nu,$$

where h is PLANCK's constant, k BOLTZMANN's constant and T the absolute temperature. Suppose that such radiation uniformly fills a ROBERTSON-WALKER model universe with scale factor $R(t)$, and that its interaction with the cosmic matter can be neglected. As the universe expands, will the radiation maintain its black-body character? If so, with what temperature law?[2]

9.9. Derive Eqs. (9.71) and (9.74) *directly* from Eqs. (9.46) and (9.47), respectively.

9.10. Derive Eqs. (9.78) and (9.80) *directly* from Eqs. (9.41) and (9.42), respectively.

9.11. Estimate, on the basis of the HUBBLE law (9.53), the linear dimensions of the '*observable*' universe. Compare your result with the value you estimate for the SCHWARZSCHILD radius of the universe.

9.12. Determine the relevant limits on the parameters ρ_0, R_0 and t_0 for a black-hole universe if the cosmological constant Λ were set identically equal to zero.

---

[1] R. K. PATHRIA, *Statistical Mechanics* (Pergamon Press, Oxford, 1972), Sec. 7.2.

[2] Such black-body radiation is, in fact, believed to fill our universe, with $T_0 \simeq 2.7 \, °\text{K}$; see, for instance, D. W. SCIAMA, in *Relativity and Gravitation*, eds. C. G. KUPER and A. PERES (Gordon and Breach, New York, 1971), pp. 283-304.

DIFFERENTIAL GEOMETRY OF A CURVED SURFACE

A.1. General considerations

To illustrate the connection between the metric ds^2 and the curvature of a given continuum, we consider in this appendix a two-dimensional curved surface embedded in a three-dimensional Euclidean space. We assume the surface to be smooth—in the sense that at each point it possesses a tangent plane. We employ rectangular Cartesian coordinates (x, y, z), with a given point P on the surface as the origin of the coordinates and the tangent plane at P as the (x, y)-plane. The equation of the surface may then be written as

$$z = F(x, y) , \qquad (A.1)$$

where the function $F(x, y)$ is supposed to be differentiable as many times as desired; clearly,

$$F(0, 0) = 0 \quad \text{and} \quad \left(\frac{\partial F}{\partial x}\right)_{x, y = 0} = \left(\frac{\partial F}{\partial y}\right)_{x, y = 0} = 0 . \qquad (A.2)$$

In the neighborhood of P, Eq. (A.1) may be expanded as a Taylor series, the first nonvanishing terms of which will be *quadratic* in x and y:

$$z = \frac{1}{2} f_{ij} x^i x^j + \ldots , \qquad (A.3)$$

where $x^1 = x$, $x^2 = y$ and $f_{ij} = (\partial^2 F/\partial x^i \, \partial x^j)_P = f_{ji}$. The derivatives f_{ij} determine the *degree of curvature* of the surface at the point P. The 'principal curvatures' of the surface are determined by the eigenvalues of the matrix (f_{ij}); let these be denoted by κ_1 and κ_2. The eigenvalue equation

$f_{ij} - \Lambda \delta_{ij} | = 0$, i.e.

$$\Lambda^2 - \Lambda(f_{11} + f_{22}) + (f_{11}f_{22} - f_{12}f_{21}) = 0 ,$$

then shows that

$$\kappa_1 + \kappa_2 = f_{11} + f_{22} \quad \text{and} \quad \kappa_1 \kappa_2 = (f_{11}f_{22} - f_{12}f_{21}) = |f_{ij}| . \quad (A.4)$$

If x^1 and x^2 were measured in the direction of the 'principal axes', we would have

$$(f_{ij}) = \begin{pmatrix} \kappa_1 & 0 \\ 0 & \kappa_2 \end{pmatrix},$$

and hence

$$z = \frac{1}{2} \kappa_1 (x^1)^2 + \frac{1}{2} \kappa_2 (x^2)^2 + \dots \quad (A.3a)$$

In a general direction, the curvature would be a *weighted mean* of the minimal and maximal values κ_1 and κ_2.

We now define the *Gaussian curvature*, or the *total curvature*, K of the surface at the point P :

$$K = \kappa_1 \kappa_2 = |f_{ij}| . \quad (A.5)$$

We shall see that K represents an *intrinsic* property of the surface, which is *not* true for the *mean curvature* $\frac{1}{2} (\kappa_1 + \kappa_2)$. For instance, in the case of a cylindrical surface $\kappa_1 = 1/R$ and $\kappa_2 = 0$ whereby $K = 0$; thus, the Gaussian curvature of a cylindrical surface is identically zero. This is indeed true, for a cylindrical surface can be unrolled into a flat surface (with no wrinkles or gaps) ! On the other hand, a spherical surface is intrinsically curved; if it is unrolled onto a plane, it will have to be torn apart and will leave gaps here and there.

We shall now attempt to relate the curvature K of the surface to its metric. For the Euclidean space, in which the given surface is embedded,

$$ds^2 = dx^i \, dx^i + (dz)^2 \quad (i = 1, 2). \quad (A.6)$$

Substituting from (A.3), whereby $dz = f_{ij} x^i \, dx^j + \dots$, we obtain

$$ds^2 \simeq (\delta_{ij} + f_{li} f_{kj} x^l x^k) \, dx^i \, dx^j. \quad (A.7)$$

If x^1 and x^2 were measured along the eigenvectors of (f_{ij}), Eq. (A.7) would have been

$$ds^2 \simeq dx^i \, dx^i + (\kappa_1 x^1 \, dx^1 + \kappa_2 x^2 \, dx^2)^2. \quad (A.7a)$$

Metric (A.7) is still in terms of the coordinates x^1 and x^2 which pertain to the three-dimensional space in which our surface is embedded. It must be expressed in terms of coordinates *intrinsic* to the surface. To do this, we look for the *geodesics* of the surface, viz. the 'lines of minimal length', given by

$$\delta \int_A^B ds = 0 , \tag{A.8}$$

with (ds) given by (A.7). By the calculus of variations, we obtain for the geodesics

$$\frac{d^2 x^i}{ds^2} + f_{ik} f_{lm} \frac{dx^l}{ds} \frac{dx^m}{ds} x^k = 0. \tag{A.9}$$

A geodesic passing through the point P will, therefore, be given by the coordinates

$$x^i(s) \simeq a^i s - \frac{1}{6} f_{ik} f_{lm} a^l a^m a^k s^3 , \tag{A.10}$$

where s is the distance measured along the geodesic, with P as the starting point, while a^i's are the direction cosines that determine the (initial) direction of the geodesic.

We now introduce, for obvious reasons, the *geodesic coordinates* y^i :

$$y^i = a^i s . \tag{A.11}$$

The y^i's are indeed intrinsic to the surface. Combining (A.10) and (A.11), we obtain

$$x^i \simeq y^i - \frac{1}{6} f_{ik} f_{lm} y^l y^m y^k, \tag{A.12}$$

which implies a *nonlinear transformation* of the coordinates. Equation (A.12) gives

$$dx^i \simeq dy^i - \frac{1}{6} f_{ik} f_{lm} y^l y^m dy^k - \frac{1}{3} f_{ik} f_{lm} y^l y^k dy^m, \tag{A.13}$$

and

$$dx^i dx^i \simeq dy^i dy^i - \frac{1}{3} f_{ik} f_{lm} y^l y^m dy^k dy^i$$

$$- \frac{2}{3} f_{ik} f_{lm} y^l y^k dy^m dy^i. \tag{A.14}$$

Substituting (A.13) and (A.14) into (A.7), we obtain

$$ds^2 \simeq dy^i \, dy^i + \frac{1}{3} \, (f_{kj} f_{li} - f_{ij} f_{lk}) \, y^k \, y^l \, dy^i \, dy^j. \tag{A.15}$$

In the non-Euclidean part of the metric (A.15), terms with $i = k$ or $j = l$ identically vanish ; the remaining terms can be written in the form

$$\frac{1}{3} \, (f_{21} f_{12} - f_{11} f_{22}) \, [(y^2 \, dy^1)^2 - 2 \, (y^1 \, y^2 \, dy^1 \, dy^2) + (y^1 \, dy^2)^2]$$

$$= -\frac{1}{3} \, K \, (y^1 \, dy^2 - y^2 \, dy^1)^2.$$

Accordingly, our metric takes the form

$$ds^2 \simeq dy^i \, dy^i - \frac{1}{3} \, K \, (y^1 \, dy^2 - y^2 \, dy^1)^2 \tag{A.16}$$

in terms of the geodesic coordinates y^1 and y^2. It will be remembered that expression (A.16) for the metric holds only in the neighbourhood of the point P and that K is the Gaussian curvature of the surface at that point. We also note that the metric obtained here is of the *Riemannian* type, viz. $ds^2 = g_{ik} \, dy^i \, dy^k$, with

$$g_{11} = 1 - \frac{1}{3} \, K \, (y^2)^2, \quad g_{22} = 1 - \frac{1}{3} \, K \, (y^1)^2 \quad \text{and} \quad g_{12} = g_{21} = +\frac{1}{3} \, K \, y^1 \, y^2.$$

We may like to adopt the *geodesic polar coordinates* (ρ, φ), in which a curve of constant φ is itself a geodesic while ρ is the distance from P measured along the geodesic. A curve of constant ρ will be a circle, with P as its centre ; in general, it will not be a geodesic. Substituting $y^1 = \rho \cos \varphi$ and $y^2 = \rho \sin \varphi$, the line element (A.16) takes the form

$$ds^2 \simeq [(d\rho)^2 + \rho^2 \, (d\varphi)^2] - \frac{1}{3} \, K \, \rho^4 \, (d\varphi)^2$$

$$= (d\rho)^2 + \rho^2 \left(1 - \frac{1}{3} \, K \, \rho^2 \right) (d\varphi)^2. \tag{A.17}$$

This gives for the perimeter p and area A of a circle (of radius ρ)

$$p \, (\rho) \simeq 2\pi\rho \left(1 - \frac{1}{3} \, K \, \rho^2 \right)^{1/2} \simeq 2\pi\rho \left(1 - \frac{1}{6} \, K \, \rho^2 \right), \tag{A.18}$$

and

$$A \, (\rho) = \int_0^\rho p \, (\rho) \, d\rho \simeq \pi\rho^2 \left(1 - \frac{1}{12} \, K \, \rho^2 \right). \tag{A.19}$$

These results are in glaring contrast with the ones pertaining to Euclidean geometry, namely $p = 2\pi\rho$ and $A = \pi\rho^2$. Deviations from the 'normal' results are governed by the Gaussian curvature K of the surface. Reversing the argument, Eqs. (A.18) and (A.19) provide a direct means of determining the curvature of a given two-dimensional surface through measurements totally intrinsic to the surface, *without making an appeal to the third dimension*! We write, in this context,

$$K = \lim_{\rho \to 0} \frac{6}{\rho^2}\left(1 - \frac{p(\rho)}{2\pi\rho}\right) = \lim_{\rho \to 0} \frac{12}{\rho^2}\left(1 - \frac{A(\rho)}{\pi\rho^2}\right). \qquad (A.20)$$

A.2.   Geometrical interpretation of curvature

We now consider a *geodesic polygon ABC* . . . , which is formed by n geodesics AB, BC, . . . Let α, β, γ, . . . denote the interior angles of the polygon; then, α' ($= \pi - \alpha$), β' ($= \pi - \beta$), γ' ($= \pi - \gamma$), . . . denote, respectively, its exterior angles. Now, let there be a vector ξ^i, at the point A, parallel to the geodesic AB. On 'parallel transport' along this geodesic, the vector ends up at the point B; being still parallel to the geodesic AB, the inclination of the vector with the geodesic BC will be $-\beta'$. Continued by 'parallel transport' along the geodesics BC, CD, . . . , it will finally end up at the point A where its inclination with the geodesic AB will be $-(\beta' + \gamma' + \ldots + \alpha')$ or, preferably, $2\pi - (\alpha' + \beta' + \gamma' + \ldots)$. Clearly, this also denotes the inclination of the transported vector ξ^i with the original vector ξ^i at the point A.

Ordinarily, when the given surface is flat the process of parallel transport of a vector along a closed contour brings the vector back to its original inclination, i.e. $\alpha' + \beta' + \gamma' + \ldots = 2\pi$ and hence

$$\alpha + \beta + \gamma + \ldots = (n - 2)\,\pi. \qquad (A.21)$$

For a curved surface, however, the transported vector turns out to be inclined to the original vector at an angle $\Delta\theta$, such that

$$\Delta\theta = \int K\,dS; \qquad (A.22)$$

here, dS is an element of area and the integration goes over the region enclosed by the contour.[1] Accordingly,

$$(\alpha' + \beta' + \gamma' + \ldots) = 2\pi - \int K\,dS \qquad (A.23)$$

---

[1] For proof, see H. P. ROBERTSON and T. W. NOONAN, *Relativity and Cosmology* (Saunders, Philadelphia, 1968); Sec. 8.12.

and hence

$$(\alpha + \beta + \gamma + \ldots) = (n - 2)\,\pi + \int K\,dS. \qquad (A.24)$$

Equation (A.22), which is known as *Peres' formula*, provides a simple geometrical interpretation for the concept of curvature. Equation (A.24), which is a direct consequence of (A.22), is equally striking. When applied to a *geodesic triangle*, it reduces to *Gauss's theorem on integral curvature*, namely

$$\alpha + \beta + \gamma = \pi + \int K\,dS. \qquad (A.25)$$

In the case of a spherical surface, $K = 1/R^2$; GAUSS's theorem on integral curvature then reduces to the theorem of 'spherical excess', namely

$$\alpha + \beta + \gamma = \pi + (S/R^2), \qquad (A.26)$$

where S is the area of the triangle and R the radius of curvature of the surface. It will be noted once again that, for determining R, one *does not* have to get out of the surface (and make an appeal to the third dimension).

A.3. Passage from a spherical surface to a plane surface and vice versa

In terms of the spherical polar coordinates (r, θ, φ), a line element on the surface of a sphere of radius R is given by

$$ds^2 = R^2\,(d\theta^2 + \sin^2\theta\,d\varphi^2). \qquad (A.27)$$

We introduce the coordinate transformation

$$r' = 2R \cot\left(\tfrac{1}{2}\,\theta\right); \quad \varphi' = \varphi, \qquad (A.28)$$

which implies

$$\theta = 2\cot^{-1}\left(\frac{r'}{2R}\right); \quad \varphi = \varphi'. \qquad (A.28a)$$

The line element (A.27) thereby takes the form

$$ds^2 = \frac{1}{[1 + (r'/2R)^2]^2}\,(dr'^2 + r'^2\,d\varphi'^2), \qquad (A.29)$$

which is closely related to the line element of a flat surface, expressed in terms of the plane polar coordinates (r', φ'). The two line elements differ through an *inhomogeneous* scale factor $[1 + (r'/2R)^2]$. Transformation (A.28) thus puts the geometry on the surface of a sphere in direct correspondence with the geometry on a plane, provided that due account is taken of the inhomogeneous scale factor.

To appreciate this relationship, we observe that transformation (A.28) is equivalent to a *stereographic projection* of a sphere on a plane; see Fig. A.1.

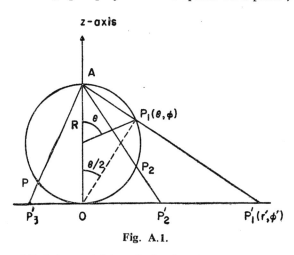

Fig. A.1.

We note that OP_1A is a right-angled triangle with $< OAP_1 = \frac{1}{2}(\pi - \theta)$. Accordingly, $\cot\left(\frac{1}{2}\theta\right) = OP_1' / OA$, which means that $OP_1' = r'$. The azimuthal angle φ determines the position of the plane AOP_1, in which lies the point P_1' as well; hence, φ' is the same as φ. Comparing line element (A.29) with its equivalent (A.27), we conclude that, on applying the inhomogeneous contraction factor $[1 + (r'/2R)^2]^{-1}$ to its line elements, the plane surface (which is originally flat) becomes *equivalent* to a spherical surface; in other words, an inhomogeneous contraction of the plane surface, which becomes more and more pronounced as one moves away from the origin, is *equivalent* to wrapping it up into the form of a spherical surface. The geometrical properties of this inhomogeneously contracted plane will be *identical* with those of a spherical surface, and the curvature of this equivalent (spherical) surface will be determined by the 'scale' employed for contraction, i.e. by the parameter R.

A circle, centred at O and passing through the point $r' = l$, should now be viewed as having a circumference equal to

$$\frac{2\pi l}{1 + (l/2R)^2},$$
(A.30)

which is zero when $l = 0$, $2\pi R$ when $l = 2R$ and again zero when $l \to \infty$. Its radius, on the other hand, is given by

$$\int_0^l \frac{dr'}{1 + (r'/2R)^2} = 2R \tan^{-1}\left(\frac{l}{2R}\right), \qquad (A.31)$$

which is zero when $l = 0$, $\frac{1}{2} \pi R$ when $l = 2R$ and πR when $l \to \infty$. Their ratio is equal to

$$2\pi \frac{(l/2R)}{\tan^{-1}\left(\dfrac{l}{2R}\right)\left\{1 + \left(\dfrac{l}{2R}\right)^2\right\}} \leqslant 2\pi ; \qquad (A.32)$$

this is very nearly equal to 2π if $l \ll R$ (i.e. *Euclidean geometry holds in the neighborhood of O, where the contraction factor $\simeq 1$*), it is 4 when $l = 2R$ and 0 when $l \to \infty$.

The curious reader may like to see for himself what happens, as a result of this transformation, to an arbitrary straight line (and to an arbitrary triangle) in the plane ; in reverse, one may like to see what happens to an arbitrary arc (in particular, that of a geodesic) on the sphere.   It will also be instructive to study the consequences of a 'parallel transport' of a vector, etc.

It is obvious that the line element (A.29) can be written as

$$ds^2 = \frac{dx'^i\, dx'^i}{[1 + (x'^i\, x'^i)/4R^2]^2} \quad (i = 1, 2). \qquad (A.33)$$

We have seen that it represents a two-dimensional surface of *constant* curvature $1/R^2$.   A generalization of (A.33) is straightforward ; one writes

$$ds^2 = \frac{dx^i\, dx^i}{\left(1 + \dfrac{K}{4}\, x^i\, x^i\right)^2} \quad (i = 1, 2, 3, \ldots, n), \qquad (A.34)$$

which represents an n-dimensional space of *constant* curvature.   The special case $n = 3$ plays a central role in relativistic cosmology, for if our three-dimensional space conforms to a metric of this form, with $K \neq 0$, we have to admit that the geometrical properties of the physical space [over distances $O\ (K^{-1/2})$] may not be Euclidean.   One of the most important consequences of introducing positive curvature is that our space, which is ordinarily regarded as *Euclidean* in character and *infinite* in size, becomes *non-Euclidean* in character and *finite* in size.   To see this, we calculate the total area of the (inhomogeneously contracted) plane ; we obtain from (A.29)

$$\int_{r'=0}^{\infty} \int_{\varphi'=0}^{2\pi} \frac{(dr')\,(r'd\varphi')}{(1 + r'^2/4R^2)^2} = 2\pi \left. \frac{-2R^2}{(1 + r'^2/4R^2)} \right|_{r'=0}^{\infty} = 4\pi R^2, \quad (A.35)$$

which is finite and, as expected, equal to the area of a spherical surface of radius R.   Similarly, the total volume of the three-dimensional space (similarly contracted) would be

$$\int_{r'=0}^{\infty} \int_{\theta'=0}^{\pi} \int_{\varphi'=0}^{2\pi} \frac{(dr')\,(r'd\theta')\,(r' \sin\theta'\,d\varphi')}{(1 + r'^2/4R^2)^3} = 4\pi \int_0^{\infty} \frac{r'^2\,dr'}{(1 + r'^2/4R^2)^3} = 2\pi^2 R^3,$$
$$(A.36)$$

which is again finite.   This calculation pertains to the *Einstein model* of the universe ; see Sec. 9.4, especially Eq. (9.36).

AUTHOR INDEX

SUBJECT INDEX

A CATALOG OF SELECTED
DOVER BOOKS
IN SCIENCE AND MATHEMATICS

A CATALOG OF SELECTED

DOVER BOOKS

IN SCIENCE AND MATHEMATICS

Astronomy

BURNHAM'S CELESTIAL HANDBOOK, Robert Burnham, Jr. Thorough guide to the stars beyond our solar system. Exhaustive treatment. Alphabetical by constellation: Andromeda to Cetus in Vol. 1; Chamaeleon to Orion in Vol. 2; and Pavo to Vulpecula in Vol. 3. Hundreds of illustrations. Index in Vol. 3. 2,000pp. 6⅛ x 9¼.
23567-X, 23568-8, 23673-0 Three-vol. set

THE EXTRATERRESTRIAL LIFE DEBATE, 1750–1900, Michael J. Crowe. First detailed, scholarly study in English of the many ideas that developed from 1750 to 1900 regarding the existence of intelligent extraterrestrial life. Examines ideas of Kant, Herschel, Voltaire, Percival Lowell, many other scientists and thinkers. 16 illustrations. 704pp. 5⅜ x 8½. 40675-X

A HISTORY OF ASTRONOMY, A. Pannekoek. Well-balanced, carefully reasoned study covers such topics as Ptolemaic theory, work of Copernicus, Kepler, Newton, Eddington's work on stars, much more. Illustrated. References. 521pp. 5⅜ x 8½.
65994-1

AMATEUR ASTRONOMER'S HANDBOOK, J. B. Sidgwick. Timeless, comprehensive coverage of telescopes, mirrors, lenses, mountings, telescope drives, micrometers, spectroscopes, more. 189 illustrations. 576pp. 5⅝ x 8¼. (Available in U.S. only.)
24034-7

STARS AND RELATIVITY, Ya. B. Zel'dovich and I. D. Novikov. Vol. 1 of *Relativistic Astrophysics* by famed Russian scientists. General relativity, properties of matter under astrophysical conditions, stars, and stellar systems. Deep physical insights, clear presentation. 1971 edition. References. 544pp. 5⅝ x 8¼. 69424-0

Chemistry

CHEMICAL MAGIC, Leonard A. Ford. Second Edition, Revised by E. Winston Grundmeier. Over 100 unusual stunts demonstrating cold fire, dust explosions, much more. Text explains scientific principles and stresses safety precautions. 128pp. 5⅜ x 8½. 67628-5

THE DEVELOPMENT OF MODERN CHEMISTRY, Aaron J. Ihde. Authoritative history of chemistry from ancient Greek theory to 20th-century innovation. Covers major chemists and their discoveries. 209 illustrations. 14 tables. Bibliographies. Indices. Appendices. 851pp. 5⅜ x 8½. 64235-6

CATALYSIS IN CHEMISTRY AND ENZYMOLOGY, William P. Jencks. Exceptionally clear coverage of mechanisms for catalysis, forces in aqueous solution, carbonyl- and acyl-group reactions, practical kinetics, more. 864pp. 5⅜ x 8½.
65460-5

Math–Geometry and Topology

ELEMENTARY CONCEPTS OF TOPOLOGY, Paul Alexandroff. Elegant, intuitive approach to topology from set-theoretic topology to Betti groups; how concepts of topology are useful in math and physics. 25 figures. 57pp. 5⅜ x 8½. 60747-X

COMBINATORIAL TOPOLOGY, P. S. Alexandrov. Clearly written, well-organized, three-part text begins by dealing with certain classic problems without using the formal techniques of homology theory and advances to the central concept, the Betti groups. Numerous detailed examples. 654pp. 5⅜ x 8½. 40179-0

EXPERIMENTS IN TOPOLOGY, Stephen Barr. Classic, lively explanation of one of the byways of mathematics. Klein bottles, Moebius strips, projective planes, map coloring, problem of the Koenigsberg bridges, much more, described with clarity and wit. 43 figures. 210pp. 5⅜ x 8½. 25933-1

CONFORMAL MAPPING ON RIEMANN SURFACES, Harvey Cohn. Lucid, insightful book presents ideal coverage of subject. 334 exercises make book perfect for self-study. 55 figures. 352pp. 5⅝ x 8¼. 64025-6

THE GEOMETRY OF RENÉ DESCARTES, René Descartes. The great work founded analytical geometry. Original French text, Descartes's own diagrams, together with definitive Smith-Latham translation. 244pp. 5⅜ x 8½. 60068-8

THE THIRTEEN BOOKS OF EUCLID'S ELEMENTS, translated with introduction and commentary by Sir Thomas L. Heath. Definitive edition. Textual and linguistic notes, mathematical analysis. 2,500 years of critical commentary. Unabridged. 1,414pp. 5⅜ x 8½. Three-vol. set.
Vol. I: 60088-2 Vol. II: 60089-0 Vol. III: 60090-4

GEOMETRY OF COMPLEX NUMBERS, Hans Schwerdtfeger. Illuminating, widely praised book on analytic geometry of circles, the Moebius transformation, and two-dimensional non-Euclidean geometries. 200pp. 5⅝ x 8¼. 63830-8

DIFFERENTIAL GEOMETRY, Heinrich W. Guggenheimer. Local differential geometry as an application of advanced calculus and linear algebra. Curvature, transformation groups, surfaces, more. Exercises. 62 figures. 378pp. 5⅜ x 8½. 63433-7

CURVATURE AND HOMOLOGY: Enlarged Edition, Samuel I. Goldberg. Revised edition examines topology of differentiable manifolds; curvature, homology of Riemannian manifolds; compact Lie groups; complex manifolds; curvature, homology of Kaehler manifolds. New Preface. Four new appendixes. 416pp. 5⅜ x 8½. 40207-X

TOPOLOGY, John G. Hocking and Gail S. Young. Superb one-year course in classical topology. Topological spaces and functions, point-set topology, much more. Examples and problems. Bibliography. Index. 384pp. 5⅝ x 8¼. 65676-4

LECTURES ON CLASSICAL DIFFERENTIAL GEOMETRY, Second Edition, Dirk J. Struik. Excellent brief introduction covers curves, theory of surfaces, fundamental equations, geometry on a surface, conformal mapping, other topics. Problems. 240pp. 5⅜ x 8½. 65609-8

Math–History of

A SHORT ACCOUNT OF THE HISTORY OF MATHEMATICS, W. W. Rouse Ball. One of clearest, most authoritative surveys from the Egyptians and Phoenicians through 19th-century figures such as Grassman, Galois, Riemann. Fourth edition. 522pp. 5⅜ x 8½. 20630-0

THE HISTORY OF THE CALCULUS AND ITS CONCEPTUAL DEVELOP-MENT, Carl B. Boyer. Origins in antiquity, medieval contributions, work of Newton, Leibniz, rigorous formulation. Treatment is verbal. 346pp. 5⅜ x 8½. 60509-4

THE HISTORICAL ROOTS OF ELEMENTARY MATHEMATICS, Lucas N. H. Bunt, Phillip S. Jones, and Jack D. Bedient. Fundamental underpinnings of modern arithmetic, algebra, geometry and number systems derived from ancient civilizations. 320pp. 5⅜ x 8½. 25563-8

A HISTORY OF MATHEMATICAL NOTATIONS, Florian Cajori. This classic study notes the first appearance of a mathematical symbol and its origin, the competition it encountered, its spread among writers in different countries, its rise to popularity, its eventual decline or ultimate survival. Original 1929 two-volume edition presented here in one volume. xxviii+820pp. 5⅜ x 8½. 67766-4

GAMES, GODS & GAMBLING: A History of Probability and Statistical Ideas, F. N. David. Episodes from the lives of Galileo, Fermat, Pascal, and others illustrate this fascinating account of the roots of mathematics. Features thought-provoking references to classics, archaeology, biography, poetry. 1962 edition. 304pp. 5⅜ x 8½. (Available in U.S. only.) 40023-9

OF MEN AND NUMBERS: The Story of the Great Mathematicians, Jane Muir. Fascinating accounts of the lives and accomplishments of history's greatest mathematical minds–Pythagoras, Descartes, Euler, Pascal, Cantor, many more. Anecdotal, illuminating. 30 diagrams. Bibliography. 256pp. 5⅜ x 8½. 28973-7

HISTORY OF MATHEMATICS, David E. Smith. Nontechnical survey from ancient Greece and Orient to late 19th century; evolution of arithmetic, geometry, trigonometry, calculating devices, algebra, the calculus. 362 illustrations. 1,355pp. 5⅜ x 8½. Two-vol. set. Vol. I: 20429-4   Vol. II: 20430-8

A CONCISE HISTORY OF MATHEMATICS, Dirk J. Struik. The best brief history of mathematics. Stresses origins and covers every major figure from ancient Near East to 19th century. 41 illustrations. 195pp. 5⅜ x 8½. 60255-9

Physics

OPTICAL RESONANCE AND TWO-LEVEL ATOMS, L. Allen and J. H. Eberly. Clear, comprehensive introduction to basic principles behind all quantum optical resonance phenomena. 53 illustrations. Preface. Index. 256pp. 5⅜ x 8½.   65533-4

ULTRASONIC ABSORPTION: An Introduction to the Theory of Sound Absorption and Dispersion in Gases, Liquids and Solids, A. B. Bhatia. Standard reference in the field provides a clear, systematically organized introductory review of fundamental concepts for advanced graduate students, research workers. Numerous diagrams. Bibliography. 440pp. 5⅜ x 8½.   64917-2

QUANTUM THEORY, David Bohm. This advanced undergraduate-level text presents the quantum theory in terms of qualitative and imaginative concepts, followed by specific applications worked out in mathematical detail. Preface. Index. 655pp. 5⅜ x 8½.   65969-0

ATOMIC PHYSICS (8th edition), Max Born. Nobel laureate's lucid treatment of kinetic theory of gases, elementary particles, nuclear atom, wave-corpuscles, atomic structure and spectral lines, much more. Over 40 appendices, bibliography. 495pp. 5⅜ x 8½.   65984-4

AN INTRODUCTION TO HAMILTONIAN OPTICS, H. A. Buchdahl. Detailed account of the Hamiltonian treatment of aberration theory in geometrical optics. Many classes of optical systems defined in terms of the symmetries they possess. Problems with detailed solutions. 1970 edition. xv + 360pp. 5⅜ x 8½.   67597-1

THIRTY YEARS THAT SHOOK PHYSICS: The Story of Quantum Theory, George Gamow. Lucid, accessible introduction to influential theory of energy and matter. Careful explanations of Dirac's anti-particles, Bohr's model of the atom, much more. 12 plates. Numerous drawings. 240pp. 5⅜ x 8½.   24895-X

ELECTRONIC STRUCTURE AND THE PROPERTIES OF SOLIDS: The Physics of the Chemical Bond, Walter A. Harrison. Innovative text offers basic understanding of the electronic structure of covalent and ionic solids, simple metals, transition metals and their compounds. Problems. 1980 edition. 582pp. 6⅛ x 9¼.
66021-4

HYDRODYNAMIC AND HYDROMAGNETIC STABILITY, S. Chandrasekhar. Lucid examination of the Rayleigh-Benard problem; clear coverage of the theory of instabilities causing convection. 704pp. 5⅜ x 8¼.   64071-X

INVESTIGATIONS ON THE THEORY OF THE BROWNIAN MOVEMENT, Albert Einstein. Five papers (1905–8) investigating dynamics of Brownian motion and evolving elementary theory. Notes by R. Fürth. 122pp. 5⅜ x 8½.   60304-0

THE PHYSICS OF WAVES, William C. Elmore and Mark A. Heald. Unique overview of classical wave theory. Acoustics, optics, electromagnetic radiation, more. Ideal as classroom text or for self-study. Problems. 477pp. 5⅜ x 8½.   64926-1

PHYSICAL PRINCIPLES OF THE QUANTUM THEORY, Werner Heisenberg. Nobel Laureate discusses quantum theory, uncertainty, wave mechanics, work of Dirac, Schroedinger, Compton, Wilson, Einstein, etc. 184pp. 5⅜ x 8½.   60113-7

ATOMIC SPECTRA AND ATOMIC STRUCTURE, Gerhard Herzberg. One of best introductions; especially for specialist in other fields. Treatment is physical rather than mathematical. 80 illustrations. 257pp. 5⅜ x 8½.   60115-3

AN INTRODUCTION TO STATISTICAL THERMODYNAMICS, Terrell L. Hill. Excellent basic text offers wide-ranging coverage of quantum statistical mechanics, systems of interacting molecules, quantum statistics, more. 523pp. 5⅜ x 8½.
65242-4

THEORETICAL PHYSICS, Georg Joos, with Ira M. Freeman. Classic overview covers essential math, mechanics, electromagnetic theory, thermodynamics, quantum mechanics, nuclear physics, other topics. First paperback edition. xxiii + 885pp. 5⅜ x 8½.   65227-0

PROBLEMS AND SOLUTIONS IN QUANTUM CHEMISTRY AND PHYSICS, Charles S. Johnson, Jr. and Lee G. Pedersen. Unusually varied problems, detailed solutions in coverage of quantum mechanics, wave mechanics, angular momentum, molecular spectroscopy, more. 280 problems plus 139 supplementary exercises. 430pp. 6½ x 9¼.   65236-X

THEORETICAL SOLID STATE PHYSICS, Vol. 1: Perfect Lattices in Equilibrium; Vol. II: Non-Equilibrium and Disorder, William Jones and Norman H. March. Monumental reference work covers fundamental theory of equilibrium properties of perfect crystalline solids, non-equilibrium properties, defects and disordered systems. Appendices. Problems. Preface. Diagrams. Index. Bibliography. Total of 1,301pp. 5⅜ x 8½. Two volumes.   Vol. I: 65015-4   Vol. II: 65016-2

A TREATISE ON ELECTRICITY AND MAGNETISM, James Clerk Maxwell. Important foundation work of modern physics. Brings to final form Maxwell's theory of electromagnetism and rigorously derives his general equations of field theory. 1,084pp. 5⅜ x 8½. Two-vol. set.   Vol. I: 60636-8   Vol. II: 60637-6

OPTICKS, Sir Isaac Newton. Newton's own experiments with spectroscopy, colors, lenses, reflection, refraction, etc., in language the layman can follow. Foreword by Albert Einstein. 532pp. 5⅜ x 8½.   60205-2

THEORY OF ELECTROMAGNETIC WAVE PROPAGATION, Charles Herach Papas. Graduate-level study discusses the Maxwell field equations, radiation from wire antennas, the Doppler effect and more. xiii + 244pp. 5⅜ x 8½.   65678-5

INTRODUCTION TO QUANTUM MECHANICS With Applications to Chemistry, Linus Pauling & E. Bright Wilson, Jr. Classic undergraduate text by Nobel Prize winner applies quantum mechanics to chemical and physical problems. Numerous tables and figures enhance the text. Chapter bibliographies. Appendices. Index. 468pp. 5⅜ x 8½.   64871-0

CATALOG OF DOVER BOOKS

METHODS OF THERMODYNAMICS, Howard Reiss. Outstanding text focuses on physical technique of thermodynamics, typical problem areas of understanding, and significance and use of thermodynamic potential. 1965 edition. 238pp. 5$\frac{3}{8}$ x 8$\frac{1}{2}$. 69445-3

TENSOR ANALYSIS FOR PHYSICISTS, J. A. Schouten. Concise exposition of the mathematical basis of tensor analysis, integrated with well-chosen physical examples of the theory. Exercises. Index. Bibliography. 289pp. 5$\frac{3}{8}$ x 8$\frac{1}{2}$. 65582-2

RELATIVITY IN ILLUSTRATIONS, Jacob T. Schwartz. Clear nontechnical treatment makes relativity more accessible than ever before. Over 60 drawings illustrate concepts more clearly than text alone. Only high school geometry needed. Bibliography. 128pp. 6$\frac{1}{8}$ x 9$\frac{1}{4}$. 25965-X

THE ELECTROMAGNETIC FIELD, Albert Shadowitz. Comprehensive undergraduate text covers basics of electric and magnetic fields, builds up to electromagnetic theory. Also related topics, including relativity. Over 900 problems. 768pp. 5$\frac{5}{8}$ x 8$\frac{1}{4}$. 65660-8

GREAT EXPERIMENTS IN PHYSICS: Firsthand Accounts from Galileo to Einstein, edited by Morris H. Shamos. 25 crucial discoveries: Newton's laws of motion, Chadwick's study of the neutron, Hertz on electromagnetic waves, more. Original accounts clearly annotated. 370pp. 5$\frac{3}{8}$ x 8$\frac{1}{2}$. 25346-5

RELATIVITY, THERMODYNAMICS AND COSMOLOGY, Richard C. Tolman. Landmark study extends thermodynamics to special, general relativity; also applications of relativistic mechanics, thermodynamics to cosmological models. 501pp. 5$\frac{3}{8}$ x 8$\frac{1}{2}$. 65383-8

LIGHT SCATTERING BY SMALL PARTICLES, H. C. van de Hulst. Comprehensive treatment including full range of useful approximation methods for researchers in chemistry, meteorology and astronomy. 44 illustrations. 470pp. 5$\frac{3}{8}$ x 8$\frac{1}{2}$. 64228-3

STATISTICAL PHYSICS, Gregory H. Wannier. Classic text combines thermodynamics, statistical mechanics and kinetic theory in one unified presentation of thermal physics. Problems with solutions. Bibliography. 532pp. 5$\frac{3}{8}$ x 8$\frac{1}{2}$. 65401-X

Paperbound unless otherwise indicated. Available at your book dealer, online at **www.doverpublications.com**, or by writing to Dept. GI, Dover Publications, Inc., 31 East 2nd Street, Mineola, NY 11501. For current price information or for free catalogues (please indicate field of interest), write to Dover Publications or log on to **www.doverpublications.com** and see every Dover book in print. Dover publishes more than 500 books each year on science, elementary and advanced mathematics, biology, music, art, literary history, social sciences, and other areas.